Theme Issue in Memory to Prof. Jiro Tsuji (1927–2022)

Theme Issue in Memory to Prof. Jiro Tsuji (1927–2022)

Guest Editors

Ewa Kowalska
Yuichi Kobayashi

Basel • Beijing • Wuhan • Barcelona • Belgrade • Novi Sad • Cluj • Manchester

Guest Editors

Ewa Kowalska
Faculty of Chemistry
Jagiellonian University
Kraków
Poland

Yuichi Kobayashi
Organization for the Strategic
Coordination of Research and
Intellectual Properties
Meiji University
Kawasaki
Japan

Editorial Office
MDPI AG
Grosspeteranlage 5
4052 Basel, Switzerland

This is a reprint of the Special Issue, published open access by the journal *Catalysts* (ISSN 2073-4344), freely accessible at: www.mdpi.com/journal/catalysts/special_issues/MS6B804M8K.

For citation purposes, cite each article independently as indicated on the article page online and using the guide below:

Lastname, A.A.; Lastname, B.B. Article Title. *Journal Name* **Year**, *Volume Number*, Page Range.

ISBN 978-3-7258-3426-6 (Hbk)
ISBN 978-3-7258-3425-9 (PDF)
https://doi.org/10.3390/books978-3-7258-3425-9

Cover image courtesy of Ewa Kowalska

© 2025 by the authors. Articles in this book are Open Access and distributed under the Creative Commons Attribution (CC BY) license. The book as a whole is distributed by MDPI under the terms and conditions of the Creative Commons Attribution-NonCommercial-NoDerivs (CC BY-NC-ND) license (https://creativecommons.org/licenses/by-nc-nd/4.0/).

Contents

About the Editors vii

Ewa Kowalska and Shuaizhi Zheng
Theme Issue in Memory to Professor Jiro Tsuji (1927–2022)
Reprinted from: *Catalysts* 2024, 14, 396, https://doi.org/10.3390/catal14070396 1

Alexander Q. Cusumano, Tianyi Zhang, William A. Goddard and Brian M. Stoltz
Origins of Enhanced Enantioselectivity in the Pd-Catalyzed Decarboxylative Allylic Alkylation of N-Benzoyl Lactams
Reprinted from: *Catalysts* 2023, 13, 1258, https://doi.org/10.3390/catal13091258 8

Yuichi Kobayashi, Takayuki Hirotsu, Yosuke Haimoto and Narihito Ogawa
Substitution of Secondary Propargylic Phosphates Using Aryl-Lithium-Based Copper Reagents
Reprinted from: *Catalysts* 2023, 13, 1084, https://doi.org/10.3390/catal13071084 14

Jianguo Hu, Hao Wan, Shengchun Wang, Hong Yi and Aiwen Lei
Electrochemical Thiocyanation/Cyclization Cascade to Access Thiocyanato-Containing Benzoxazines
Reprinted from: *Catalysts* 2023, 13, 631, https://doi.org/10.3390/catal13030631 28

Sylwia Ostrowska, Lorenzo Palio, Agnieszka Czapik, Subhrajyoti Bhandary, Marcin Kwit and Kristof Van Hecke et al.
A Second-Generation Palladacycle Architecture Bearing a N-Heterocyclic Carbene and Its Catalytic Behavior in Buchwald–Hartwig Amination Catalysis
Reprinted from: *Catalysts* 2023, 13, 559, https://doi.org/10.3390/catal13030559 38

Kazuya Ito, Takayuki Doi and Hirokazu Tsukamoto
De Novo Synthesis of Polysubstituted 3-Hydroxypyridines Via "Anti-Wacker"-Type Cyclization
Reprinted from: *Catalysts* 2023, 13, 319, https://doi.org/10.3390/catal13020319 51

Zhicheng Bao, Chaoqiang Wu and Jianbo Wang
Palladium-Catalyzed Three-Component Coupling of Benzynes, Benzylic/Allylic Bromides and 1,1-Bis[(pinacolato)boryl]methane
Reprinted from: *Catalysts* 2023, 13, 126, https://doi.org/10.3390/catal13010126 69

Masaya Kumashiro, Kosuke Ohsawa and Takayuki Doi
Photocatalyzed Oxidative Decarboxylation Forming Aminovinylcysteine Containing Peptides
Reprinted from: *Catalysts* 2022, 12, 1615, https://doi.org/10.3390/catal12121615 77

Simon V. Sieger, Ilja Lubins and Bernhard Breit
Rhodium-Catalyzed Dynamic Kinetic Resolution of Racemic Internal Allenes towards Chiral Allylated Triazoles and Tetrazoles
Reprinted from: *Catalysts* 2022, 12, 1209, https://doi.org/10.3390/catal12101209 96

Xiao Wang, Ming-Zhu Lu and Teck-Peng Loh
Transition-Metal-Catalyzed C–C Bond Macrocyclization via Intramolecular C–H Bond Activation
Reprinted from: *Catalysts* 2023, 13, 438, https://doi.org/10.3390/catal13020438 107

Yuji Nishii and Masahiro Miura
Construction of Benzo-Fused Polycyclic Heteroaromatic Compounds through Palladium-Catalyzed Intramolecular C-H/C-H Biaryl Coupling
Reprinted from: *Catalysts* 2022, 13, 12, https://doi.org/10.3390/catal13010012 130

Yuichi Kobayashi
Coupling Reactions on Secondary Allylic, Propargylic, and Alkyl Carbons Using Organoborates/Ni and RMgX/Cu Reagents
Reprinted from: *Catalysts* **2023**, *13*, 132, https://doi.org/10.3390/catal13010132 . . . . **140**

About the Editors

Ewa Kowalska

Ewa Kowalska received her PhD with honors in chemical technology from Gdansk University of Technology in 2004. After completing JSPS (2005–2007) and GCOE (2007–2009) postdoctoral fellowships in Japan, and Marie Sklodowska-Curie (MSC) fellowships in France (2002–2003) and Germany (2009–2012), she worked at the Institute for Catalysis (ICAT), Hokkaido University, as an associate professor and a leader of the research cluster on plasmonic photocatalysis (2012–2023). In 2022, she was appointed professor at the Faculty of Chemistry, Jagiellonian University. Her research interests focus on environmental protection, AOPs, solar energy conversion, plasmonic photocatalysis, nanoscience, "green" chemistry, and vis-responsive materials for the degradation of chemical and microbiological pollutants. She has published over 120 papers, has led research grants from different funding agencies (e.g., Bill & Melinda Gates Foundation, CONCERT Japan, European Commission (MSC), NCN, NAWA), and has served as editor of *Materials Science in Semiconductor Processing* (MSSP), *Micro & Nano Letters* (MNL), and *Energy, Water and Air Catalysis Research* (EWA Cat. Res.).

Yuichi Kobayashi

Yuichi KOBAYASHI, an emeritus professor of Tokyo Institute of Technology, received his Ph. D. from Tokyo Institute of Technology in 1981 under the supervision by Prof. J. Tsuji; worked as a post-doctor for Professor Gilbert Stork at Columbia University, New York (1981 to 1982); and joined Fumie Sato's laboratory at Tokyo Institute of Technology in 1982 as an assistant professor. Since then, he spent 37 years at the university for teaching and research until his retirement in March, 2019. During this period, he was promoted to an associate professor in 1988, became an independent researcher in 1995, and advanced to a full professor in Dec., 2010. In April, 2019, he moved to Meiji University, Organization for the Strategic Coordination of Research and Intellectual Properties and has been there since (March, 2025). His research interests include organic synthesis using transition metal catalysts and the development of reactions for chiral C–C bond formation. He has published more than 300 papers. Targets include cinchona alkaloids such as quinine, phoslactomycins, fatty acid metabolites known as resolvins and leukotrienes, isoprostanes, epi-jasmonates, tetrahydrocannabinoids, etc. Chiral C–C bond forming reactions include the substitution of 4-cyclopent-1,3-diol derivatives, allylic substitutions, propargylic substitution, and Cu-catalyzed substitution at allylic, propargylic, and saturated secondary carbons. He received the Young Chemist Award of the Chemical Society of Japan in 1988 and the LSU Neuroscience Center of Excellence Dean's Award Lecture at Louisiana State University, New Orleans, in 2019.

Editorial

Theme Issue in Memory to Professor Jiro Tsuji (1927–2022)

Ewa Kowalska [1,*] and Shuaizhi Zheng [2,*]

1. Faculty of Chemistry, Jagiellonian University, 30-387 Krakow, Poland
2. School of Materials Science and Engineering, Xiangtan University, Xiangtan 411105, China
* Correspondence: ewa.k.kowalska@uj.edu.pl (E.K.); xishuai423@hotmail.com (S.Z.)

The importance of catalysis is obvious and unquestionable, especially bearing in mind that about 90% of all commercially produced chemical products involve catalysts at some step of their manufacture [1]. The history of the catalysis study has been presented in many fruitful review papers [1–7]. Considering the keywords "catalysis" and "history", more than one thousand papers could be found (e.g., 1300 in the Web of Science on 16 April 2024). Interestingly, the special aspects on the development of catalysis research have been presented in respect not only to the different research fields of catalysis (e.g., organotextile catalysis [8], heterogeneous catalysis [9,10], homogeneous catalysis [11,12], electrocatalysis [13], single-atom catalysis [14], phase transfer [15], solid hydrogen storage [16], plasma catalysis [17], catalytic combustion [18], minerals [19], small molecules [20], transition metals [21], computational study/machine learning [22,23], enzyme [24], polymerization [25], water oxidation [26], photoredox catalysis [27,28], photocatalysis [29], artificial photosynthesis [30], fluid catalytic cracking [31], gasoline automobile catalysis [32,33], nanozymes [34], "the origin of life" [35], applied/industrial catalysis [1,36,37], etc.) and particular catalysts (e.g., Ziegler–Natta catalysts [38]), but also to the specific countries (e.g., Korea [39], Brazil [40–42], Portugal [43], Italy [44], and India [45]), companies (e.g., Ciba–Geigy [46], Roche [47], CF-industries [48], Dow Chemical Company [49], Eastman Chemical Company [50], Leuna Factory [51], and Volkswagen [52]), world regions (Latin America [53]), institutes and consortiums (e.g., Dalian Institute of Chemical Physics [54] and CIRCC in Italy [55]), catalysis societies and their meetings (e.g., The Catalysis Society of Japan (CATSJ) [56], The Swiss Industrial Biocatalysis Consortium (SIBC) [57], and "50 Years of German Catalysis Meetings" [58]), and even the collaboration and friendships between countries (Japan–China [59] and France–Venezuela [60]), etc.

Among various papers, the development of catalysis research in the field of organic chemistry has been well documented [61], as shown by a significant increase in the number of published papers (Figure 1). Many different aspects of organic chemistry have been discussed, including single-atom catalysis [62], hydrogenation [63], C-H functionalization [64], enantioselective catalysis [65], methylation [66], nucleophilic phosphinocatalysis [67], chemistry of alkynes [12], isocyanide-based multicomponent reactions [28], aldol reaction [68], Fischer–Tropsch (FT) synthesis [69,70] (and even the reactors [71]), microwave-assisted synthesis [42], Diels–Alder reaction [72], Suzuki–Miyaura cross-couplings [73], synthesis of nitrile [74], formose reaction [75], π-allylpalladium chemistry [76], palladium-based catalysts [77–80], and nickel catalysts [79].

Obviously the studies on catalysis by famous scientists, including the works by Lavlavel G. Ionescu [81], Nikolai Zelinsky [82,83], Heinz Heinemann [84], Paul Sabatier [85], Jean Baptiste Senderens [85], Mikhail Kucherov [86], William Crowell Bray [87], Slobodan Anic [88], Luigi Casale [89], and Gilbert Stork [90,91], have also been acknowledged.

This Special Issue of *Catalysts* is dedicated to Professor Jiro Tsuji, who passed away on 1 April 2022. Jiro Tsuji was born on 11 May 1927, in Shiga, Japan. He received a Bachelor of Science degree from Kyoto University, Japan, in 1951, a M.Sc. degree from Baylor University, USA, in 1957, and a Ph.D. from Columbia University, USA, in 1960 under the supervision

Citation: Kowalska, E.; Zheng, S. Theme Issue in Memory to Professor Jiro Tsuji (1927–2022). *Catalysts* **2024**, *14*, 396. https://doi.org/10.3390/catal14070396

Received: 9 June 2024
Accepted: 18 June 2024
Published: 21 June 2024

Copyright: © 2024 by the authors. Licensee MDPI, Basel, Switzerland. This article is an open access article distributed under the terms and conditions of the Creative Commons Attribution (CC BY) license (https://creativecommons.org/licenses/by/4.0/).

of Gilbert Stork. Tsuji returned to Japan in 1962 to work as a research associate at Toray Industries, where he began his career in transition metal-catalyzed reactions. In 1974, he moved to the Tokyo Institute of Technology and served as a professor until retirement age of 60 in 1988. Thereafter, he was a professor at Okayama University of Science, Japan, from 1988 to 1996, and then at Kurashiki University of Science and the Arts from 1996 to 1999.

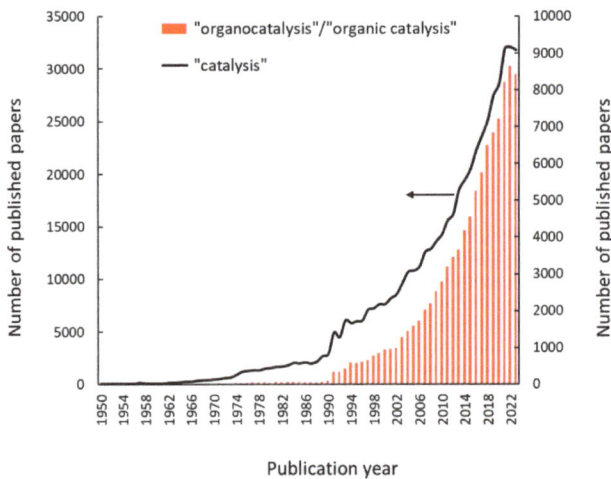

Figure 1. Number of papers published annually on catalysis and organocatalysis (searched in the Web of Science using "catalysis" and "organocatalysis"/"organic catalysis", 27 March 2024).

Professor Jiro Tsuji pioneered the discovery of many transition metal-catalyzed reactions and showed a general idea of developing and applying these reactions in organic synthesis. Among the well-known reactions are several types of Pd-catalyzed ones, such as (i) the substitution of allylic substrates based on the reaction of π-allyl palladium with carbon nucleophiles, discovered in 1965; (ii) reactions of allyl β-keto esters, resulting in allylation, olefin formation, and reduction; (iii) reactions of propargylic substrates; and (iv) the formation of methyl ketones from 1-olefins, based on the Wacker process of ethylene. It is noteworthy that the olefin formation is used as the key step in the industrial synthesis of jasmonate [92]. Other reactions catalyzed by Pd, Ru, and Cu are carbonylation of olefins, dienes, acetylenes, and allyl compounds; decarbonylation of acid chloride and aldehydes; oxidative decomposition of catechol to muconic acid; etc.

Professor Jiro Tsuji focused on the carbon–carbon bond-forming reaction from the very beginning of his research. According to Tsuji, the importance of organic synthesis was taught by professor Gilbert Stork (who he acknowledged in two review papers [90,91]) when he was in the doctoral program at Columbia University.

The significance of the reactions found by Tsuji has been proven by their widespread adoption in academic and industrial laboratories. Consequently, it is not surprising that Tsuji has been honored with the Chemical Society of Japan Award in 1981, the Japanese Medal of Honor with Purple Ribbon in 1994, the Japan Academy Prize in 2004, and the Tetrahedron Prize in 2014. He received the title of honorary professor at Tokyo Institute of Technology on 21 July 2011.

It must be underlined that two reactions were named after him, i.e., the Tsuji–Trost reaction and the Tsuji–Wilkinson decarbonylation reaction. The former (also called the Trost allylic alkylation or allylic alkylation) is a substitution reaction with a palladium catalyst, involving a substrate that contains a leaving group in an allylic position. This work (already mentioned above) was first pioneered by professor Tsuji in 1965 [93] and, then, continued by professor Barry Trost in 1973 with the introduction of phosphine ligands [94]. The latter is a method for the decarbonylation of aldehydes and some acyl chlorides. The name has recognized professor Tsuji, whose team reported as first the use of RhCl(PPh$_3$)$_3$ catalyst (Wilkinson) for these reactions.

The scientific achievements of professor Jiro Tsuji have resulted in more than 250 scientific papers, cited in more than 8,000 works, and reaching almost 13,000 citations (according to the Web of Science, on 23 April 2024). Obviously, the papers with the highest interests of the scientific community present palladium catalysts in organic synthesis [95–101]. However, also other catalysts have been successfully applied, and the data reported have been highly cited, including osmium tetroxide [102], metal complexes (e.g., rhodium, ruthenium, molybdenum, nickel, palladium, cuprous, titanium, and aluminum [103–108]), and other metals (copper [109,110], bismuth [111], and iron [110]).

This Special Issue of *Catalysts*, in memory of prof. Jiro Tsuji, was announced to acknowledge the magnificent impact of his study on others. In total, eleven papers were published, including three reviews, six research papers, and two communications. It must be pointed out that six of them have been selected as feature papers. The review papers focus on (i) the construction of structurally intriguing π-extended polycyclic heteroaromatics through catalytic coupling reactions [112]; (ii) the coupling reactions using organoborates/Ni and RMgX/Cu reagents [113]; and (iii) the transition metal-catalyzed ring-closing metathesis and coupling reactions [114].

The communications by Sieger et al. and Cusumano et al. present their findings on the Rh-catalyzed addition reaction of nitrogen-containing heterocycles to internal allenes [115] and the origins of enantioselectivity in the Pd-catalyzed decarboxylative allylic alkylation of N-benzoyl lactams [116], respectively.

Among research papers, besides classical organic synthesis approaches, other so-called "green" methods have also been discussed, i.e., using photocatalysis and electrochemistry. The photocatalytic formation of (2S,3S)-S-[(Z)-aminovinyl]-3-methyl-D-cysteine (AviMeCys) has been proposed by Kumashiro et al. [117]. In the case of electrochemistry, oxidative cyclization of ortho-vinyl aniline has been proposed by Hu et al. [118].

The favorite topic of professor Tsuji, i.e., palladium catalysis, has been discussed by Bao et al. [119], presenting a three-component cross-coupling reaction of 2-(trimethylsilyl)phenyl trifluoromethanesulfonate, benzylic/allylic bromides, and 1,1-bis[(pinacolato)boryl]methane. An interesting approach has been proposed by Ito et al. for the preparation of polysubstituted 3-hydroxypyridines from amino acids, propargyl alcohols, and arylboronic acids [120]. Ostrowska et al. proposed the synthetic protocol for palladacycle complexes using a mild base and an environmentally desirable solvent [121]. Finally, the substitution of secondary propargylic phosphates has been carried out by Kobayashi et al. with the use of aryl-lithium-based copper reagents [122].

To conclude, the significant contribution of professor Jiro Tsuji in the fields of organic chemistry and catalysis is unquestionable. It is thought that the research started by Tsuji will inspire others further in the development of new, environmentally friendly synthesis methods for world sustainability.

Finally, guest editors of this Special Issue thank all authors for their valuable contributions, without which this Special Issue would not have been possible. Moreover, we would like to express our sincerest thanks also to the editorial team of *Catalysts* for their kind support, advice, and fast responses.

Acknowledgments: The fruitful comments from Yuichi Kobayashi (guest editor of this Special Issue) are highly acknowledged.

Conflicts of Interest: The authors declare no conflicts of interest.

References

1. Armor, J.N. A history of industrial catalysis. *Catal. Today* **2011**, *163*, 3–9. [CrossRef]
2. Piumetti, M. A brief history of the science of catalysis—I: From the early concepts to single-site heterogeneous catalysts. *Chim. Oggi-Chem. Today* **2014**, *32*, 22–27.
3. da Rocha, M.G.; Nakagaki, S.; Machado, G.S. Revisiting the History of Catalysis: Elizabeth Fulhame's Contributions and Other Historical Aspects. *Rev. Virtual Quim.* **2023**, *15*, 283–294. [CrossRef]

4. Lindström, B.; Pettersson, L.J. A brief history of catalysis. *CatTech* **2003**, *7*, 130–138. [CrossRef]
5. Ben Kilani, C.; Batis, H.; Chastrette, M. Development of the Ideas Concerning Catalysis at the Beginning of the XIXth Century. *L'Actualité Chim.* **2001**, 44–50, ISSN: 0151-9093. Available online: https://www.webofscience.com/wos/woscc/full-record/WOS:000170674000007?SID=EUW1ED0FD05YO5GfFVkDU5dTu6Uku (accessed on 20 June 2024).
6. Somorjai, G.A.; McCrea, K. Roadmap for catalysis science in the 21st century: A personal view of building the future on past and present accomplishments. *Appl. Catal. A-Gen.* **2001**, *222*, 3–18. [CrossRef]
7. Ertl, G.; Gloyna, T. Catalysis—From philosopher's stone to Wilhelm Ostwald. *Z. Phys. Chem.* **2003**, *217*, 1207–1219. [CrossRef]
8. Lee, J.W.; Mayer-Gall, T.; Opwis, K.; Song, C.E.; Gutmann, J.S.; List, B. Organotextile Catalysis. *Science* **2013**, *341*, 1225–1229. [CrossRef]
9. Di Monte, R.; Kaspar, J. Heterogeneous environmental catalysis—A gentle art: CeO_2-ZrO_2 mixed oxides as a case history. *Catal. Today* **2005**, *100*, 27–35. [CrossRef]
10. Fukuoka, A.; Kobayashi, H. Valorization of Cellulose and Chitin into Valuable Chemicals by Heterogeneous Catalysis. *J. JPN Petrol. Inst.* **2023**, *66*, 48–56. [CrossRef]
11. Klein, A.; Goldfuss, B.; van der Vlugt, J.I. From Mechanisms in Homogeneous Metal Catalysis to Applications in Chemical Synthesis. *Inorganics* **2018**, *6*, 19. [CrossRef]
12. Temkin, O.N. "Golden Age" of Homogeneous Catalysis Chemistry of Alkynes: Dimerization and Oligomerization of Alkynes. *Kinet. Catal.* **2019**, *60*, 689–732. [CrossRef]
13. Zang, W.J.; Kou, Z.K.; Pennycook, S.J.; Wang, J. Heterogeneous Single Atom Electrocatalysis, Where "Singles" Are "Married". *Adv. Energy Mater.* **2020**, *10*, 1903181. [CrossRef]
14. Zhou, Y.; Jiang, Y.; Ji, Y.X.; Lang, R.; Fang, Y.X.; Wu, C.D. The Opportunities and Challenges in Single-Atom Catalysis. *ChemCatChem* **2023**, *15*, e202201176. [CrossRef]
15. Makosza, M.; Fedorynski, M. Thirty years of phase transfer catalysis. *Pol. J. Chem.* **1996**, *70*, 1093–1110.
16. Salman, M.S.; Pratthana, C.; Lai, Q.W.; Wang, T.; Rambhujun, N.; Srivastava, K.; Aguey-Zinsou, K.F. Catalysis in Solid Hydrogen Storage: Recent Advances, Challenges, and Perspectives. *Energy Technol.* **2022**, *10*, 2200433. [CrossRef]
17. Whitehead, J.C. Plasma-catalysis: The known knowns, the known unknowns and the unknown unknowns. *J. Phys. D Appl. Phys.* **2016**, *49*, 243001. [CrossRef]
18. Arai, H.; Fukuzawa, H. Research and development on high temperature catalytic combustion. *Catal. Today* **1995**, *26*, 217–221. [CrossRef]
19. Zhong, M.; Huang, H.P.; Xu, P.C.; Hu, J. Catalysis of Minerals in Pyrolysis Experiments. *Minerals* **2023**, *13*, 515. [CrossRef]
20. Cai, M.; Zhang, R.Z.; Yang, C.M.; Luo, S.Z. Bio-inspired Small Molecular Catalysis. *Chin. J. Chem.* **2023**, *41*, 548–559. [CrossRef]
21. Yan, N.; Xiao, C.X.; Kou, Y. Transition metal nanoparticle catalysis in green solvents. *Coord. Chem. Rev.* **2010**, *254*, 1179–1218. [CrossRef]
22. Besora, M.; Maseras, F. Microkinetic modeling in homogeneous catalysis. *Wiley Interdiscip. Rev. Comput. Stat.* **2018**, *8*, e1372. [CrossRef]
23. Chen, D.X.; Shang, C.; Liu, Z.P. Machine-learning atomic simulation for heterogeneous catalysis. *NPJ Comput. Mater.* **2023**, *9*, 2. [CrossRef]
24. Scrutton, N.S. Unravelling the complexity of enzyme catalysis. *FEBS J.* **2023**, *290*, 2204–2207. [CrossRef] [PubMed]
25. Mülhaupt, R. Catalytic polymerization and post polymerization catalysis fifty years after the discovery of Ziegler's catalysts. *Macromol. Chem. Phys.* **2003**, *204*, 289–327. [CrossRef]
26. Reith, L.; Lienau, K.; Triana, C.A.; Siol, S.; Patzke, G.R. Preparative History vs Driving Force in Water Oxidation Catalysis: Parameter Space Studies of Cobalt Spinels. *Acs Omega* **2019**, *4*, 15444–15456. [CrossRef]
27. Teply, F. Visible-light photoredox catalysis with $[Ru(bpy)_3]^{2+}$: General principles and the twentieth-century roots. *Phys. Sci. Rev.* **2020**, *5*, 20170171.
28. Russo, C.; Brunelli, F.; Tron, G.C.; Giustiniano, M. Isocyanide-Based Multicomponent Reactions Promoted by Visible Light Photoredox Catalysis. *Chem.-Eur. J.* **2023**, *29*, e202203150. [CrossRef]
29. Wang, K.L.; Janczarek, M.; Wei, Z.S.; Raja-Mogan, T.; Endo-Kimura, M.; Khedr, T.M.; Ohtani, B.; Kowalska, E. Morphology- and crystalline composition-governed activity of titania-based photocatalysts: Overview and perspective. *Catalysts* **2019**, *9*, 1054. [CrossRef]
30. Zhang, B.B.; Sun, L.C. Artificial photosynthesis: Opportunities and challenges of molecular catalysts. *Chem. Soc. Rev.* **2019**, *48*, 2216–2264. [CrossRef]
31. Vogt, E.T.C.; Weckhuysen, B.M. Fluid catalytic cracking: Recent developments on the grand old lady of zeolite catalysis. *Chem. Soc. Rev.* **2015**, *44*, 7342–7370. [CrossRef]
32. Farrauto, R.J.; Deeba, M.; Alerasool, S. Gasoline automobile catalysis and its historical journey to cleaner air. *Nat. Catal.* **2019**, *2*, 603–613. [CrossRef]
33. Shelef, M.; McCabe, R.W. Twenty-five years after introduction of automotive catalysts: What next? *Catal. Today* **2000**, *62*, 35–50. [CrossRef]
34. Zandieh, M.; Liu, J.W. Nanozymes: Definition, Activity, and Mechanisms. *Adv. Mater.* **2024**, *36*, 2211041. [CrossRef] [PubMed]
35. de Graaf, R.; De Decker, Y.; Sojo, V.; Hudson, R. Quantifying Catalysis at the Origin of Life. *Chemistry* **2023**, *29*, e202301447. [CrossRef] [PubMed]

36. Adams, C. Applied Catalysis: A Predictive Socioeconomic History. *Top. Catal.* **2009**, *52*, 924–934. [CrossRef]
37. Kochloefl, K. Development of industrial solid catalysts. *Chem. Eng. Technol.* **2001**, *24*, 229–234. [CrossRef]
38. Thakur, A.; Chammingkwan, P.; Wada, T.; Onishi, R.; Kamimura, W.; Seenivasan, K.; Terano, M.; Taniike, T. Solution-state NMR study of organic components of industrial Ziegler-Natta catalysts: Effect of by-products on catalyst performance. *Appl. Catal. A-Gen.* **2021**, *611*, 117971. [CrossRef]
39. Moon, S.H.; Lee, W.Y.; Kim, Y.G. Catalysis in Korea. *Catal. Surv. Asia* **2004**, *8*, 225–229. [CrossRef]
40. Dupont, J. The catalysis in Brazil: A history of success in the last 25 years. *Quim. Nova* **2002**, *25*, 12–13. [CrossRef]
41. Bernardo-Gusmaoa, K.; Pergher, S.B.C.; dos Santos, E.N. A Panorama of Catalysis in Brazil in the Last 40 Years. *Quim. Nova* **2017**, *40*, 650–655.
42. de Souza, R.; Miranda, L. Microwave Assisted Organic Synthesis: A History of Success in Brazil. *Quim. Nova* **2011**, *34*, 497–506.
43. Pombeiro, A.J.L.; Burke, A.J. Virtual Collection of Portuguese Catalysis. *Chemcatchem* **2018**, *10*, 2712–2716. [CrossRef]
44. Ferriani, S.; Lazerson, M.H.; Lorenzoni, G. Anchor entrepreneurship and industry catalysis: The rise of the Italian Biomedical Valley. *Res. Policy* **2020**, *49*, 104045. [CrossRef]
45. Sahoo, J.; Panda, J.; Sahoo, G. Unravelling the Development of Non-Covalent Organocatalysis in India. *Synlett* **2023**, *34*, 729–758. [CrossRef]
46. Bader, R.R.; Baumeister, P.; Blaser, H.U. Catalysis at Ciba-Geigy. *Chimia* **1996**, *50*, 99–105. [CrossRef]
47. Roessler, F. Catalysis in the industrial production of pharmaceuticals and fine chemicals. *Chimia* **1996**, *50*, 106–109. [CrossRef]
48. Steininger, B. Ammonia synthesis on the banks of the Mississippi: A molecular-planetary technology. *Anthr. Rev.* **2021**, *8*, 262–279. [CrossRef]
49. Chum, P.S.; Swogger, K.W. Olefin polymer technologies-History and recent progress at The Dow Chemical Company. *Prog. Polym. Sci.* **2008**, *33*, 797–819. [CrossRef]
50. Puckette, T.A. Hydroformylation Catalysis at Eastman Chemical: Generations of Catalysts. *Top. Catal.* **2012**, *55*, 421–425. [CrossRef]
51. Becker, K. Catalysts of the Leuna Factory 1960–1986: From Empirie to Science Part IV Hydrocracking Catalysts. *Chem. Ing. Tech.* **2016**, *88*, 1050–1067. [CrossRef]
52. König, A.; Held, W.; Richter, T. Lean-burn catalysts from the perspective of a car manufacturer.: Early work at Volkswagen Research. *Top. Catal.* **2004**, *28*, 99–103. [CrossRef]
53. Ramirez-Corredores, M.M. Catalysis research in Latin America. *Appl. Catal. A-Gen.* **2000**, *197*, 3–9. [CrossRef]
54. Lin, L.W.; Liang, D.B.; Wang, Q.X.; Cai, G.Y. Research and development of catalytic processes for petroleum and natural gas conversions in the Dalian Institute of Chemical Physics. *Catal. Today* **1999**, *51*, 59–72. [CrossRef]
55. Aresta, M. The Contribution of CIRCC Partners to the Birth and Growth of CO_2 Chemistry. *Eur. J. Inorg.* **2022**, *2022*, e202200321. [CrossRef]
56. Iwasawa, Y.; Iwamoto, M.; Deguchi, T.; Kubota, Y.; Machida, M.; Yamashita, H. The Catalysis Society of Japan (CATSJ): History and Activities. *Angew. Chem. Int. Ed.* **2008**, *47*, 9180–9185. [CrossRef]
57. Hanlon, S. The Swiss Industrial Biocatalysis Consortium (SIBC): Past, Present and Future. *Chimia* **2020**, *74*, 342–344. [CrossRef]
58. Völter, J.; Schlögl, R. 50 Years of German Catalysis Meetings: From Twin Roots to a Joint Success Story. *Chemcatchem* **2017**, *9*, 527–532. [CrossRef]
59. Oyama, S.T.; Xin, Q.; Xiong, G.X.; Shen, W.J.; Xu, J.; Yin, H.M.; Yuan, Y.Z.; Liu, H.C.; Zheng, H.D. History of the Dalian Institute of Chemical Physics and the Friendship between China and japan in catalysis. *Chin. J. Catal.* **2019**, *40*, 1591–1614. [CrossRef]
60. Arvanitis, R.; Vessuri, H. Cooperation between France and Venezuela in the field of catalysis. *Int. Soc. Sci. J.* **2001**, *53*, 201–217. [CrossRef]
61. Chen, D.Y.K. A Personal Perspective on Organic Synthesis: Past, Present, and Future. *Isr. J. Chem.* **2018**, *58*, 85–93. [CrossRef]
62. Li, W.H.; Yang, J.R.; Wang, D.S.; Li, Y.D. Striding the threshold of an atom era of organic synthesis by single-atom catalysis. *Chem* **2022**, *8*, 119–140. [CrossRef]
63. Pritchard, J.; Filonenko, G.A.; van Putten, R.; Hensen, E.J.M.; Pidko, E.A. Heterogeneous and homogeneous catalysis for the hydrogenation of carboxylic acid derivatives: History, advances and future directions. *Chem. Soc. Rev.* **2015**, *44*, 3808–3833. [CrossRef] [PubMed]
64. Piou, T.; Rovis, T. Electronic and Steric Tuning of a Prototypical Piano Stool Complex: Rh(III) Catalysis for C-H Functionalization. *Acc. Chem. Res.* **2018**, *51*, 170–180. [CrossRef]
65. Blaser, H.U.; Spindler, F. Enantioselective catalysis for agrochemicals. The case histories of (S)–metolachlor, (R)–metalaxyl and clozylacon. *Top. Catal.* **1997**, *4*, 275–282. [CrossRef]
66. Zhou, J.; Liu, Z.C.; Wang, Y.D.; Kong, D.J.; Xie, Z.K. Shape selective catalysis in methylation of toluene: Development, challenges and perspectives. *Front. Chem. Sci. Eng* **2018**, *12*, 103–112. [CrossRef]
67. San, K.; Venkatesh, T.; Kwon, O. Nucleophilic Phosphine Catalysis: The Untold Story. *Asian J. Org. Chem.* **2021**, *10*, 2699–2708.
68. Beutner, G.L.; Denmark, S.E. Lewis Base Catalysis of the Mukaiyama Directed Aldol Reaction: 40 Years of Inspiration and Advances. *Angew. Chem. Int. Ed.* **2013**, *52*, 9086–9096. [CrossRef] [PubMed]
69. Schulz, H. Short history and present trends of Fischer-Tropsch synthesis. *Appl. Catal. A-Gen.* **1999**, *186*, 3–12. [CrossRef]
70. Hazra, S.; Elias, A.J. One Hundred Years of the Fischer-Tropsch Reaction. *Resonance* **2023**, *28*, 1875–1889. [CrossRef]
71. Guettel, R.; Kunz, U.; Turek, T. Reactors for Fischer-Tropsch synthesis. *Chem. Eng. Technol.* **2008**, *31*, 746–754. [CrossRef]

72. Oikawa, H.; Tokiwano, T. Enzymatic catalysis of the Diels-Alder reaction in the biosynthesis of natural products. *Nat. Prod. Rep.* **2004**, *21*, 321–352. [CrossRef] [PubMed]
73. Ahmed, A.; Mushtaq, I.; Chinnam, S. Suzuki-Miyaura cross-couplings for alkyl boron reagent: Recent developments—A review. *Futur. J. Pharm. Sci.* **2023**, *9*, 67. [CrossRef]
74. Trifirò, F. Some history on the new ways of synthesis of nitriles. *Catal. Today* **2021**, *363*, 10–14. [CrossRef]
75. Raos, N. Formose Reaction—The Holy Grail of Chemists. *Kem. Ind.* **2018**, *67*, 127–134. [CrossRef]
76. Tsuji, J. The birth and development of π-allylpalladium chemistry. *J. Synth. Org. Chem. Jpn.* **1999**, *57*, 1036–1050. [CrossRef]
77. Tsuji, J. 25 Years in the Organic-Chemistry of Palladium. *J. Organomet. Chem.* **1986**, *300*, 281–305. [CrossRef]
78. Tsuji, J. Organopalladium chemistry in the '60s and '70s. *New J. Chem.* **2000**, *24*, 127–135. [CrossRef]
79. Tsuji, J. Remarkable advances in palladium and nickel catalyzed reactions: Increasing impact on organic synthesis. *J. Synth. Org. Chem.* **2001**, *59*, 607–616. [CrossRef]
80. Tsuji, J. Dawn of organopalladium chemistry in the early 1960s and a retrospective overview of the research on palladium-catalyzed reactions. *Tetrahedron* **2015**, *71*, 6330–6348. [CrossRef]
81. Martinez, A.D.; Kid, B.J. A Tribute to Prof. Dr. Lavinel G. Ionescu on His 70th Birthday. *Periódico Tchê Química* **2017**, *14*, 144–161.
82. Beloglazkina, E.K.; Bogatova, T.V.; Nenajdenko, V.G. Nikolay Zelinsky (1861–1953): Mendeleev's Protege, a Brilliant Scientist, and the Top Soviet Chemist of the Stalin Era. *Angew. Chem. Int. Ed.* **2020**, *59*, 20744–20752. [CrossRef] [PubMed]
83. Bogatova, T.V.; Beloglazkina, E.K.; Nenaidenko, V.G.N.D. Zelinskii: To the 160th Anniversary of Birth. *Russ. J. Org. Chem.* **2021**, *57*, 1191–1211. [CrossRef]
84. Somorjai, G.A. Heinz Heinemann. The Berkeley Years (1978–1993). *Catal. Lett.* **2009**, *133*, 232–233. [CrossRef]
85. Couderc, F.; Ong-Meang, V. Paul Sabatier and Father Jean Baptiste Senderens, distant witnesses of a "French positive secularism". *Comptes Rendus Chim.* **2011**, *14*, 516–523. [CrossRef]
86. Rulev, A.Y.; Ponomarev, D.A. Mikhail Kucherov: "The Experiment Confirmed my Hypothesis". *Angew. Chem. Int. Ed.* **2019**, *58*, 7914–7920. [CrossRef] [PubMed]
87. Cervellati, R. William Crowell Bray and the discovery of the first periodic homogeneous reaction in 1921. *React. Kinet. Mech. Catal.* **2022**, *135*, 1139–1146. [CrossRef]
88. Schmitz, G. Historical overview of the oscillating reactions. Contribution of Professor Slobodan Anic. *React. Kinet. Mech. Catal.* **2016**, *118*, 5–13. [CrossRef]
89. Travis, A.S. Luigi Casale's enterprise: Pioneer of global catalytic high-pressure industrial chemistry. *Catal. Today* **2022**, *387*, 4–8. [CrossRef]
90. Tsuji, J. Memory of Professor Gilbert Stork. *J. Synth. Org. Chem.* **2021**, *79*, 157–159. [CrossRef]
91. Tsuji, J. Contributions of Stork, Gilbert in Heterocyclic Chemistry. *Heterocycles* **1987**, *25*, 1–6. [CrossRef]
92. Kataoka, H.; Yamada, T.; Goto, K.; Tsuji, J. An efficiemt synthetic method of methyl (±)-jasmonate. *Tetrahedron* **1987**, *43*, 4107–4112. [CrossRef]
93. Tsuji, J.; Takahashi, H.; Morikawa, M. Organic syntheses by means of noble metal compounds XVII. Reaction of π-allylpalladium chloride with nucleophiles. *Tetrahedron Lett.* **1965**, *6*, 4387–4388. [CrossRef]
94. Trost, B.M.; Fullerton, T.J. New synthetic reactions. Allylic alkylation. *J. Am. Chem. Soc.* **1973**, *95*, 292–294. [CrossRef]
95. Tsuji, J. Synthetic Applications of The Palladium-Catalyzed Oxidation of Olefins to Ketones. *Synthesis-Stuttgart* **1984**, *1984*, 369–384. [CrossRef]
96. Tsuji, J. New General Synthetic Methods Involving Pi-Allylpalladium Complexes as Intermediates and Neutral Reaction Conditions. *Tetrahedron* **1986**, *42*, 4361–4401. [CrossRef]
97. Tsuji, J.; Minami, I. New Synthetic Reactions of Allyl Alkyl Carbonates, Allyl Beta-Keto Carboxylates, and Allyl Vinylic Carbonates Catalyzed by Palladium Complexes. *Acc. Chem. Res.* **1987**, *20*, 140–145. [CrossRef]
98. Tsuji, J.; Mandai, T. Palladium-catalyzed reactions of propargylic compounds in organic synthesis. *Angew. Chem. Int. Ed.* **1995**, *34*, 2589–2612. [CrossRef]
99. Tsuji, J.; Kataoka, H.; Kobayashi, Y. Regioselective 1,4-Addition of Nucleophiles to 1,3-Diene Mono-Epoxides Catalyzed by Palladium Complex. *Tetrahedron Lett.* **1981**, *22*, 2575–2578. [CrossRef]
100. Shimizu, I.; Yamada, T.; Tsuji, J. Palladium-Catalyzed Rearrangement of Allylic Esters of Acetoacetic Acid to Give Gamma,Delta-Unsaturated Methyl Ketones. *Tetrahedron Lett.* **1980**, *21*, 3199–3202. [CrossRef]
101. Tsuji, J.; Shimizu, I.; Yamamoto, K. Convenient General Synthetic Method for 1,4- and 1,5-Diketones by Palladium Catalyzed Oxidation of Alpha-Allyl and Alpha-3-Butenyl Ketones. *Tetrahedron Lett.* **1976**, *17*, 2975–2976. [CrossRef]
102. Minato, M.; Yamamoto, K.; Tsuji, J. Osmium Tetraoxide Catalyzed Vicinal Hydroxylation of Higher Olefins by Using Hexacyanoferrate(Iii) Ion as A Cooxidant. *J. Org. Chem.* **1990**, *55*, 766–768. [CrossRef]
103. Minami, I.; Shimizu, I.; Tsuji, J. Reactions of Allylic Carbonates Catalyzed by Palladium, Rhodium, Ruthenium, Molybdenum, And Nickel-Complexes—Allylation of Carbonucleophiles and Decarboxylation-Dehydrogenation. *J. Organomet. Chem.* **1985**, *296*, 269–280. [CrossRef]
104. Tsuji, J.; Takayanagi, H. Organic Synthesis by Means of Metal-Complexes. XIII. Efficient, Nonenzymatic Oxidation of Catechol with Molecular-Oxygen Activated by Cuprous Chloride to CIS, CIS-Muconate as Model Reaction For Pyrocatechase. *J. Am. Chem. Soc.* **1974**, *96*, 7349–7350. [CrossRef]

105. Tsuji, J.; Minami, I.; Shimizu, I. Allyation of Carbonucleophiles with Allylic Carbonates under Neutral Conditions Catalyzed by Rhodium Complexes. *Tetrahedron Lett.* **1984**, *25*, 5157–5160. [CrossRef]
106. Yamamoto, K.; Yoshitake, J.; Qui, N.T.; Tsuji, J. Simple Synthesis of Alpha,Beta-Unsaturated Aldehydes by Reaction of Vinylsilanes with Dichloromethyl Methyl-Ether Promoted by Titanium(Iv) Chloride—Application to Synthesis of Ethyl 12-Oxo-10(E)-Dodecenoate And Nuciferal. *Chem. Lett.* **1978**, 859–862. [CrossRef]
107. Tsuji, J.; Yamada, T.; Kaito, M.; Mandai, T. Efficient Regioselective Aldol Condensation of Methyl Ketones Promoted by Organo-Aluminum Compounds, and Its Application to Muscone Synthesis. *Bull. Chem. Soc. Jpn.* **1980**, *53*, 1417–1420. [CrossRef]
108. Tsuji, J.; Takayanagi, H. Cuprous Ion-Catalyzed Oxidative Cleavage of Aromatic Ortho-Diamines by Oxygen—(Z,Z)-2,4-Hexadienedinitrile. *Org. Synth.* **1988**, *50*, 662–663.
109. Nagashima, H.; Seki, K.; Ozaki, N.; Wakamatsu, H.; Itoh, K.; Tomo, Y.; Tsuji, J. Transition-Metal-Catalyzed Radical Cyclization—Copper-Catalyzed Cyclization of Allyl Trichloroacetates to Trichlorinated Gamma-Lactones. *J. Org. Chem.* **1990**, *55*, 985–990. [CrossRef]
110. Tsuji, J.; Sakai, K.; Nemoto, H.; Nagashima, H. Iron and Copper Catalyzed Reaction of Benzylamine with Carbon-Tetrachloride—Facile Formation of 2,4,5-Triphenylimidazoline Derivatives. *J. Mol. Catal.* **1983**, *18*, 169–176. [CrossRef]
111. Minato, M.; Tsuji, J. Allylation of Aldehydes in an Aqueous 2-Phase System by Electrochemically Regenerated Bismuth Metal. *Chem. Lett.* **1988**, *17*, 2049–2052. [CrossRef]
112. Nishii, Y.; Miura, M. Construction of Benzo-Fused Polycyclic Heteroaromatic Compounds through Palladium-Catalyzed Intramolecular C-H/C-H Biaryl Coupling. *Catalysts* **2023**, *13*, 12. [CrossRef]
113. Kobayashi, Y. Coupling Reactions on Secondary Allylic, Propargylic, and Alkyl Carbons Using Organoborates/Ni and RMgX/Cu Reagents. *Catalysts* **2023**, *13*, 132. [CrossRef]
114. Wang, X.; Lu, M.-Z.; Loh, T.-P. Transition-Metal-Catalyzed C–C Bond Macrocyclization via Intramolecular C–H Bond Activation. *Catalysts* **2023**, *13*, 438. [CrossRef]
115. Sieger, S.V.; Lubins, I.; Breit, B. Rhodium-Catalyzed Dynamic Kinetic Resolution of Racemic Internal Allenes towards Chiral Allylated Triazoles and Tetrazoles. *Catalysts* **2022**, *12*, 1209. [CrossRef]
116. Cusumano, A.Q.; Zhang, T.; Goddard, W.A.; Stoltz, B.M. Origins of Enhanced Enantioselectivity in the Pd-Catalyzed Decarboxylative Allylic Alkylation of N-Benzoyl Lactams. *Catalysts* **2023**, *13*, 1258. [CrossRef] [PubMed]
117. Kumashiro, M.; Ohsawa, K.; Doi, T. Photocatalyzed Oxidative Decarboxylation Forming Aminovinylcysteine Containing Peptides. *Catalysts* **2022**, *12*, 1615. [CrossRef]
118. Hu, J.; Wan, H.; Wang, S.; Yi, H.; Lei, A. Electrochemical Thiocyanation/Cyclization Cascade to Access Thiocyanato-Containing Benzoxazines. *Catalysts* **2023**, *13*, 631. [CrossRef]
119. Bao, Z.; Wu, C.; Wang, J. Palladium-Catalyzed Three-Component Coupling of Benzynes, Benzylic/Allylic Bromides and 1,1-Bis[(pinacolato)boryl]methane. *Catalysts* **2023**, *13*, 126. [CrossRef]
120. Ito, K.; Doi, T.; Tsukamoto, H. De Novo Synthesis of Polysubstituted 3-Hydroxypyridines Via "Anti-Wacker"-Type Cyclization. *Catalysts* **2023**, *13*, 319. [CrossRef]
121. Ostrowska, S.; Palio, L.; Czapik, A.; Bhandary, S.; Kwit, M.; Van Hecke, K.; Nolan, S.P. A Second-Generation Palladacycle Architecture Bearing a N-Heterocyclic Carbene and Its Catalytic Behavior in Buchwald–Hartwig Amination Catalysis. *Catalysts* **2023**, *13*, 559. [CrossRef]
122. Kobayashi, Y.; Hirotsu, T.; Haimoto, Y.; Ogawa, N. Substitution of Secondary Propargylic Phosphates Using Aryl-Lithium-Based Copper Reagents. *Catalysts* **2023**, *13*, 1084. [CrossRef]

Disclaimer/Publisher's Note: The statements, opinions and data contained in all publications are solely those of the individual author(s) and contributor(s) and not of MDPI and/or the editor(s). MDPI and/or the editor(s) disclaim responsibility for any injury to people or property resulting from any ideas, methods, instructions or products referred to in the content.

Communication

Origins of Enhanced Enantioselectivity in the Pd-Catalyzed Decarboxylative Allylic Alkylation of *N*-Benzoyl Lactams [†]

Alexander Q. Cusumano [1], Tianyi Zhang [1], William A. Goddard III [2,*] and Brian M. Stoltz [1,*]

[1] The Warren and Katharine Schlinger Laboratory for Chemistry and Chemical Engineering, Division of Chemistry and Chemical Engineering, California Institute of Technology, Pasadena, CA 91125, USA; alex.cusumano@gmail.com (A.Q.C.); tz6009@princeton.edu (T.Z.)
[2] Materials and Process Simulation Center, Beckman Institute, California Institute of Technology, Pasadena, CA 91125, USA
* Correspondence: wag@caltech.edu (W.A.G.III); stoltz@caltech.edu (B.M.S.)
[†] Dedicated to the memory of the late Professor Jiro Tsuji (1927–2022), a true pioneer, scholar, scientist, and friend.

Abstract: We explore the origins of the marked improvement in enantioselectivity in the inner-sphere (PHOX)Pd-catalyzed allylic alkylation of *N*-benzoyl lactam nucleophiles over their carbocyclic counterparts. We employ density functional theory calculations to aid in the interpretation of experimental results. Ultimately, we propose that the enhancement in enantioselectivity arises primarily from noncovalent interactions between the substrate and ligand rather than secondary substrate chelation, as previously hypothesized.

Keywords: $C(sp^3)$–$C(sp^3)$ cross-coupling; asymmetric catalysis; computation; allylic alkylation

Citation: Cusumano, A.Q.; Zhang, T.; Goddard, W.A., III; Stoltz, B.M. Origins of Enhanced Enantioselectivity in the Pd-Catalyzed Decarboxylative Allylic Alkylation of *N*-Benzoyl Lactams. *Catalysts* **2023**, *13*, 1258. https://doi.org/10.3390/catal13091258

Academic Editors: Ewa Kowalska and Yuichi Kobayashi

Received: 20 July 2023
Revised: 22 August 2023
Accepted: 25 August 2023
Published: 30 August 2023

Copyright: © 2023 by the authors. Licensee MDPI, Basel, Switzerland. This article is an open access article distributed under the terms and conditions of the Creative Commons Attribution (CC BY) license (https://creativecommons.org/licenses/by/4.0/).

1. Introduction

The Pd-catalyzed decarboxylative asymmetric allylic alkylation of *hard* enolate nucleophiles is a proven tactic for the formation of all-carbon quaternary stereogenic centers [1,2]. Employing chiral *tert*-butyl phosphinooxazoline (*t*-BuPHOX) ligands renders the transformation asymmetric, with an enantiodetermining inner-sphere reductive elimination [3–9] (Figure 1A). Despite extensive ligand optimization efforts, enantioenrichment of carbocyclic ketone products (**2**), derived from b-ketoesters (**1**) or enol carbonates, are generally limited to 80–90% ee. In contrast, *N*-benzoyl lactams (**3**) undergo the analogous transformation with markedly higher levels of enantioselectivity, often ≥99% ee (**4**) [10]. This represents a substantial increase in the effective difference in barrier height between diastereomeric enantiodetermining transition states ($\Delta\Delta G^{\ddagger}$), from *ca.* 1.6 to >3.3 kcal/mol (Effective $\Delta\Delta G^{\ddagger}$ calculated from Eyring equation at 40 °C. Note the effective $\Delta\Delta G^{\ddagger}$ may not directly correspond to the free energy difference between only the two lowest diastereomeric transition states if multiple low energy conformeric transition states are present) (Figure 1B). Compared to their carbocyclic counterparts, the lactam substrate class would afford a potentially more electron-rich Pd-enolate, which may serve to reinforce a highly selective inner-sphere mechanism over a poorly selective outer-sphere process [7]. However, we also posited that the presence of an adjacent Lewis basic carbonyl group may enable additional interactions with the metal center. To independently examine each of these variables, we explored the a- and b-enaminone substrate classes (Figure 1C). Of note, a-enaminones (**5**) with a-heteroatom chelating groups retain the high levels of enantioselectivity of the *N*-benzoyl lactam substrate class [11], while b-enaminones (**6**) featuring more electron-rich enolates but lacking the ability to engage in hypothesized secondary interactions afford products in <90% ee [12]. These results suggest that the a-heteroatom-containing fragment of the substrate appears to play a key part in improving enantioselectivity. Here, we employ computational tools to elucidate this role, ultimately deepening our understanding of the origins of enantioselectivity in the inner-sphere allylic alkylation reaction.

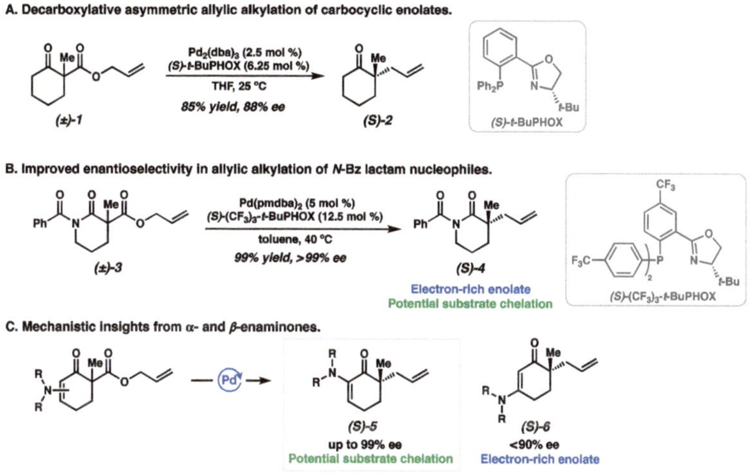

Figure 1. (**A**) The decarboxylative asymmetric allylic alkylation of cyclic ketone nucleophiles. (**B**) Allylic alkylation of *N*-benzoyl lactams. (**C**) Mechanistic insights from a- and b-enaminone substrate classes.

2. Results and Discussion

The enantiodetermining C–C bond formation in the (PHOX)Pd-catalyzed asymmetric allylic alkylation occurs via a seven-centered pericyclic transition state [7,13]. Enantioselectivity arises from preferential exposure of the *Re* face of the prochiral enolate ligand to the h^1-allyl terminus (**TS1**) (Figure 2A). Bond formation from the *Si* face (**TS2**) is disfavored due to steric incursions between the carbocyclic enolate backbone and the ligand scaffold. The dramatic improvement in enantioenrichment of *N*-benzoyl lactam products suggests an enhanced favorability of the analogous *Re* transition states over their *Si* counterparts. As similar levels of enantioselectivity were observed with a-enaminones (Figure 1C), we posited that such a-heteroatom-containing motifs may reinforce the *Re* facial preference through axial chelation with the Pd^{II} center in the reductive elimination transition state (Figure 2B, right). To further probe this hypothesis, we turned to computations.

Beginning with enolate **7**, derived from carbocyclic substrate **1**, we find a 2.0 kcal/mol preference for **TS1** over **TS2** at the revDOD-PBEP86-NL/def2-TZVPP/SMD(PhMe)//r^{2}SCAN-D4/def2-TZVP[Pd], def2-SVP level of theory—in accord with our prior studies (Figure 3) [7]. Maintaining a similar steric profile while perturbing enolate electronics with N–H lactam-derived enolate **8** did not significantly alter $\Delta\Delta G^‡$. Accounting for distribution across all conformers, enantiomeric excesses of 89% and 90% are computed, respectively (experimentally, the corresponding N–H lactam is not compatible in the transformation).

We then explored the effect of *N*-substitution on the relative free energies. *N*-benzoyl substitution affords two low-energy transition states from the favored *Si* face—one conformer with the flanking carbonyl of the benzoyl group oriented away from (**TS5**) and another toward (**TS6**) the metal center (Figure 4). **TS6** is reminiscent of our chelating heteroatom hypothesis (Figure 2B). However, **TS5** is computed to be favored over **TS6** by 5.2 and 4.7 kcal/mol with (S)-*t*-BuPHOX and (S)-(CF$_3$)$_3$-*t*-BuPHOX ligands, respectively. With regard to enantioselectivity, **TS5** is favored over the lowest energy *Si* face transition states (**TS7**) by 2.3 and 3.2 kcal/mol with the (S)-*t*-BuPHOX and (S)-(CF$_3$)$_3$-*t*-BuPHOX ligands, respectively. Computed enantiomeric excesses of 95% and 99% are found when accounting for all transition state conformers. The additional increase in $\Delta\Delta G^‡$ of 0.9 kcal/mol with incorporation of *p*-CF$_3$ groups may arise from increasing favorable electrostatic interactions between the benzoyl and PHOX ligand arene quadrupoles in **TS5b**. We note the electron-poor (S)-(CF$_3$)$_3$-t-BuPHOX ligand is also crucial in promoting the inner-sphere mechanism discussed herein over less selective outer-sphere pathways.

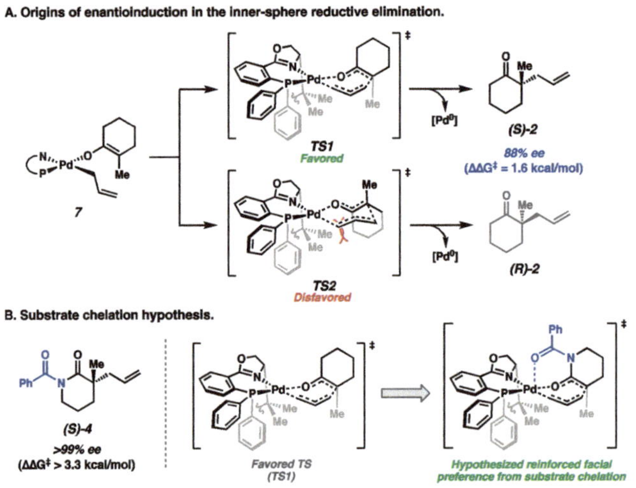

Figure 2. (**A**) Enantioinduction via inner-sphere reductive elimination. (**B**) Initial hypothesis for enhanced enantioenrichment of *N*-benzoyl lactam substrates as compared to cyclohexanones.

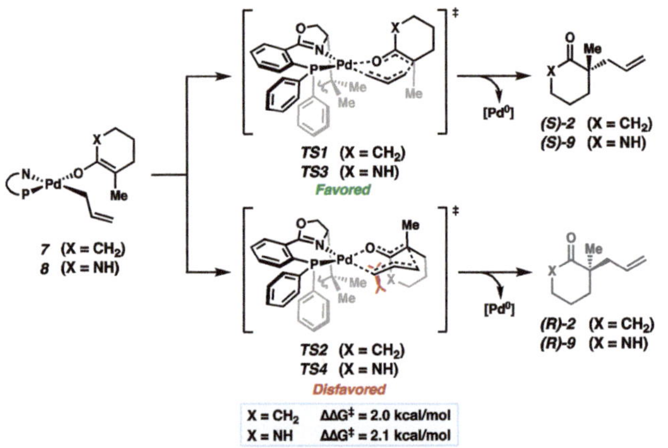

Figure 3. Minor effect of alteration in enolate electronics.

While the computed trends in enantiomeric excess are in accord with experimental values, the energetic preference for **TS5** over **TS6** mandates a re-evaluation of our initial hypothesis regarding axial chelation. **TS6a** and **TS6b** feature axial Pd–O distances of 2.64 and 2.63 Å (compared to equatorial Pd–O distances of 2.20 and 2.19 Å), highlighting the lack of strong axial binding of the carbonyl oxygen. While the s-donating oxygen lone pair is repelled by the occupied axially-oriented 4d(z^2) orbital of the d^8 Pd^{II} center, mixing with the empty 5p(z) orbital may contribute to a partial s bonding interaction (for further discussion on such 3-center 4-electron bonding arrays in d^8 complexes, see [14]). Geometric constraints inhibit π bonding interactions with the carbonyl group. We also suspect an electrostatic contribution to the weak Pd–O axial binding. However, four-coordinate **TS5a** and **TS5b** are the favored conformation of the *Si* transition states by a considerable margin.

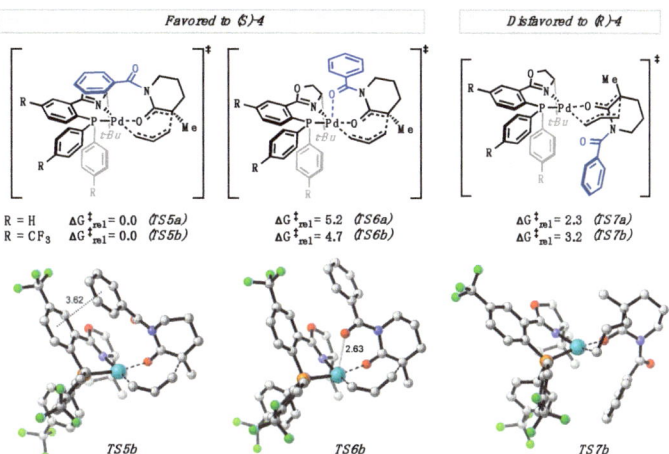

Figure 4. C–C bond-forming transition states for the *N*-benzoyl lactam substrates.

A similar trend is observed with *N*-acetyl lactam transition states (**TS5c** and **TS6c**), highlighting that the preference for four-coordinate transition states is not a conformational artifact of the benzoyl arene (Figure 5). Additionally, *N*-carbamate groups (Boc, CBz, and Fmoc) that are more Lewis basic lead to reduced enantioselectivities of 73–87% ee. Hence, axial chelation to the square planar Pd center in the reductive elimination does not appear to enhance enantioselectivity.

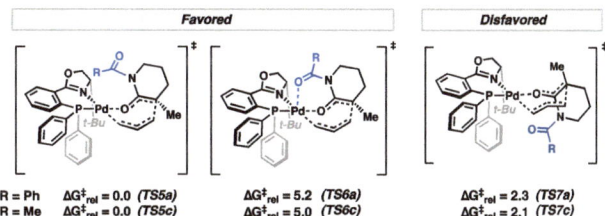

Figure 5. Comparison of relative barriers between diastereomeric C–C bond-forming transition states in both *N*-Bz and *N*-Ac lactams.

In lieu of secondary substrate chelation to Pd, we posit that the success of the benzoyl group lies in its ability to adopt a parallel orientation to the open face of the PHOX backbone in **TS5**. However, the *tert*-butyl group occupies this site in the *Si* transition state (**TS7**). Hence, this low-energy orientation is not accessible, and a large energetic penalty is incurred in C–C bond formation from the *Si* face. The results presented herein suggest that the improved enantioselectivity observed in the *N*-Bz class of substrates is noncovalent in nature. While no evidence of secondary substrate chelation is found for *N*-Bz lactams, such interactions may prevail in other substrate classes. The detailed investigation of these systems will be reported in due course.

3. Computational Details

All quantum mechanics calculations were carried out with the ORCA program (version 5) [15]. The r^{2}SCAN functional [16] paired with D4 dispersion corrections [17], henceforth referred to as r^{2}SCAN-D4, was employed for geometry optimizations and harmonic frequency calculations. Similar geometries were obtained across a variety of density functionals. For geometry optimization and harmonic frequency calculations, Pd is described by the def2-TZVP basis set [18] and the ECP28MWB small-core (18 explicit valence electrons) quasi-relativistic pseudopotential [19], while C, H, and P are assigned the def2-SVP basis. Diffuse functions are added to O, N, and F (ma-def2-SVP). All Hessians were computed analytically. Stationary points are characterized by the correct number of imaginary vibrational modes (zero for minima and one for saddle points). Cartesian coordinates of all optimized structures are included as ".xyz" files and are available online in a compressed zip file format (see Supplementary Materials).

Electronic energies are further refined with single-point calculations employing the revDOD-PBEP86-NL double hybrid functional with non-local dispersion corrections [20] and the def2-TZVPP basis set on all atoms (with the ECP28MWB pseudopotential for Pd) with additional diffuse functions on O, N, and F (ma-def2-TZVPP). Solvation was accounted for with the SMD solvation model for toluene. Similar results were obtained from single-point calculations employing the range-separated hybrid wB97M-V functional [21]. To check for basis set superposition error, single-point calculations of select transition states (**TS5**, **TS6**, and **TS7**) were carried out with revDOD-PBEP86-NL and wB97M-V functionals paired with the quadruple-z quality (ma-)def2-QZVPP basis set. Similar results are obtained; hence, we recommend the more computationally tractable triple-z quality basis set for this application. Final Gibbs free energies were obtained by applying thermodynamic corrections obtained at the optimization level of theory to these refined electronic energies. Thermodynamic corrections from harmonic frequency calculations employ the quasi-rigid rotor harmonic oscillator approach to correct the breakdown of the harmonic oscillator approximation at low vibrational frequencies [22].

All stereochemical perturbations (*Re/Si*, chair/boat, axial/equatorial) and conformations (carbonyl distal, carbonyl proximal) are considered for each reaction pathway. Computed enantiomeric excess accounts for contributions from all considered transition states weighted by their final relative Gibbs free energies at 40 °C. All quantum mechanical data are included online in the supplementary Excel file (see Supplementary Materials).

Supplementary Materials: The following supporting information can be downloaded at: https://www.mdpi.com/article/10.3390/catal13091258/s1, all computed quantum mechanics energies and coordinates of computed transition states.

Author Contributions: Conceptualization, computations, manuscript preparation, A.Q.C.; Conceptualization, computations, editing, T.Z.; conceptualization, supervision, editing, B.M.S. and W.A.G.III. All authors have read and agreed to the published version of the manuscript.

Funding: The NIH-NIGMS (R35GM145239), Heritage Medical Research Investigators Program, and Caltech are thanked for their support of our research program. We thank the Caltech High Performance Computing Center for support. W.A.G. thanks the NSF (CBET-2005250) for support.

Data Availability Statement: The authors confirm that the data supporting the findings of this study are available within the article and its Supplementary Materials.

Acknowledgments: We further thank Zachary P. Sercel for insightful discussion and assistance.

Conflicts of Interest: The authors declare no conflict of interest.

References

1. Liu, Y.; Han, S.-J.; Liu, W.-B.; Stoltz, B.M. Catalytic Enantioselective Construction of Quaternary Stereocenters: Assembly of Key Building Blocks for the Synthesis of Biologically Active Molecules. *Acc. Chem. Res.* **2015**, *48*, 740–751. [CrossRef]
2. Trost, B.M.; Crawley, M.L. Asymmetric Transition-Metal-Catalyzed Allylic Alkylations: Applications in Total Synthesis. *Chem. Rev.* **2003**, *103*, 2921–2944. [CrossRef]
3. Behenna, D.C.; Stoltz, B.M. The Enantioselective Tsuji Allylation. *J. Am. Chem. Soc.* **2004**, *126*, 15044–15045. [CrossRef]
4. Mohr, J.T.; Behenna, D.C.; Harned, A.M.; Stoltz, B.M. Deracemization of Quaternary Stereocenters by Pd-Catalyzed Enantioconvergent Decarboxylative Allylation of Racemic β-Ketoesters. *Angew. Chem. Int. Ed.* **2005**, *44*, 6924–6927. [CrossRef]
5. Keith, J.A.; Behenna, D.C.; Mohr, J.T.; Ma, S.; Marinescu, S.C.; Oxgaard, J.; Stoltz, B.M.; Goddard, W.A. The Inner-Sphere Process in the Enantioselective Tsuji Allylation Reaction with (*S*)-*t*-Bu-Phosphinooxazoline Ligands. *J. Am. Chem. Soc.* **2007**, *129*, 11876–11877. [CrossRef]

6. Keith, J.A.; Behenna, D.C.; Sherden, N.; Mohr, J.T.; Ma, S.; Marinescu, S.C.; Nielsen, R.J.; Oxgaard, J.; Stoltz, B.M.; Goddard, W.A. The Reaction Mechanism of the Enantioselective Tsuji Allylation: Inner-Sphere and Outer-Sphere Pathways, Internal Rearrangements, and Asymmetric C–C Bond Formation. *J. Am. Chem. Soc.* **2012**, *134*, 19050–19060. [CrossRef]
7. Cusumano, A.Q.; Stoltz, B.M.; Goddard, W.A. Reaction Mechanism, Origins of Enantioselectivity, and Reactivity Trends in Asymmetric Allylic Alkylation: A Comprehensive Quantum Mechanics Investigation of a $C(sp^3)$–$C(sp^3)$ Cross-Coupling. *J. Am. Chem. Soc.* **2020**, *142*, 13917–13933. [CrossRef]
8. McPherson, K.E.; Croatt, M.P.; Morehead, A.T.; Sargent, A.L. DFT Mechanistic Investigation of an Enantioselective Tsuji–Trost Allylation Reaction. *Organometallics* **2018**, *37*, 3791–3802. [CrossRef]
9. Sherden, N.H.; Behenna, D.C.; Virgil, S.C.; Stoltz, B.M. Unusual Allylpalladium Carboxylate Complexes: Identification of the Resting State of Catalytic Enantioselective Decarboxylative Allylic Alkylation Reactions of Ketones. *Angew. Chem. Int. Ed.* **2009**, *48*, 6840–6843. [CrossRef]
10. Behenna, D.C.; Liu, Y.; Yurino, T.; Kim, J.; White, D.E.; Virgil, S.C.; Stoltz, B.M. Enantioselective Construction of Quaternary N-Heterocycles by Palladium-Catalysed Decarboxylative Allylic Alkylation of Lactams. *Nat. Chem.* **2012**, *4*, 130–133. [CrossRef]
11. Duquette, D.C.; Cusumano, A.Q.; Lefoulon, L.; Moore, J.T.; Stoltz, B.M. Probing Trends in Enantioinduction via Substrate Design: Palladium-Catalyzed Decarboxylative Allylic Alkylation of α-Enaminones. *Org. Lett.* **2020**, *22*, 4966–4969. [CrossRef]
12. Bennett, N.B.; Duquette, D.C.; Kim, J.; Liu, W.-B.; Marziale, A.N.; Behenna, D.C.; Virgil, S.C.; Stoltz, B.M. Expanding Insight into Asymmetric Palladium-Catalyzed Allylic Alkylation of N-Heterocyclic Molecules and Cyclic Ketones. *Chem. Eur. J.* **2013**, *19*, 4414–4418. [CrossRef]
13. Cusumano, A.Q.; Goddard, W.A.I.; Stoltz, B.M. The Transition Metal Catalyzed [π2s + π2s + σ2s + σ2s] Pericyclic Reaction: Woodward–Hoffmann Rules, Aromaticity, and Electron Flow. *J. Am. Chem. Soc.* **2020**, *142*, 19033–19039. [CrossRef]
14. Aullón, G.; Alvarez, S. Axial Bonding Capabilities of Square Planar d^8-ML_4 Complexes. Theoretical Study and Structural Correlations. *Inorg. Chem.* **1996**, *35*, 3137–3144. [CrossRef]
15. Neese, F. Software Update: The ORCA Program System—Version 5.0. *Wiley Interdiscip. Rev. Comput. Mol. Sci.* **2022**, *12*, e1606. [CrossRef]
16. Ehlert, S.; Huniar, U.; Ning, J.; Furness, J.W.; Sun, J.; Kaplan, A.D.; Perdew, J.P.; Brandenburg, J.G. r2SCAN-D4: Dispersion Corrected Meta-Generalized Gradient Approximation for General Chemical Applications. *J. Chem. Phys.* **2021**, *154*, 061101. [CrossRef]
17. Caldeweyher, E.; Ehlert, S.; Hansen, A.; Neugebauer, H.; Spicher, S.; Bannwarth, C.; Grimme, S. A Generally Applicable Atomic-Charge Dependent London Dispersion Correction. *J. Chem. Phys.* **2019**, *150*, 154122. [CrossRef]
18. Weigend, F.; Ahlrichs, R. Balanced Basis Sets of Split Valence, Triple Zeta Valence and Quadruple Zeta Valence Quality for H to Rn: Design and Assessment of Accuracy. *Phys. Chem. Chem. Phys.* **2005**, *7*, 3297–3305. [CrossRef]
19. Peterson, K.A.; Figgen, D.; Goll, E.; Stoll, H.; Dolg, M. Systematically Convergent Basis Sets with Relativistic Pseudopotentials. II. Small-Core Pseudopotentials and Correlation Consistent Basis Sets for the Post-*d* Group 16–18 Elements. *J. Chem. Phys.* **2003**, *119*, 11113–11123. [CrossRef]
20. Santra, G.; Sylvetsky, N.; Martin, J.M.L. Minimally Empirical Double-Hybrid Functionals Trained against the GMTKN55 Database: revDSD-PBEP86-D4, revDOD-PBE-D4, and DOD-SCAN-D4. *J. Phys. Chem. A* **2019**, *123*, 5129–5143. [CrossRef]
21. Mardirossian, N.; Head-Gordon, M. ωB97M-V: A Combinatorially Optimized, Range-Separated Hybrid, Meta-GGA Density Functional with VV10 Nonlocal Correlation. *J. Chem. Phys.* **2016**, *144*, 214110. [CrossRef] [PubMed]
22. Grimme, S. Supramolecular Binding Thermodynamics by Dispersion-Corrected Density Functional Theory. *Chem. Eur. J.* **2012**, *18*, 9955–9964. [CrossRef]

Disclaimer/Publisher's Note: The statements, opinions and data contained in all publications are solely those of the individual author(s) and contributor(s) and not of MDPI and/or the editor(s). MDPI and/or the editor(s) disclaim responsibility for any injury to people or property resulting from any ideas, methods, instructions or products referred to in the content.

Article

Substitution of Secondary Propargylic Phosphates Using Aryl-Lithium-Based Copper Reagents †

Yuichi Kobayashi [1,2,*], Takayuki Hirotsu [1], Yosuke Haimoto [3] and Narihito Ogawa [3]

[1] Department of Bioengineering, Tokyo Institute of Technology, 4259 Nagatsuta-cho, Midori-ku, Yokohama 226-8501, Japan
[2] Organization for the Strategic Coordination of Research and Intellectual Properties, Meiji University, 1-1-1 Higashimita, Tama-ku, Kawasaki 214-8571, Japan
[3] Department of Applied Chemistry, Meiji University, 1-1-1 Higashimita, Tama-ku, Kawasaki 214-8571, Japan
* Correspondence: ykobayas@bio.titech.ac.jp
† This paper is dedicated to late Professor Jiro Tsuji.

Abstract: The substitution of secondary propargylic phosphates ROP(O)(OEt)$_2$ where R = [Ph(CH$_2$)$_2$]C(H)(C≡CTMS)] Ph(CH$_2$)$_2$CH(OP(O)(OEt)$_2$)(C≡CTMS) with copper reagents derived from PhLi and copper salts such as CuCl, CuCN, and Cu(acac)$_2$ was studied to establish an ArLi-based reagent system. Among the reagents prepared, PhLi/CuCl (2:1) showed 98% α regioselectivity (rs), while PhLi/Cu(acac)$_2$ was γ selective (>99% rs). PhLi prepared in situ from PhI and PhBr by Li-halogen exchange with *t*-BuLi was also used for the α selective substitution. A study using the (*S*)-phosphate disclosed 99% enantiospecificity (es) and the inversion of the stereochemistry. The substitution of five phosphates with substituted aryl reagents produced the corresponding propargylic products with high rs and es values. Similar reactivity and selectivity were observed with 2-furyl and 2-thienyl reagents, which were prepared via direct lithiation with *n*-BuLi.

Keywords: aryl lithium; substitution; coupling; secondary carbon; propargylic; regioselectivity; enantiospecificity; copper

Citation: Kobayashi, Y.; Hirotsu, T.; Haimoto, Y.; Ogawa, N. Substitution of Secondary Propargylic Phosphates Using Aryl-Lithium-Based Copper Reagents. *Catalysts* **2023**, *13*, 1084. https://doi.org/10.3390/catal13071084

Academic Editor: Victorio Cadierno

Received: 16 June 2023
Revised: 4 July 2023
Accepted: 7 July 2023
Published: 10 July 2023

Copyright: © 2023 by the authors. Licensee MDPI, Basel, Switzerland. This article is an open access article distributed under the terms and conditions of the Creative Commons Attribution (CC BY) license (https://creativecommons.org/licenses/by/4.0/).

1. Introduction

Coupling reaction of secondary alcohol derivatives with organometallic reagents is a promising method to construct chiral centers on secondary carbons. Thus far, alkyl, allylic, and propargylic alcohol derivatives have been investigated as substrates [1–3]. Among them, alkyl substrates are less reactive than the other substrates that are activated by the double or triple bond. Consequently, only a few reagent/catalyst/leaving group systems have been published [4–7]. Among these, the highly reactive system developed by us consists of the PySO$_3$ leaving group, Grignard reagents, and Cu(OTf)$_2$ as a catalyst [7]. In contrast, various reagents/catalyst systems have been developed for the allylic coupling at secondary carbons [8–10]. For organic synthesis, regiocontrol between α and γ carbons and stereocontrol in it, as well as convenience in reagent/catalyst preparation, are highly important. The copper-catalyzed allylic substitution of allylic picolinates with alkyl and aryl Grignard reagents developed by us meets these requirements and proceeds with high S$_N$2' selectivity, furnishing tertiary carbons [11,12] and quaternary carbons [13,14].

The α-substitution using enantioenriched secondary propargylic alcohol derivatives has also been studied to develop ammonium salt/ArMgBr/Cu cat. [15], salicylate/ArMgCl/Cu cat. [16], bromide/heteroaryl cuprate [17], and sulfonate/TMSCF$_3$/Cu cat./KF [18]. The α-substitution using racemic substrates was reported as well [19–22]. Furthermore, asymmetric version using chiral ligands was developed by Fu [23–25] and Nishibayashi [26,27]. Propargylic reactions have been studied by Professor Tsuji after his palladium-catalyzed reactions using allylic alcohol derivatives [28]. In consideration of his way of the research development and the experimental convenience mentioned above, we explored

Cu-catalyzed propargylic substitution with aryl Grignard reagents on the basis of our allylic substitution. We found the α selective substitution of phosphate **1** with ArMgBr by using CuCN and CuBr·Me$_2$S to give acetylenes **2** (Scheme 1) and the γ regioselective reaction with Cu(acac)$_2$ [29]. The substitution was successfully applied for the synthesis of biologically active compounds [29–32]. With these results in mind, we focused our attention on aryl lithium compounds (ArLi), with the expectation that several preparations of ArLi would expand the scope of the reagents. Herein, we present the results along this line (Scheme 1).

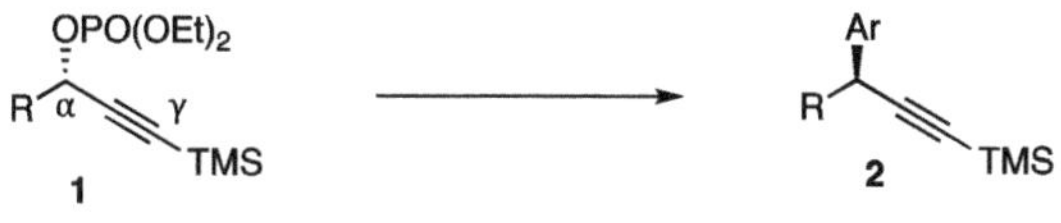

previous work	ArMgBr with CuCN or CuBr·Me$_2$S
present work	ArLi with CuCl

Scheme 1. Propargylic substitution with aryl copper reagents.

2. Results and Discussion

Racemic phosphates **1a–1c** and **7** and enantiomerically enriched phosphates (*S*)-**1**, -**7**, and -**8** were prepared according to the previous methods [29]. The substitution of racemic propargylic phosphate **1a** with copper reagents derived from PhLi and copper salts was studied first (Table 1). Ratios of acetylene **2aa**, the regioisomer **3aa**, and **1a** (if recovered) were determined by ^{1}H NMR spectroscopy and are summarized in entries. Alcohol **4a** was not formed. The reaction of **1a** with PhLi, without any copper salt, gave a mixture of unidentified products and recovered **1a** (Entry 1). The reaction with a phenylcopper reagent prepared from PhLi and CuCN (formal structure PhCu·LiCN) was slow, and a roughly 1:1 mixture of **2aa** and regioisomer **3aa** was produced after 17 h (Entry 2). In contrast, α selective reaction was realized with phenyl cuprate, and the desired product, **2aa**, was obtained with 94% regioselectivity (rs) (Entry 3), whereas the catalytic use of CuCN retarded the reaction. Moderate γ selectivity was observed using reagents derived with CuBr·Me$_2$S and with CuBr$_2$ (Entries 4 and 5). Phenyl copper derived from CuCl was also γ selective (Entry 6). Fortunately, phenyl cuprate derived from CuCl disclosed high α selectivity, which was higher than that achieved using the CuCN-derived phenyl cuprate (Entry 7 vs. Entry 3). Products **2aa** and **3aa**, in a 98:2 ratio, were isolated in a 76% yield after chromatography on silica gel (Entry 7).

The substitution conditions used in Entry 7 were applied to PhLi, which was prepared from iodo- and bromobenzene by Li-halogen exchange with *t*-BuLi. An X-shape flask [33] with two bottoms was used for our convenience. Briefly, CuCl was placed in one bottom, and lithiation of PhI in Et$_2$O was carried out in the other bottom at 0 °C for 30 min. The solution was diluted with THF and mixed with CuCl by tilting the flask. Racemic phosphate **1a** was added to the resulting reagent, and the mixture was stirred at 0 °C for 1 h to afford **2aa** in an 80% yield with 98% rs (Entry 8). Bromobenzene was converted to the phenyl reagent in a similar way, and **2aa** was obtained in a 73% yield with 97% rs (Entry 9). These results indicate that LiI and LiBr produced by the lithiation affected neither the reactivity nor the rs.

In addition, a reagent derived from PhLi, Cu(acac)$_2$, and MgBr$_2$ was found to be γ regioselective (Entry 10). Without MgBr$_2$, a mixture of unidentified products and the unreacted phosphate was obtained.

Table 1. Substitution of phosphate 1a with PhLi-based reagents.

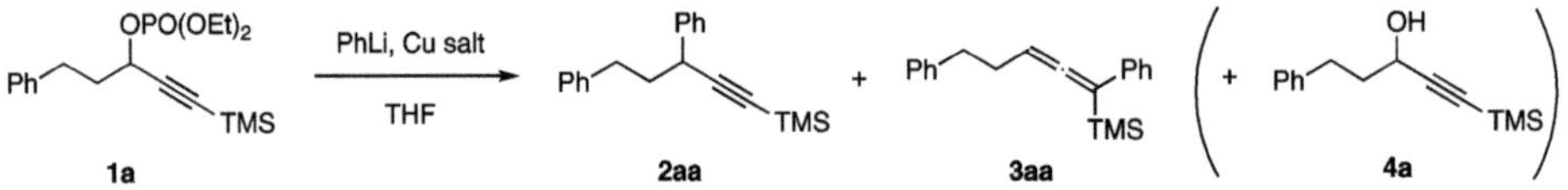

Entry	Ph Reagent	Cu Salt	PhLi [1]/Cu Salt	Temp.	Time	2aa/3aa/1a [2]	Yield (%)
1	PhLi	–	–	rt	6 h	– [3]	–
2	PhLi	CuCN	2.5:3	rt	17 h	44:56:0	nd [4]
3	PhLi	CuCN	3:1.5	rt	2 h	94:6:0	nd [4]
4	PhLi, MgBr$_2$ [5]	CuBr·Me$_2$S	3:3	0 °C	2 h	13:87:0	nd [4]
5	PhLi	CuBr$_2$	3:1.5	0 °C	2 h	7:45:48	nd [4]
6	PhLi	CuCl	2.5:3	0 °C	7 h	17:83:0	nd [4]
7	PhLi	CuCl	3:1.5	0 °C	2 h	98:2:0	76%
8	PhI, t-BuLi [6]	CuCl	3:1.5	0 °C	1 h	98:2:0	80%
9	PhBr, t-BuLi [6]	CuCl	3:1.5	0 °C	2 h	97:3:0	73%
10	PhLi, MgBr$_2$ [5]	Cu(acac)$_2$	3.6:2:5	0 °C	4 h	1:>99:0	85%

[1] PhLi in c-hexane and Et$_2$O from Kanto, Japan, was used in Entries 1–7 and 10. [2] Determined by ^{1}H NMR. [3] A mixture of 1a and unidentified products. [4] nd: not determined. [5] Added MgBr$_2$ in Entries 4 (6 equiv.) and 10 (5 equiv.), respectively. [6] Lithiation in Et$_2$O.

The reaction conditions used in Entry 7 were applied to enantiomerically enriched phosphate (S)-1a of 98% ee to determine enantiospecificity (es) [34] and a stereochemical course (Scheme 2). A slightly higher equiv. of PhLi (3.2 equiv.) was used to ensure the formation of the cuprate. The reaction gave (S)-2aa in a 70% yield, and the HPLC analysis of the product on chiral stationary (abbreviated as chiral HPLC) disclosed 97% ee, which was calculated to be 99% es. Furthermore, the (S) configuration was assigned to the product by comparing the retention times of the derived phenylacetylene (S)-5aa with the reported data [29] (Scheme 3). These results indicate the inversion of the stereochemistry with marginal racemization.

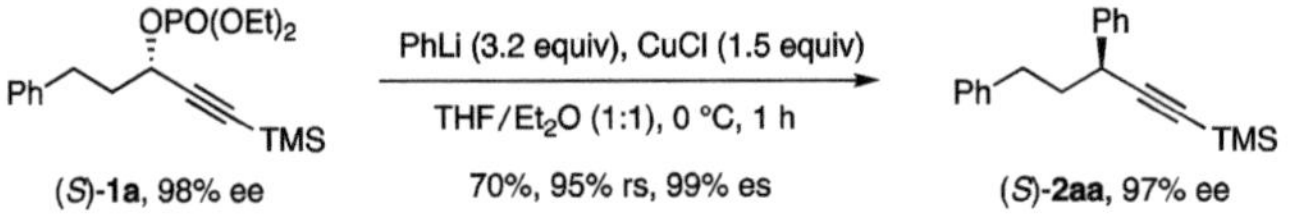

Scheme 2. Stereochemistry of the propargylic substitution.

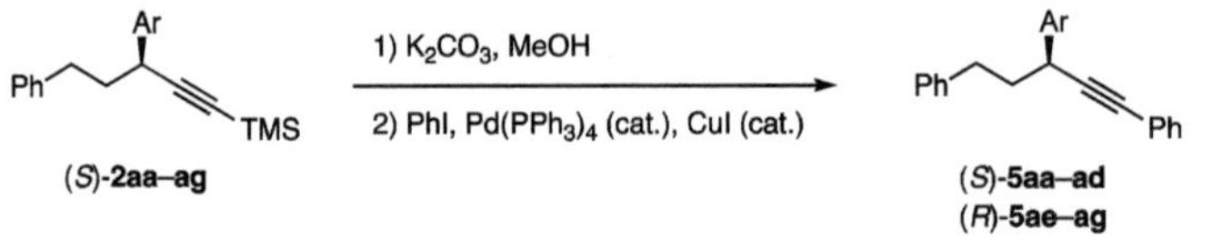

Scheme 3. Conversion to phenylacetylenes for determination of es and the absolute configuration. Ar for substrates and products: aa, Ph; ab, 4-MeC$_6$H$_4$; ac, 2-MeC$_6$H$_4$; ad, 4-MeOC$_6$H$_4$; ae, 2-MeOC$_6$H$_4$; af, 2-furyl; ag, 2-thienyl.

Several reagents prepared via the Li-Br exchange with t-BuLi or the direct lithiation with n-BuLi were subjected to the substitution with (S)-1a (97–98% ee) at 0 °C for 1 h. The isolated yields, rs, and es are delineated in Scheme 4.

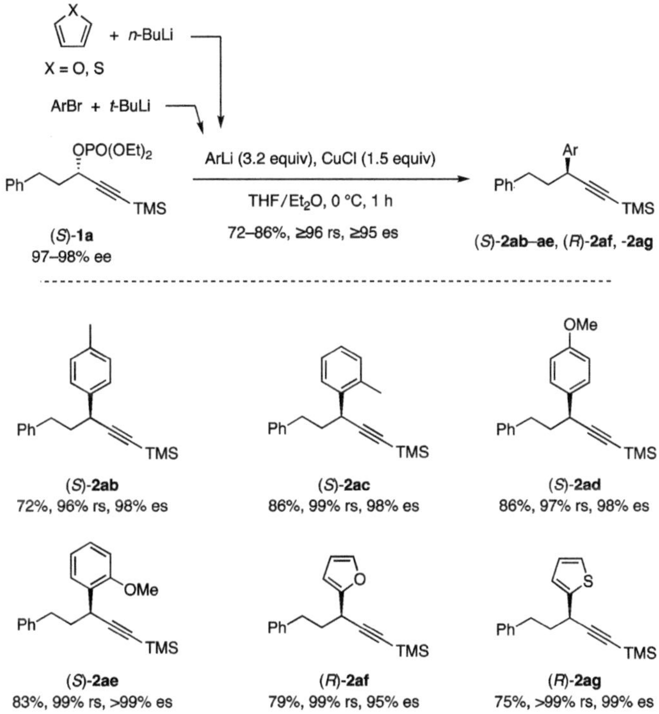

Scheme 4. Propargylic substitution with ArLi-based copper reagents.

The products were converted to phenylacetylenes by the method shown in Scheme 3, and each es of the phenyl derivatives was determined by chiral HPLC analysis. The configurations of (S)-**2ab**–(S)-**2ae** were established by comparing the retention times with those reported [29]. The same configuration was assigned to the products derived from furan and thiophene by analogy. The (R) stereochemistry was assigned by the priority rule. Two tolyl reagents prepared from 4- and 2-bromotoluenes via lithiation afforded (S)-**2ab** and -**2ac** in 72% and 86% yields, respectively, with 98% es for both. Similarly, 4- and 2-bromoanisoles produced (S)-**2ad** and -**2ae**, respectively, in good yields with high es values. Notably, the 2-substituted reagents afforded a higher rs than the 4-substituted reagents. A similar increase in the rs is presented later (**2be** vs. **2ba** in Scheme 5). The substitution of (S)-**1a** with the 2-furyl reagent prepared via the direct lithiation with n-BuLi resulted in high selectivity in rs and es, affording (R)-**2af** in a 79% yield. In combination with the oxidative conversion of the furan ring to the 2-butene-1,4-dione moiety, the present coupling would be useful in organic synthesis [35–37]. Similarly, the 2-thienyl reagent prepared by the direct lithiation gave (R)-**2ag** with high selectivity. Previously, furyl and thienyl copper reagents have been reported to be less nucleophilic for 1,4-addition [38–40], whereas high reactivity is reported for the propargylic substitution [17]. The reactivity of furyl and thienyl reagents in this study was sufficiently high and comparable to that of the aryl reagents shown in Scheme 4.

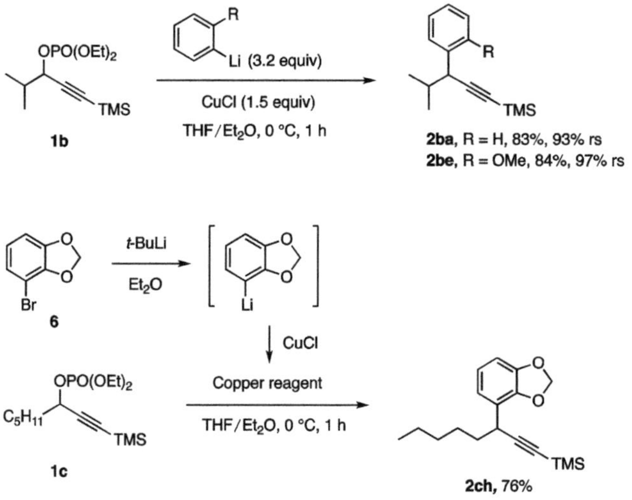

Scheme 5. Further study of propargylic substitution.

Substrates **1b** and **1c** were next subjected to the substitution (Scheme 5). Despite the high steric congestion in **1b**, the reaction proceeded well, although the rs decreased slightly to 93%. In contrast, a high rs (97% rs) was recorded with the 2-MeOC$_6$H$_4$ reagent to produce **2be** in an 84% yield. A similar ortho effect is mentioned above in the substitution reactions, giving (*S*)-**2ac** and -**2ae** (Scheme 4).

The reaction of **1c** to produce **2ch** was examined next (Scheme 5) to demonstrate the advantage of the present reaction using the ArLi over the previous method, which used the Grignard reagent [29]. 4-Bromo-1,3-benzodioxole (**6**) was converted to the copper reagent via the Li-Br exchange, and the reaction with phosphate **1c** proceeded smoothly to afford **2ch** in a 76% yield. No signal for the allenylic regioisomer was identified in the ^{1}H NMR spectrum probably due to the ortho effect observed above. Bromide **6** is convertible to the corresponding Grignard reagent [41], which would produce **2ch** by using the original method [29]. However, a part of the Grignard solution will be lost by titration, and the remaining solution after the use will be discarded; thus, the usage efficiency of bromide **6** via the Grignard reagent would not be high. In contrast, bromide **6** is converted quantitatively to the lithium reagent by the Li-Br exchange. This reaction would be an example to show that the exchange would especially be convenient for the substitution reaction using bromide and iodide that are prepared by a multistep procedure.

To evaluate the contribution of the TMS group on the regioselectivity, propargylic phosphates possessing phenylacetylenic and pentylacetylenic moieties were subjected to the substitution reaction (Scheme 6). The reaction of phenylacetylenic phosphate **7** with the PhLi-base reagent afforded **5aa** with a 92% rs. The rs was somewhat low compared to that of the reaction using TMS-acetylenic phosphate **1a** (98% rs in Table 1, Entry 7). In contrast, a high rs of >99% was observed for the substitution with the 2-MeOC$_6$H$_4$ reagent, giving (*R*)-**5ae** from (*S*)-**7**. Pentylacetylenic phosphate (*S*)-**8** produced (*S*)-**9** with an 83% rs. These results suggested that the rs of pentylacetylenic and other alkyl acetylenic phosphates would be low compared to that of TMS-acetylenic phosphates; thus, no further studies were examined.

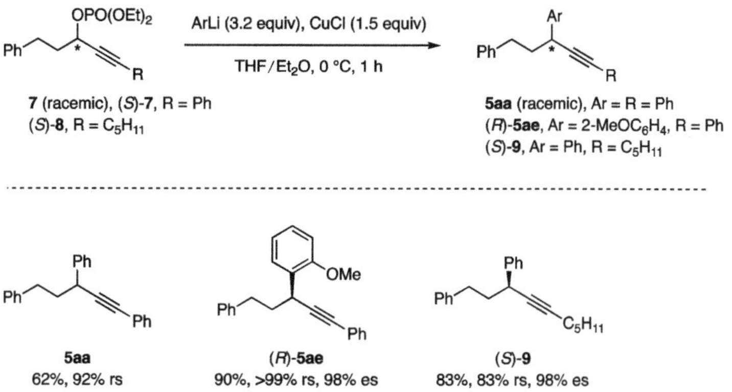

Scheme 6. Propargylic substitution of phenyl and alkyl acetylenic phosphates.

3. Materials and Methods

3.1. General

The ^{1}H (300 or 400 MHz) and ^{13}C NMR (75 or 100 MHz) spectroscopic data were recorded in CDCl$_3$, using Me$_4$Si (δ = 0 ppm) and the centerline of the triplet (δ = 77.1 ppm), respectively, as internal standards. Signal patterns are indicated as br s (broad singlet), s (singlet), d (doublet), t (triplet), q (quartet), and m (multiplet). Coupling constants (J) are given in Hertz (Hz). The ^{13}C–APT data (APT: attached proton test) are added to ^{13}C chemical shifts with minus (for C and CH$_2$) and plus (for CH and CH$_3$) signs. The solvents that were distilled prior to use are THF (from Na/benzophenone), Et$_2$O (from Na/benzophenone), and CH$_2$Cl$_2$ (from CaH$_2$). After the reaction was quenched, the organic extracts were concentrated by using an evaporator. The silica gel used for chromatography was purchased (Merck (Tokyo, Japan), silica gel 60; KANTO, silica gel 60N). PhLi in c-hexane/Et$_2$O, t-BuLi in n-pentane, and n-BuLi in hexane were purchased from Kanto, Japan, while CuCl and Cu(acac)$_2$ were obtained from Tokyo Chemical Industry (TCI) (Tokyo, Japan) and used without purification.

According to the previous method [29], racemic phosphates 1a–c and 7 and enantiomerically enriched phosphates (S)-1a (97–98% ee), (S)-7 (96% ee), and (S)-8 (95% ee) were synthesized. The synthesis of (S)-1a, -7, and -8 is described in the Supplementary Materials. 4-Bromo-1,3-benzodioxole (6) was prepared according to the literature method [41,42].

Ratios of acetylenes and allenes produced by the coupling reactions were determined by integration of the diagnostic signals in ^{1}H NMR spectra and converted to the regioselectivity (rs). HPLC analyses of the derived phenylacetylenes were performed using chiral stationary columns (abbreviated as chiral HPLC) to determine the enantiomeric ratios and absolute configurations. The enantiomer ratios were converted to the enantiospecificity (es) according to the following equation: (% ee of the product) × 100/(% ee of the phosphate) [34].

3.2. Representative Procedures of the Coupling Reaction

3.2.1. Method A Using Commercial PhLi (Table 1, Entry 7)

To an ice-cold suspension of CuCl (21.9 mg, 0.221 mmol) in Et$_2$O (1 mL) and THF (0.5 mL) was added PhLi (1.13 M in c-hexane/Et$_2$O, 0.40 mL, 0.452 mmol) dropwise. The mixture was stirred at 0 °C for 30 min, and then a solution of racemic phosphate 1a (54.5 mg, 0.148 mmol) in THF (1 mL) was added. The mixture was stirred for 2 h and diluted with saturated NH$_4$Cl and a small amount of 28% NH$_4$OH. The resulting mixture was extracted with EtOAc twice. The combined extracts were washed with brine, dried over MgSO$_4$, and concentrated to leave an oil, which was purified by chromatography on silica gel to afford acetylene 2aa (33.1 mg, 76% yield, 98% rs by ^{1}H NMR analysis).

3.2.2. Method B Using PhLi Derived from PhI and t-BuLi (Table 1, Entry 8)

To an ice-cold solution of PhI (0.060 mL, 0.538 mmol) in Et$_2$O (0.7 mL) was added t-BuLi (1.59 M in pentane, 0.56 mL, 0.89 mmol). After 30 min, THF (0.7 mL) and CuCl (21.1 mg, 0.213 mmol) were added to the solution. The mixture was stirred for 30 min at 0 °C, and then a solution of racemic phosphate **1a** (54.2 mg, 0.147 mmol) in THF (0.7 mL) was added. The resulting mixture was stirred at 0 °C for 1 h and diluted with saturated NH$_4$Cl and a small amount of 28% NH$_4$OH. Extraction of the products and purification were carried out as described in Method A to afford **2aa** (34.6 mg, 80% yield, 98% rs by ^{1}H NMR analysis).

3.2.3. Method C Using PhLi Derived from PhBr and t-BuLi (Table 1, Entry 9)

To a solution of PhBr (0.050 mL, 0.478 mmol) in Et$_2$O (1.5 mL) at −15 °C was added t-BuLi (1.59 M in pentane, 0.61 mL, 0.97 mmol). After 30 min, THF (0.4 mL) and CuCl (21.6 mg, 0.218 mmol) were added to the solution. The mixture was stirred at 0 °C for 1 h, and then a THF solution of racemic phosphate **1a** (59.1 mg, 0.160 mmol) was added. The resulting mixture was stirred at 0 °C for 2 h and diluted with saturated NH$_4$Cl and a small amount of 28% NH$_4$OH. Extraction of the products and purification were carried out as described in Method A to afford **2aa** (34.2 mg, 73% yield, 97% rs by ^{1}H NMR analysis).

3.3. Representative Procedure for the Conversion of TMS-Acetylenes to Phenylacetylenes

To a solution of acetylene (S)-**2aa** (184 mg, 0.630 mmol) in MeOH (1.3 mL) was added K$_2$CO$_3$ (104 mg, 0.753 mmol). The mixture was stirred at room temperature for 2 h, diluted with Et$_2$O, and filtered through a pad of Celite. The filtrate was concentrated to afford an oil, which was purified by chromatography on silica gel with hexane/EtOAc for the next reaction.

To a solution of the above acetylene, PhI (0.084 mL, 0.753 mmol), t-BuNH$_2$ (0.66 mL, 6.23 mmol), and Pd(PPh$_3$)$_4$ (73.0 mg, 0.0632 mmol) in benzene (6 mL) was added CuI (36.0 mg, 0.189 mmol). The mixture was stirred at room temperature for 14 h and diluted with saturated NH$_4$Cl. The resulting mixture was extracted with EtOAc twice. The combined organic layers were washed with brine, dried over MgSO$_4$, and concentrated to afford an oil, which was purified by chromatography on silica gel with hexane/EtOAc to afford acetylene (S)-**5aa** (146 mg, 78% yield).

3.4. Experiments and Characterization of the Products

3.4.1. Synthesis of (S)-(3,5-Diphenylpent-1-yn-1-yl)trimethylsilane [(S)-**2aa**], Its Conversion to Ph-Acetylene (S)-**5aa**, and Chiral HPLC Analysis

According to Method A, PhLi (1.13 M in c-hexane/Et$_2$O, 0.46 mL, 0.52 mmol) was mixed with CuCl (23.9 mg, 0.241 mmol) in Et$_2$O (0.8 mL) and THF (0.4 mL) at 0 °C for 30 min. A solution of (S)-**1a** (98% ee, 60.0 mg, 0.163 mmol) in THF (0.5 mL) was added to the copper reagent, and the mixture was stirred at 0 °C for 1 h to afford (S)-**2aa** (33.5 mg, 70% yield): 95% rs; 99% es by chiral HPLC analysis using Chiralcel OJ-H, hexane/i-PrOH (99.9:0.1), 0.2 mL/min, 25 °C, and t_R/min = 37.3 (minor) and 44.3 (major); ^{1}H NMR (300 MHz, CDCl$_3$) δ 0.21 (s, 9 H), 1.99–2.08 (m, 2 H), 2.76 (dt, J = 3.3, 8.5 Hz, 2 H), 3.65 (t, J = 7.2 Hz, 1 H), and 7.16–7.37 (m, 10 H); and ^{13}C NMR (75 MHz, CDCl$_3$) δ 0.3 (+), 33.5 (−), 38.2 (+), 40.3 (−), 87.9 (−), 108.1 (−), 126.0 (+), 126.8 (+), 127.5 (+), 128.4 (+), 128.5 (+), 128.6 (+), 141.6 (−), and 141.8 (−). The ^{1}H, ^{13}C, and ^{13}C-APT NMR spectra were consistent with those reported [29]. The absolute configuration of (S)-**2aa** was determined by the chiral HPLC analysis of the derived phenylacetylene (S)-**5aa** (see below).

The procedure was described as a representative example (vide supra): ^{1}H NMR (300 MHz, CDCl$_3$) δ 2.08–2.22 (m, 2 H), 2.85 (dt, J = 3.0, 7.4, 2 H), 3.84 (t, J = 7.4 Hz, 1 H), and 7.15–7.51 (m, 15 H); and ^{13}C NMR (75 MHz, CDCl$_3$) δ 33.7 (−), 37.9 (+), 40.2 (−), 83.9 (−), 91.2 (−), 123.8 (−), 126.0 (+), 126.9 (+), 127.6 (+), 127.9 (+), 128.3 (+), 128.5 (+), 128.6 (+), 131.7 (+), 141.7 (−), and 141.9 (−). The spectra were consistent with those reported [29].

Chiral HPLC analysis using Chiralcel OD-H, hexane/i-PrOH (99.5:0.5), 0.3 mL/min, 25 °C, t_R/min = 23.2 (major), and 25.5 (minor): 98% es; (S)-configuration by comparing the relative t_R values with the published values [29]: t_R/min = 35.3 for (S)-isomer and 38.7 for (R)-isomer.

3.4.2. Synthesis of (S)-Trimethyl[5-phenyl-3-(p-tolyl)pent-1-yn-1-yl]silane [(S)-**2ab**], Its Conversion to Ph-Acetylene (S)-**5ab**, and Chiral HPLC Analysis

According to Method C, lithiation of 4-bromotoluene (0.060 mL, 0.490 mmol) in Et$_2$O (0.7 mL) with t-BuLi (1.59 M, 0.56 mL, 0.890 mmol) (0 °C for 30 min) was followed by the addition of THF (0.35 mL) and a reaction with CuCl (19.9 mg, 0.201 mmol) (0 °C for 15 min). A solution of (S)-**1a** (98% ee, 50.2 mg, 0.136 mmol) in THF (0.35 mL) was added to the copper reagent, and the mixture was stirred at 0 °C for 1 h to give (S)-**2ab** (30.2 mg, 72% yield): 96% rs; 98% es by chiral HPLC analysis of the corresponding Ph-acetylene (S)-**5ab** (vide infra); ^{1}H NMR (300 MHz, CDCl$_3$) δ 0.20 (s, 9 H), 1.97–2.07 (m, 2 H), 2.33 (s, 3 H), 2.75 (dt, J = 2.7, 7.5 Hz, 2 H), 3.61 (t, J = 7.5 Hz, 1 H), and 7.10–7.32 (m, 9 H); and ^{13}C NMR (75 MHz, CDCl$_3$) δ 0.3 (+), 21.1 (+), 33.5 (−), 37.8 (+), 40.2 (−), 87.6 (−), 108.4 (−), 125.9 (+), 127.4 (+), 128.4 (+), 128.6 (+), 129.2 (+), 136.3 (−), 138.6 (−), and 141.8 (−). The ^{1}H, ^{13}C, and ^{13}C-APT NMR spectra were consistent with those reported [29].

According to the representative procedure, the reaction of (S)-**2ab** (39.7 mg, 0.130 mmol) with K$_2$CO$_3$ (25.0 mg, 0.181 mmol) in MeOH (1.3 mL) at rt for 3 h afforded the desilylated acetylene, and the subsequent coupling reaction with PhI (0.020 mL, 0.179 mmol) in benzene (1.1 mL), using t-BuNH$_2$ (0.130 mL, 1.23 mmol), Pd(PPh$_3$)$_4$ (18.8 mg, 0.0163 mmol), and CuI (22.5 mg, 0.118 mmol), at rt for 14 h gave acetylene (S)-**5ab** (31.9 mg, 79% yield): ^{1}H NMR (300 MHz, CDCl$_3$) δ 2.04–2.22 (m, 2 H), 2.34 (s, 3 H), 2.80–2.87 (m, 2 H), 3.81 (dd, J = 7.8, 6.6 Hz, 1 H), 7.13–7.48 (m, 12 H), and 7.45–7.50 (m, 2 H). The ^{1}H spectrum was consistent with that reported [29].

Chiral HPLC analysis using Chiralcel OD-H, hexane/i-PrOH (99.8:0.2), 0.3 mL/min, 25 °C, t_R/min = 28.5 (major), and 35.0 (minor): 98% es; (S)-configuration by comparing the relative t_R values with the published data [29]: t_R/min = 32.6 for (S)-isomer and 40.9 for (R)-isomer.

3.4.3. Synthesis of (S)-Trimethyl[5-phenyl-3-(o-tolyl)pent-1-yn-1-yl]silane [(S)-**2ac**], Its Conversion to Ph-Acetylene (S)-**5ac**, and Chiral HPLC Analysis

According to Method C, lithiation of 2-bromotoluene (0.080 mL, 0.664 mmol) in Et$_2$O (1 mL) with t-BuLi (1.59 M, 0.84 mL, 1.34 mmol) (0 °C for 30 min) was followed by the addition of THF (0.5 mL) and a reaction with CuCl (29.3 mg, 0.296 mmol) (0 °C for 15 min). A solution of (S)-**1a** (98% ee, 76.3 mg, 0.207 mmol) in THF (1 mL) was added to the copper reagent, and the mixture was stirred at 0 °C for 1 h to afford (S)-**2ac** (54.4 mg, 86% yield): 99% rs; 98% es by chiral HPLC analysis of the corresponding Ph-acetylene (R)-**5ac** (vide infra); ^{1}H NMR (300 MHz, CDCl$_3$) δ 0.21 (s, 9 H), 1.89–2.04 (m, 2 H), 2.18 (s, 3 H), 2.72–2.94 (m, 2 H), 3.79 (dd, J = 8.8, 5.8 Hz, 1 H), 7.07–7.34 (m, 8 H), and 7.52 (d, J = 7.5 Hz, 1 H); ^{13}C NMR (75 MHz, CDCl$_3$) δ 0.3 (+), 19.0 (+), 33.8 (−), 34.7 (+), 38.7 (−), 87.3 (−), 108.4 (−), 126.0 (+), 126.3 (+), 126.7 (+), 127.6 (+), 128.4 (+), 128.6 (+), 130.5 (+), 134.9 (−), 139.8 (−), and 141.7 (−). The ^{1}H, ^{13}C, and ^{13}C-APT NMR spectra were consistent with those reported [29].

According to the representative procedure, the reaction of (S)-**2ac** (54.4 mg, 0.177 mmol) with K$_2$CO$_3$ (34.6 mg, 0.250 mmol) in MeOH (1.8 mL) at rt for 3 h gave the desilylated acetylene, and the subsequent coupling reaction with PhI (0.030 mL, 0.269 mmol) in benzene (1.5 mL), using t-BuNH$_2$ (0.260 mL, 2.45 mmol), Pd(PPh$_3$)$_4$ (23.1 mg, 0.020 mmol), and CuI (41.7 mg, 0.219 mmol), at rt for 14 h afforded acetylene (R)-**5ac** (47.0 mg, 85% yield): ^{1}H NMR (300 MHz, CDCl$_3$) δ 1.97–2.18 (m, 2 H), 2.24 (s, 3 H), 2.81–3.03 (m, 2 H), 3.99 (dd, J = 9.3, 5.1 Hz, 1 H), 7.11–7.35 (m, 11 H), 7.43–7.50 (m, 2 H), and 7.59 (d, J = 6.9 Hz, 1 H). The ^{1}H spectrum was consistent with the reported data [29].

Chiral HPLC analysis using Chiralcel OD-H, hexane/i-PrOH (99.8:0.2), 0.3 mL/min, 25 °C, and t_R/min = 31.6 (minor) and 35.5 (major): 98% es; (R)-configuration by comparing the relative t_R values with the published data [29]: t_R/min = 33.5 for (S)-isomer and 39.6 for (R)-isomer.

3.4.4. Synthesis of (S)-[3-(4-Methoxyphenyl)-5-phenylpent-1-yn-1-yl]trimethylsilane [(S)-2ad], Its Conversion to Ph-Acetylene (S)-5ad, and Chiral HPLC Analysis

According to Method C, the lithiation of 4-bromoanisole (0.060 mL, 0.479 mmol) in Et_2O (0.7 mL) with t-BuLi (1.59 M in pentane, 0.58 mL, 0.922 mmol) (0 °C for 30 min) was followed by the addition of THF (0.4 mL) and a reaction with CuCl (21.6 mg, 0.218 mmol) (0 °C for 30 min). A solution of (S)-1a (98% ee, 53.2 mg, 0.144 mmol) in THF (0.4 mL) was added to the copper reagent, and the mixture was stirred at 0 °C for 1 h to give (S)-2ad (45.9 mg, 86% yield): 97% rs; 98% es by chiral HPLC analysis of the corresponding Ph-acetylene (vide infra); ^{1}H NMR (300 MHz, $CDCl_3$) δ 0.20 (s, 9 H), 1.96–2.06 (m, 2 H), 2.70–2.78 (m, 2 H), 3.60 (t, J = 7.1 Hz, 1 H), 3.79 (s, 3 H), 6.85 (d, J = 9.0 Hz, 2 H), and 7.16–7.32 (m, 7 H); ^{13}C NMR (75 MHz, $CDCl_3$) δ 0.3 (+), 33.5 (−), 37.4 (+), 40.3 (−), 55.4 (+), 87.6 (−), 108.5 (−), 113.9 (+), 125.9 (+), 128.4 (+), 128.5 (+), 128.6 (+), 133.7 (−), 141.8 (−), and 158.5 (−). The ^{1}H, ^{13}C, and ^{13}C-APT NMR spectra were consistent with those reported [29].

According to the representative procedure, the reaction of (S)-2ad (40.8 mg, 0.127 mmol) with K_2CO_3 (26.5 mg, 0.192 mmol) in MeOH (1.3 mL) at rt for 3 h gave the desilylated acetylene, and the subsequent coupling reaction with PhI (0.020 mL, 0.179 mmol) in benzene (1.1 mL), using t-$BuNH_2$ (0.140 mL, 1.32 mmol), $Pd(PPh_3)_4$ (27.4 mg, 0.0237 mmol), and CuI (16.4 mg, 0.0861 mmol), at rt for 14 h afforded acetylene (S)-5ad (29.4 mg, 71% yield): ^{1}H NMR (300 MHz, $CDCl_3$) δ 2.05–2.21 (m, 2 H), 2.78–2.87 (m, 2 H), 3.80 (s, 3 H), 3.74–3.87 (m, 1 H), 6.88 (d, J = 9.0 Hz, 2 H), 7.16–7.36 (m, 10 H), and 7.43–7.50 (m, 2 H). The ^{1}H spectrum was consistent with that reported [29].

Chiral HPLC analysis using Chiralcel OD-H, hexane/i-PrOH (99.5:0.5), 0.3 mL/min, 25 °C, and t_R/min = 34.2 (major) and 44.9 (minor): 98% es; (S)-configuration by comparing the relative t_R values with the published data [29]: t_R/min = 38.5 for (S)-isomer and 51.2 for (R)-isomer.

3.4.5. Synthesis of (S)-[3-(2-Methoxyphenyl)-5-phenylpent-1-yn-1-yl]trimethylsilane [(S)-2ae], Its Conversion to Ph-Acetylene (S)-5ae, and Chiral HPLC Analysis

According to Method C, lithiation of 2-bromoanisole (0.070 mL, 0.569 mmol) in Et_2O (0.9 mL) with t-BuLi (1.59 M, 0.69 mL, 1.10 mmol) (0 °C for 30 min) was followed by the addition of THF (0.4 mL) and a reaction with CuCl (24.9 mg, 0.252 mmol) (0 °C for 30 min). A solution of (S)-1a (98% ee, 62.9 mg, 0.171 mmol) in THF (0.5 mL) was added to the copper reagent, and the mixture was stirred at 0 °C for 1 h to produce (S)-2ae (45.8 mg, 83% yield): 99% rs; >99% es by chiral HPLC analysis of the corresponding Ph-acetylene (R)-5ae (vide infra); ^{1}H NMR (300 MHz, $CDCl_3$) δ 0.21 (s, 9 H), 1.83–2.11 (m, 2 H), 2.68–2.88 (m, 2 H), 3.76 (s, 3 H), 4.13 (dd, J = 8.7, 5.1 Hz, 1 H), 6.82 (d, J = 8.1 Hz, 1 H), 6.96 (t, J = 7.4 Hz, 1 H), 7.13–7.30 (m, 6 H), and 7.60 (dd, J = 7.6, 1.6 Hz, 1 H); and ^{13}C NMR (75 MHz, $CDCl_3$) δ 0.3 (+), 31.9 (+), 33.7 (−), 38.2 (−), 55.4 (+), 87.1 (−), 108.7 (−), 110.4 (+), 120.7 (+), 125.8 (+), 127.9 (+), 128.3 (+), 128.6 (+), 129.9 (−), 142.2 (−), and 156.2 (−). The ^{1}H, ^{13}C, and ^{13}C-APT NMR spectra were consistent with those reported [29].

According to the representative procedure, the reaction of (S)-2ae (37.1 mg, 0.0983 mmol) with K_2CO_3 (27.3 mg, 0.198 mmol) in MeOH (1.3 mL) at rt for 3 h afforded the desilylated acetylene, and the subsequent coupling reaction with PhI (0.020 mL, 0.179 mmol) in benzene (1.1 mL), using t-$BuNH_2$ (0.14 mL, 1.32 mmol), $Pd(PPh_3)_4$ (24.1 mg, 0.021 mmol), and CuI (11.5 mg, 0.060 mmol), at rt for 14 h gave acetylene (R)-5ae (22.1 mg, 69% yield): ^{1}H NMR (300 MHz, $CDCl_3$) δ 1.96–2.21 (m, 2 H), 2.77–2.97 (m, 2 H), 3.80 (s, 3 H), 4.33 (dd, J = 8.8, 5.3 Hz, 1 H), 6.85 (dd, J = 8.1, 1.2 Hz, 1 H), 6.97 (dt, J = 1.1, 7.2 Hz, 1 H), 7.13–7.35 (m, 9 H), 7.45–7.51 (m, 2 H), and 7.64 (dd, J = 7.8, 1.5 Hz, 1 H); ^{13}C NMR (75 MHz, $CDCl_3$) δ 31.4 (+), 33.8 (−), 38.2 (−), 55.4 (+), 83.1 (−), 91.8 (−), 110.5 (+), 120.7 (+), 124.0 (−), 125.8 (+), 127.7 (+), 127.9 (+), 128.29 (+), 128.31 (+), 128.59 (+), 128.62 (+), 130.2 (−), 131.8 (+), 142.1 (−), and 156.2 (−). The ^{1}H, ^{13}C, and ^{13}C-APT NMR spectra were consistent with those reported [29].

Chiral HPLC analysis using Chiralcel OD-H, hexane/i-PrOH (99.5:0.5), 0.3 mL/min, 25 °C, and t_R/min = 24.2 (major) and 28.0 (minor): >99% es; (R)-configuration by comparing the relative t_R values with the published data [29]: t_R/min = 35.1 for (R)-isomer and 40.0 for (S)-isomer.

3.4.6. Synthesis of (R)-[3-(Furan-2-yl)-5-phenylpent-1-yn-1-yl]trimethylsilane [(R)-**2af**], Its Conversion to Ph-Acetylene (S)-**5af**, and Chiral HPLC Analysis

To an ice-cold solution of furan (0.040 mL, 0.552 mmol) in THF (0.4 mL) was added n-BuLi (1.65 M, 0.29 mL, 0.479 mmol) dropwise. The solution was stirred at 0 °C for 30 min. According to Method C, CuCl (20.5 mg, 0.207 mmol) was added to the solution, and, after 15 min of stirring at 0 °C, a solution of phosphate (S)-**1a** (97% ee, 54.7 mg, 0.148 mmol) in THF (0.4 mL) was then added. The mixture was stirred at 0 °C for 1 h to afford (R)-**2af** (33.0 mg, 79% yield): 99% rs; 95% es by chiral HPLC analysis of the corresponding Ph-acetylene (R)-**5af** (vide infra); ^{1}H NMR (400 MHz, CDCl$_3$) δ 0.20 (s, 9 H), 2.02–2.22 (m, 2 H), 2.77 (t, J = 8.0 Hz, 2 H), 3.75 (dd, J = 8.2, 5.8 Hz, 1 H), 6.21 (d, J = 3.2 Hz, 1 H), 6.30 (dd, J = 3.0, 2.2 Hz, 1 H), and 7.16–7.35 (m, 5 H); ^{13}C NMR (100 MHz, CDCl$_3$) δ 0.2 (+), 32.0 (+), 33.1 (−), 36.3 (−), 87.4 (−), 105.2 (−), 106.0 (+), 110.3 (+), 126.0 (+), 128.5 (+), 128.6 (+), 141.6 (−), 141.7 (+), and 154.1 (−); and HRMS (FD) calcd for C$_{18}$H$_{26}$OSi [M]$^+$ 282.14399, found 282.14264.

According to the representative procedure, the reaction of (R)-**2af** (24.8 mg, 0.0878 mmol) with K$_2$CO$_3$ (22.7 mg, 0.164 mmol) in MeOH (0.9 mL) at rt for 3 h gave the desilylated acetylene, and the subsequent coupling reaction with PhI (0.020 mL, 0.179 mmol) in benzene (0.8 mL), using t-BuNH$_2$ (0.10 mL, 0.94 mmol), Pd(PPh$_3$)$_4$ (13.5 mg, 0.0117 mmol), and CuI (8.6 mg, 0.045 mmol), at rt for 16 h afforded Ph-acetylene (R)-**5af** (18.0 mg, 73% yield): ^{1}H NMR (300 MHz, CDCl$_3$) δ 2.12–2.33 (m, 2 H), 2.85 (t, J = 7.8 Hz, 2 H), 3.86 (dd, J = 8.2, 5.8 Hz, 1 H), 6.28 (dt, J = 3.3, 0.8 Hz, 1 H), 6.33 (dd, J = 3.3, 2.1 Hz, 1 H), 7.16–7.34 (m, 8 H), 7.36 (dd, J = 2.1, 0.9 Hz, 1 H), and 7.43–7.50 (m, 2 H).

Chiral HPLC analysis using Chiralpak AD-H, hexane/i-PrOH (99.5:0.5), 0.3 mL/min, 25 °C, and t_R/min = 22.9 (minor) and 26.5 (major): 95% es; (R)-configuration was assigned by analogy with that shown in Scheme 4.

3.4.7. Synthesis of (R)-Trimethyl[5-phenyl-3-(thiophen-2-yl)pent-1-yn-1-yl]silane [(R)-**2ag**], Its Conversion to Ph-Acetylene (S)-**5ag**, and Chiral HPLC Analysis

To an ice-cold solution of thiophene (0.050 mL, 0.636 mmol) in THF (0.4 mL) was added n-BuLi (1.65 M, 0.31 mL, 0.51 mmol) dropwise. The solution was stirred at 0 °C for 30 min. According to Method C, CuCl (22.9 mg, 0.231 mmol) was added to the solution, and, after 15 min of stirring at 0 °C, a solution of phosphate (S)-**1a** (98% ee, 57.4 mg, 0.156 mmol) in THF (0.4 mL) was added. The mixture was stirred at 0 °C for 1 h to afford (R)-**2ag** (34.8 mg, 75% yield): >99% rs; 99% es by chiral HPLC analysis of the corresponding Ph-acetylene (vide infra); ^{1}H NMR (300 MHz, CDCl$_3$) δ 0.21 (s, 9 H), 2.09–2.17 (m, 2 H), 2.77–2.83 (m, 2 H), 3.93 (t, J = 7.2 Hz, 1 H), 6.93 (dd, J = 5.0, 3.8 Hz, 1 H), 6.96–6.98 (m, 1 H), and 7.16–7.32 (m, 6 H); ^{13}C NMR (100 MHz, CDCl$_3$) δ 0.2 (+), 33.3 (−), 33.5 (+), 40.2 (−), 87.9 (−), 107.1 (−), 124.0 (+), 124.6 (+), 126.1 (+), 126.7 (+), 128.5 (+), 128.7 (+), 141.6 (−), and 145.2 (−); and HRMS (FD) calcd for C$_{18}$H$_{22}$SSi [M]$^+$ 298.12115, found 298.11989.

According to the representative procedure, the reaction of (R)-**2ag** (34.8 mg, 0.117 mmol) with K$_2$CO$_3$ (26.1 mg, 0.189 mmol) in MeOH (1.3 mL) at rt for 3 h afforded the desilylated acetylene, and the subsequent coupling reaction with PhI (0.020 mL, 0.179 mmol) in benzene (1 mL), using t-BuNH$_2$ (0.13 mL, 1.23 mmol), Pd(PPh$_3$)$_4$ (15.8 mg, 0.0137 mmol), and CuI (7.1 mg, 0.0373 mmol), at rt for 16 h produced Ph-acetylene (R)-**5ag** (23.5 mg, 67% yield): ^{1}H NMR (400 MHz, CDCl$_3$) δ 2.24 (dt, J = 8.4, 7.2, 2 H), 2.88 (dd, J = 9.3, 6.6 Hz, 2 H), 4.13 (t, J = 7.2 Hz, 1 H), 6.96 (dd, J = 5.0, 3.4 Hz, 1 H), 7.02–7.05 (m, 1 H), 7.17–7.36 (m, 9 H), and 7.44–7.52 (m, 2 H).

Chiral HPLC analysis using Chiralpak AD-H, hexane/i-PrOH (99.5:0.5), 0.3 mL/min, and 25 °C, t_R/min = 26.1 (minor) and 28.7 (major): 99% es; (R)-configuration was assigned by analogy with that shown in Scheme 4.

3.4.8. Synthesis of Trimethyl(4-methyl-3-phenylpent-1-yn-1-yl)silane (**2ba**)

According to Method A, PhLi (1.13 M, 0.57, mL, 0.644 mmol) and, after 15 min at 0 °C, a solution of **1b** (56.4 mg, 0.184 mmol) in THF (0.5 mL) was added to CuCl (25.7 mg, 0.260 mmol) in THF (0.4 mL) and Et$_2$O (0.9 mL). The mixture was stirred at 0 °C for 1 h to produce **2ba** (35.0 mg, 83% yield): 93% rs; and ^{1}H NMR (300 MHz, CDCl$_3$) δ 0.19 (s, 9 H), 0.90 (d, *J* = 6.6 Hz, 3 H), 0.98 (d, *J* = 6.9 Hz, 3 H), 1.88–2.00 (m, 1 H), 3.54 (d, *J* = 5.7 Hz, 1 H), and 7.18–7.33 (m, 5 H). The ^{1}H NMR spectrum was consistent with that reported [29].

3.4.9. Synthesis of [3-(2-Methoxyphenyl)-4-methylpent-1-yn-1-yl]trimethylsilane (**2be**)

According to Method C, lithiation of 2-bromoanisole (0.080 mL, 0.650 mmol) in Et$_2$O (1 mL) with *t*-BuLi (1.59 M, 0.77 mL, 1.22 mmol) (0 °C for 30 min) was followed by the addition of THF (0.5 mL) and a reaction with CuCl (27.1 mg, 0.274 mmol) (0 °C for 15 min). A solution of **1b** (58.1 mg, 0.190 mmol) in THF (0.5 mL) was added to the copper reagent, and the mixture was stirred at 0 °C for 1 h to produce **2be** (41.3 mg, 84% yield): 97% rs; ^{1}H NMR (300 MHz, CDCl$_3$) δ 0.18 (s, 9 H), 0.86 (d, *J* = 6.9 Hz, 3 H), 1.01 (d, *J* = 6.9 Hz, 3 H), 1.89–2.02 (m, 1 H), 3.80 (s, 3 H), 4.06 (d, *J* = 5.1 Hz, 1 H), 6.83 (d, *J* = 7.6 Hz, 1 H), 6.94 (t, *J* = 7.6 Hz, 1 H), 7.20 (dt, *J* = 1.8, 7.6 Hz, 1 H), and 7.49 (dd, *J* = 7.6, 1.8 Hz, 1 H); and ^{13}C NMR (75 MHz, CDCl$_3$) δ 0.3 (+), 18.1 (+), 21.5 (+), 32.6 (+), 39.0 (+), 55.4 (+), 87.5 (−), 107.5 (−), 110.3 (+), 120.3 (+), 127.6 (+), 129.4 (−), 129.5 (+), and 156.4 (−). The ^{1}H, ^{13}C, and ^{13}C–APT NMR spectra were consistent with the reported data [29].

3.4.10. Synthesis of [3-(Benzo[d][1,3]dioxol-4-yl)oct-1-yn-1-yl]trimethylsilane (**2ch**)

According to Method C, lithiation of 2-bromocatechol derivative **6** (139.0 mg, 0.691 mmol) in Et$_2$O (1 mL) with *t*-BuLi (1.59 M, 0.82 mL, 1.30 mmol) (0 °C for 30 min) was followed by the addition of THF (0.5 mL) and a reaction with CuCl (28.5 mg, 0.288 mmol) (0 °C for 15 min). A solution of **1c** (68.1 mg, 0.204 mmol) in THF (0.5 mL) was added to the copper reagent, and the mixture was stirred at 0 °C for 1 h to afford **2ch** (51.3 mg, 76% yield): ^{1}H NMR (400 MHz, CDCl$_3$) δ 0.17 (s, 9 H), 0.88 (t, *J* = 6.8, 3 H), 1.23–1.52 (m, 6 H), 1.65–1.79 (m, 2 H), 3.81 (dd, *J* = 7.6, 6.4 Hz, 1 H), 5.905 (d, *J* = 1.4 Hz, 1 H), 5.940 (d, *J* = 1.4 Hz, 1 H), 6.71 (dd, *J* = 8.0, 1.4 Hz, 1 H), 6.81 (t, *J* = 8.0 Hz, 1 H), and 6.98 (dd, *J* = 8.0, 0.8 Hz, 1 H); ^{13}C NMR (100 MHz, CDCl$_3$) δ 0.2 (+), 14.1 (+), 22.6 (−), 26.9 (−), 31.4 (−), 32.8 (+), 36.4 (−), 86.9 (−), 100.7 (−), 107.1 (+), 107.5 (−), 121.1 (+), 121.6 (+), 123.7 (−), 144.5 (−), and 147.1 (−); and HRMS (FD) calcd for C$_{18}$H$_{26}$O$_2$Si [M]$^+$ 302.17021, found 302.17080.

3.4.11. Synthesis of Pent-1-yne-1,3,5-triyltribenzene (**5aa**)

According to Method A, PhLi (1.13 M, 0.48 mL, 0.542 mmol) and, after 15 min at 0 °C, a solution of **7** (57.6 mg, 0.155 mmol) in THF (0.4 mL) were added to CuCl (20.5 mg, 0.207 mmol) in THF (0.4 mL) and Et$_2$O (0.8 mL). The mixture was stirred at 0 °C for 1 h to produce **5aa** (27.9 mg, 62% yield, 92% rs). The ^{1}H NMR spectrum was consistent with that derived from TMS-acetylene **2aa**.

3.4.12. Synthesis of (*R*)-[3-(2-Methoxyphenyl)pent-1-yne-1,5-diyl]dibenzene [(*R*)-**5ae**]

According to Method C, lithiation of 2-bromoanisole (0.060 mL, 0.488 mmol) in Et$_2$O (0.7 mL) with *t*-BuLi (1.59 M, 0.57 mL, 0.906 mmol) (0 °C for 30 min) was followed by the addition of THF (0.35 mL) and a reaction with CuCl (19.8 mg, 0.200 mmol) (0 °C for 15 min). A solution of (*S*)-**7** (96% ee, 52.2 mg, 0.140 mmol) in THF (0.35 mL) was added to the copper reagent, and the mixture was stirred at 0 °C for 1 h to afford (*R*)-**5ae** (41.1 mg, 90% yield, >99% rs, 98% es by chiral HPLC analysis). The ^{1}H NMR spectrum was consistent with that reported [29]. (*R*)-Configuration was determined by comparing the relative t_R values with those of (*R*)-**5ae**, which was derived from (*S*)-**2ae** (vide supra).

3.4.13. Synthesis of (*S*)-Dec-4-yne-1,3-diyldibenzene [(*S*)-**9**]

According to Method A, PhLi (1.13 M, 0.54 mL, 0.61 mmol) and, after 15 min at 0 °C, a solution of (*S*)-**8** (95% ee, 64.3 mg, 0.175 mmol) in THF (0.45 mL) were added to CuCl

(25.4 mg, 0.257 mmol) in THF (0.45 mL) and Et$_2$O (0.9 mL). The mixture was stirred at 0 °C for 1 h to give (S)-9 (42.5 mg, 83% yield): 83% rs; 98% es by chiral HPLC analysis, using Chiralcel OJ-H, hexane/i-PrOH (99.9:0.1), 0.3 mL/min, 25 °C, t_R/min = 38.1 (minor), and 44.7 (major); ^{1}H NMR (300 MHz, CDCl$_3$) δ 0.91 (t, J = 7.0 Hz, 3 H), 1.24–1.52 (m, 4 H), 1.52–1.63 (m, 2 H), 2.01 (q, J = 8.0 Hz, 2 H), 2.26 (dt, J = 2.3, 6.6 Hz, 2 H), 2.68–2.86 (m, 2 H), 3.60 (tt, J = 6.9, 2.1 Hz, 1 H), and 7.14–7.39 (m, 10 H); and ^{13}C NMR (75 MHz, CDCl$_3$) δ 14.1 (+), 18.9 (−), 22.3 (−), 28.9 (−), 31.2 (−), 33.7 (−), 37.4 (+), 40.6 (−), 81.4 (−), 83.9 (−), 125.9 (+), 126.6 (+), 127.5 (+), 128.40 (+), 128.45 (+), 128.6 (+), 142.0 (−), and 142.8 (−). The ^{1}H, ^{13}C, and ^{13}C–APT NMR spectra were consistent with the reported data [29]. (S)-Configuration was assigned by analogy with that shown in Scheme 4.

3.5. Synthesis of (1,5-Diphenylpenta-1,2-dien-1-yl)trimethylsilane (3aa) (Table 1, Entry 10)

To an ice-cold mixture of Cu(acac)$_2$ (87.3 mg, 0.334 mmol) and MgBr$_2$ in THF (0.20 M, 4.2 mL, 0.84 mmol) was added PhLi (1.13 M, 0.54 mL, 0.61 mmol) dropwise. The mixture was stirred at 0 °C for 1 h, and phosphate 1a (61.5 mg, 0.167 mmol) in THF (1.7 mL) was added. The reaction was carried out at 0 °C for 4 h and diluted with saturated NH$_4$Cl and 28% NH$_4$OH. The product was extracted with EtOAc twice. The combined extracts were dried over MgSO$_4$ and concentrated. The residual oil was purified by chromatography on silica gel with hexane/EtOAc to afford allene 3aa (41.6 mg, 85% yield): >99% rs; ^{1}H NMR (300 MHz, CDCl$_3$) δ 0.22 (s, 9 H), 2.33–2.50 (m, 2 H), 2.68–2.86 (m, 2 H), 5.19 (t, J = 6.8 Hz, 1 H), and 7.12–7.32 (m, 10 H); and ^{13}C NMR (75 MHz, CDCl$_3$) δ −0.2 (+), 30.3 (−), 36.0 (−), 86.7 (+), 100.6 (−), 126.0 (+), 126.1 (+), 127.7 (+), 128.43 (+), 128.45 (+), 128.6 (+), 137.9 (−), 141.9 (−), and 208.3 (−). The ^{1}H, ^{13}C, and ^{13}C–APT NMR spectra were consistent with those reported [29].

4. Conclusions

Aryl-lithium-based copper reagents were developed for the propargylic substitution of TMS-acetylenic phosphates. Aryl lithium compounds prepared in situ by the lithium-halogen exchange with *t*-BuLi were converted to highly regioselective and enantiospecific aryl reagents. 2-Furyl and 2-thienyl reagents prepared via the direct lithiation with *n*-BuLi were successful as well. The present propargylic substitution with several preparations of aryl lithiums would be a useful reaction in organic synthesis.

Supplementary Materials: The following supporting information can be downloaded at https://www.mdpi.com/article/10.3390/catal13071084/s1, general information of experiments, remaining procedures, ^{1}H, ^{13}C, and ^{13}C-APT NMR spectra [29].

Author Contributions: Conceptualization, Y.K.; investigation and data curation, T.H., Y.H. and N.O.; writing, Y.K. All authors have read and agreed to the published version of the manuscript.

Funding: This work was supported by KAKENHI Grant No. 20K05501.

Data Availability Statement: All experimental data are contained in the article and Supplementary Materials.

Conflicts of Interest: The authors declare no conflict of interest.

References

1. Kobayashi, Y. Alkyl Pyirdinesulfonates and Allylic Pyridinecarboxylates, New Boosters for the Substitution at Secondary Carbons. *Heterocycles* **2020**, *100*, 499–546. [CrossRef]
2. Kobayashi, Y.; Shimoda, M. Substitution of Allylic Picolinates with Various Copper Reagents and Synthetic Applications. In *Cutting-Edge of Organic Synthesis and Chemical Biology of Bioactive Molecules*; Kobayashi, Y., Ed.; Springer Nature: Singapore, 2019; Chapter 7; pp. 145–169. ISBN 978-981-13-6243-9 for hardcover. Available online: https://www.springer.com/gp/book/9789811362439 (accessed on 14 June 2019).
3. Kobayashi, Y. Coupling Reactions on Secondary Allylic, Propargylic, and Alkyl Carbons using Organoborates/Ni and RMgX/Cu Reagents. *Catalysts* **2023**, *13*, 132. [CrossRef]

4. Burns, D.H.; Miller, J.D.; Chan, H.-K.; Delaney, M.O. Scope and Utility of a New Soluble Copper Catalyst [CuBr·LiSPh·LiBr·THF]: A Comparison with Other Copper Catalysts in Their Ability to Couple One Equivalent of a Grignard Reagent with an Alkyl Sulfonate. *J. Am. Chem. Soc.* **1997**, *119*, 2125–2133. [CrossRef]
5. Yang, C.-T.; Zhang, Z.-Q.; Liang, J.; Liu, J.-H.; Lu, X.-Y.; Chen, H.-H.; Liu, L. Copper-Catalyzed Cross-Coupling of Nonactivated Secondary Alkyl Halides and Tosylates with Secondary Alkyl Grignard Reagents. *J. Am. Chem. Soc.* **2012**, *134*, 11124–11127. [CrossRef] [PubMed]
6. Xu, S.; Oda, A.; Bobinski, T.; Li, H.; Matsueda, Y.; Negishi, E. Highly Efficient, Convergent, and Enantioselective Synthesis of Phthioceranic Acid. *Angew. Chem. Int. Ed.* **2015**, *54*, 9319–9322. [CrossRef]
7. Shinohara, R.; Morita, M.; Ogawa, N.; Kobayashi, Y. Use of the 2-Pyridinesulfonyloxy Leaving Group for the Fast Copper-Catalyzed Coupling Reaction at Secondary Alkyl Carbons with Grignard Reagents. *Org. Lett.* **2019**, *21*, 3247–3251. [CrossRef] [PubMed]
8. Breit, B.; Schmidt, Y. Directed Reactions of Organocopper Reagents. *Chem. Rev.* **2008**, *108*, 2928–2951. [CrossRef]
9. Prakash, J.; Marek, I. Enantioselective synthesis of all-carbon quaternary stereogenic centers in acyclic systems. *Chem. Commun.* **2011**, *47*, 4593–4623. [CrossRef]
10. Ohmiya, H.; Sawamura, M. Copper Catalyzed Allylic Substitution and Conjugate Addition with Alkylboranes. *J. Synth. Org. Chem. Jpn.* **2014**, *72*, 1207–1217. [CrossRef]
11. Kiyotsuka, Y.; Acharya, H.P.; Katayama, Y.; Hyodo, T.; Kobayashi, Y. Picolinoxy Group, a New Leaving Group for anti S_N2' Selective Allylic Substitution with Aryl Anions Based on Grignard Reagents. *Org. Lett.* **2008**, *10*, 1719–1722. [CrossRef]
12. Kiyotsuka, Y.; Katayama, Y.; Acharya, H.P.; Hyodo, T.; Kobayashi, Y. New General Method for Regio- and Stereoselective Allylic Substitution with Aryl and Alkenyl Coppers Derived from Grignard Reagents. *J. Org. Chem.* **2009**, *74*, 1939–1951. [CrossRef] [PubMed]
13. Kaneko, Y.; Kiyotsuka, Y.; Acharya, H.P.; Kobayashi, Y. Construction of a quaternary carbon at the carbonyl carbon of the cyclohexane ring. *Chem. Commun.* **2010**, *46*, 5482–5484. [CrossRef]
14. Feng, C.; Kobayashi, Y. Allylic Substitution for Construction of a Chiral Quaternary Carbon Possessing an Aryl Group. *J. Org. Chem.* **2013**, *78*, 3755–3766. [CrossRef] [PubMed]
15. Guisán-Ceinos, M.; Martín-Heras, V.; Tortosa, M. Regio- and Stereospecific Copper-Catalyzed Substitution Reaction of Propargylic Ammonium Salts with Aryl Grignard Reagents. *J. Am. Chem. Soc.* **2017**, *139*, 8448–8451. [CrossRef] [PubMed]
16. Jiang, Y.; Ma, Y.; Ma, E.; Li, Z. Copper-Catalyzed Selective Cross-Couplings of Propargylic Ethers with Aryl Grignard Reagents. *Asian J. Org. Chem.* **2019**, *8*, 1834–1837. [CrossRef]
17. Trost, B.M.; Debien, L. Re-orienting coupling of organocuprates with propargyl electrophiles from SN20 to SN2 with stereocontrol. *Chem. Sci.* **2016**, *7*, 4985–4989. [CrossRef] [PubMed]
18. Gao, X.; Xiao, Y.-L.; Wan, X.; Zhang, X. Copper-Catalyzed Highly Stereoselective Trifluoromethylation and Difluoroalkylation of Secondary Propargyl Sulfonates. *Angew. Chem. Int. Ed.* **2018**, *57*, 3187–3191. [CrossRef]
19. Ma, S.; Wang, G. Regioselectivity Control by a Ligand Switch in the Coupling Reaction Involving Allenic/Propargylic Palladium Species. *Angew. Chem. Int. Ed.* **2003**, *42*, 4215–4217. [CrossRef]
20. Domingo-Legarda, P.; Soler-Yanes, R.; Quirós-López, M.T.; Buñuel, E.; Cárdenas, D.J. Iron-Catalyzed Coupling of Propargyl Bromides and Alkyl Grignard Reagents. *Eur. J. Org. Chem.* **2018**, *2018*, 4900–4904. [CrossRef]
21. Manjón-Mata, I.; Quirós, M.T.; Buñuel, E.; Cárdenas, D.J. Regioselective Iron-Catalysed Cross-Coupling Reaction of Aryl Propargylic Bromides and Aryl Grignard Reagents. *Adv. Synth. Catal.* **2020**, *362*, 146–151. [CrossRef]
22. Tsuji, H.; Kawatsura, M. Transition-Metal-Catalyzed Propargylic Substitution of Propargylic Alcohol Derivatives Bearing an Internal Alkyne Group. *Asian J. Org. Chem.* **2020**, *9*, 1924–1941. [CrossRef]
23. Smith, S.W.; Fu, G.C. Nickel-Catalyzed Negishi Cross-Couplings of Secondary Nucleophiles with Secondary Propargylic Electrophiles at Room Temperature. *Angew. Chem. Int. Ed.* **2008**, *47*, 9334–9336. [CrossRef] [PubMed]
24. Schley, N.D.; Fu, G.C. Nickel-Catalyzed Negishi Arylations of Propargylic Bromides: A Mechanistic Investigation. *J. Am. Chem. Soc.* **2014**, *136*, 16588–16593. [CrossRef]
25. Oelke, A.J.; Sun, J.; Fu, G.C. Nickel-Catalyzed Enantioselective Cross-Couplings of Racemic Secondary Electrophiles That Bear an Oxygen Leaving Group. *J. Am. Chem. Soc.* **2012**, *134*, 2966–2969. [CrossRef] [PubMed]
26. Nishibayashi, Y. Development of Asymmetric Propargylic Substitution Reactions Using Transition Metal Catalysts. *Chem. Lett.* **2021**, *50*, 1282–1288. [CrossRef]
27. Sakata, K.; Nishibayashi, Y. Mechanism and reactivity of catalytic propargylic substitution reactions via metal–allenylidene intermediates: A theoretical perspective. *Catal. Sci. Technol.* **2018**, *8*, 12–25. [CrossRef]
28. Tsuji, J.; Mandai, T. Palladium-Catalyzed Reactions of Propargylic Compounds in Organic Synthesis. *Angew. Chem. Int. Ed. Engl.* **1996**, *34*, 2589–2612. [CrossRef]
29. Kobayashi, Y.; Takashima, Y.; Motoyama, Y.; Isogawa, Y.; Katagiri, K.; Tsuboi, A.; Ogawa, N. α- and γ-Regiocontrol and Enantiospecificity in the Copper-catalyzed Substitution Reaction of Propargylic Phosphates with Grignard Reagents. *Chem. Eur. J.* **2021**, *27*, 3779–3785. [CrossRef]
30. Ogawa, N.; Uematsu, C.; Kobayashi, Y. Stereoselective Synthesis of (−)-Heliannuol E by α-Selective Propargyl Substitution. *Synlett* **2021**, *32*, 2071–2074. [CrossRef]

31. Takashima, Y.; Isogawa, Y.; Tsuboi, A.; Ogawa, N.; Kobayashi, Y. Synthesis of a TNF inhibitor, flurbiprofen and an *i*-Pr analogue in enantioenriched forms by copper catalyzed propargylic substitution with Grignard reagents. *Org. Biomol. Chem.* **2021**, *19*, 9906–9909. [CrossRef]
32. Kobayashi, Y.; Hirotsu, T. Synthesis of (*S*)-Nyasol through the Copper-catalyzed Propargylic Substitution. *Synlett* **2023**, *34*, 159–162. [CrossRef]
33. Kobayashi, Y.; Kiyotsuka, Y.; Sugihara, Y.; Wada, K. Installation of the imidazole ring on chiral substrates via allylic substitution. *Tetrahedron* **2015**, *71*, 6481–6487. [CrossRef]
34. Denmark, S.E.; Vogler, T. Synthesis and Reactivity of Enantiomerically Enriched Thiiranium Ions. *Chem. Eur. J.* **2009**, *15*, 11737–11745. [CrossRef] [PubMed]
35. Makarov, A.S.; Uchuskin, M.G.; Trushkov, I.V. Furan Oxidation Reactions in the Total Synthesis of Natural Products. *Synthesis* **2018**, *50*, 3059–3086. [CrossRef]
36. Kusakabe, M.; Kitano, Y.; Kobayashi, Y.; Sato, F. Preparation of optically active 2-furylcarbinols by kinetic resolution using the Sharpless reagent and their application in organic synthesis. *J. Org. Chem.* **1989**, *54*, 2085–2091. [CrossRef]
37. Kobayashi, Y.; Nakano, M.; Kumar, G.B.; Kishihara, K. Efficient Conditions for Conversion of 2-Substituted Furans into 4-Oxygenated 2-Enoic Acids and Its Application to Synthesis of (+)-Aspicilin, (+)-Patulolide A, and (−)-Pyrenophorin. *J. Org. Chem.* **1998**, *63*, 7505–7515. [CrossRef]
38. Ng, J.S.; Behling, J.R.; Campbell, A.L.; Nguyen, D.; Lipshutz, B. Reactions of higher order cyanocuprates derived from 2-lithiated furans: Scope, limitations, and synthetic utility. *Tetrahedron Lett.* **1988**, *29*, 3045–3048. [CrossRef]
39. Lipshutz, B.H.; Koerner, M.; Parker, D.A. 2-thienyl(cyano)copper lithium. A lower order, "cuprate in a bottle" precursor to higher order reagents. *Tetrahedron Lett.* **1987**, *28*, 945–948. [CrossRef]
40. Lipshutz, B.H.; Kozlowski, J.A.; Parker, D.A.; Nguyen, S.L.; McCarthy, K.E. More highly mixed, higher order cyanocuprates "RT(2-thienyl)Cu(CN)Li$_2$". Efficient reagents which promote selective ligand transfer. *J. Organomet. Chem.* **1985**, *285*, 437–447. [CrossRef]
41. Kobayashi, Y.; Lalitnorasate, P.; Kaneko, Y.; Kiyotsuka, Y.; Endo, Y. Synthesis of ACAT inhibitors through substitution using allylic picolinate and copper reagent. *Tetrahedron Lett.* **2010**, *51*, 6018–6021. [CrossRef]
42. Hansen, T.V.; Skattebøl, L. One-pot synthesis of substituted catechols from the corresponding phenols. *Tetrahedron Lett.* **2005**, *46*, 3357–3358. [CrossRef]

Disclaimer/Publisher's Note: The statements, opinions and data contained in all publications are solely those of the individual author(s) and contributor(s) and not of MDPI and/or the editor(s). MDPI and/or the editor(s) disclaim responsibility for any injury to people or property resulting from any ideas, methods, instructions or products referred to in the content.

Article

Electrochemical Thiocyanation/Cyclization Cascade to Access Thiocyanato-Containing Benzoxazines

Jianguo Hu [1,2,†], Hao Wan [1,†], Shengchun Wang [3], Hong Yi [3,*] and Aiwen Lei [1,3,*]

[1] National Research Center for Carbohydrate Synthesis, Jiangxi Normal University, Nanchang 330022, China
[2] Academician Workstation, Jiangxi University of Chinese Medicine, Nanchang 330004, China
[3] The Institute for Advanced Studies (IAS), College of Chemistry and Molecular Sciences, Wuhan University, Wuhan 430072, China
* Correspondence: hong.yi@whu.edu.cn (H.Y.); aiwenlei@whu.edu.cn (A.L.)
† These authors contributed equally to this work.

Abstract: Due to the importance of SCN-containing heteroarenes, developing novel and green synthetic protocols for the synthesis of SCN-containing compounds has drawn much attention over the last decades. We reported here an electrochemical oxidative cyclization of *ortho*-vinyl aniline to access various SCN-containing benzoxazines. Mild conditions, an extra catalyst-free and oxidant-free system, and good tolerance for air highlight the application potential of this method.

Keywords: electrochemistry; cascade cyclization; thiocyanation; difunctionalization

1. Introduction

Due to the unique physiological activities of heteroarenes, heterocyclic compounds are widely present in natural products, pharmaceuticals, pesticides, and materials [1–4]. Among these valuable heterocycles, benzoxazine has also served as a key skeleton in polymers, contributing to their outstanding characteristics [5,6]. Therefore, the construction and modification of benzoxazines have drawn much attention from synthetic chemists and material scientists.

To date, the flourishing development of radical chemistry has provided attractive protocols to access heterocycles via cascade routes [7–13]. Utilizing radicals as functional reagents, the complicated heterocycles could be effectively obtained under mild conditions. In this context, the radical-induced cyclization cascade process is a considerable path for synthesizing benzoxazines (Scheme 1A). Recently, several breakthroughs have been achieved in such processes. In 2015, Ji and co-workers developed a Cu-catalyzed system for cascade cyclization using nitrile as radical precursors [14]. Two years later, Zhao reported a similar catalytic condition in which alkane was used as radical precursors [15]. Additionally, the radical cascade cyclization was also tolerated with S-centered radicals. In 2019, Li developed an Ag-induced reaction to obtain benzoxazines in which sulfonyl radicals served as a key [16]. Recently, Liang discovered a Mn(OAc)$_3$-promoted sulfonation-cyclization cascade via the SO$_3^-$ radical [17]. Without the assistance of transition-metal, Guo developed a K$_2$S$_2$O$_8$-induced strategy to achieve radical thiocyanooxygenation [18]. Despite of these advances, the heat condition, the use of transition-metal and/or sacrificial oxidant promote the development of alternative methods. Photoredox chemistry provide a mild route to radical cyclization [19,20]. Xiao and colleagues developed an oxytrifluoromethylation of *N*-allylamides to access CF$_3$-containing oxazolines and benzoxazines with Ru-photocatalysts [21]. In 2016, Fu and co-authors reported a photo-induced oxydifluoromethylation of olefinic amides via a difluoromethyl radical method [22]. Three years later, Sun used bromomethyl cyanides as radical precursors to synthesize 4-cyanoethylated benzoxazines by photo-induction [23]. However, the using of expensive catalysts may limit their further application. Overall, developing a practical and green method with bulk radical precursors is in demand for cascade cyclization to synthesize benzoxazines.

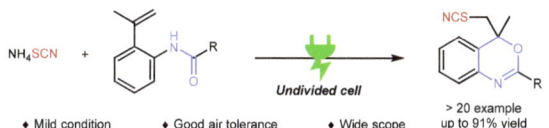

Scheme 1. Recent advances in cyclization cascade to access benzoxazines. (**A**) Advances in radical cascade cyclization. (**B**) Outline of this work: electrochemical thiocyanation/cyclization cascade to access thiocyanato-containing benzoxazines.

Over the last decade, electrochemical organic synthesis has been regarded as a sustainable technology in which electrons serve as redox reagents [24–27]. Especially, benefiting from diverse derivatizations of the thiocyanic group, electrochemical alkene thiocyanation has undergone vigorous development [28]. For example, the aryl thiocyanate generated by electrochemistry can be effectively transformed to other valuable chemicals, including trifluoromethyl thioether, alkyl thioethers, and tetrazole [29]. Since the wide application of ammonium thiocyanates [30–32], constructing thiocyanato-containing benzoxazines via an *S*-centered radical process is a considerable route [33]. Recently, we have developed an efficient electrochemical method to oxidize the olefinic amides to construct the derivatives of benzoxazines and iminoisobenzofurans [34,35]. Based on these advances, we reported here an electrochemical thiocyanation/cyclization cascade to construct benzoxazine under mild conditions (Scheme 1B). The merit of this method was demonstrated by its extra catalyst-free and oxidant-free conditions. While we were preparing this paper, Huang and coworkers reported a similar work that an electrochemical oxythiocyanation of *ortho*-olefinic amides enables the synthesis of thiocyanated benzoxazines [36].

2. Results

2.1. Condition Optimization

Initial condition optimization was examined with *N*-(2-(prop-1-en-2-yl)phenyl)benzamide **1a** as radical acceptor and ammonium thiocyanate **2** as radical precursor (Table 1). After a series of efforts, the optimized condition was established with a carbon rod as the anode, Pt as the cathode, 0.5 M CH_3CN as the solvent, and 1 equivalent H_2SO_4 as the acid. Under a 15 mA electrolysis with 3 h, the desired product **3a** was obtained in 91% isolated yield (entry 1). Without H_2SO_4, this organic transformation was realized in a low yield (entry 2). When trifluoroacetic acid (TFA) was used as the acid, the desired transformation was achieved smoothly in 74% GC yield (entry 3). Using H_2O or 2,2,2-trifluoroethanol (TFE) instead of H_2SO_4, reaction yields obviously decreased (entries 4–5). Moreover, this electrochemical transformation performed worse with other solvents, such as THF, DMSO, and EtOH (entries 6–8). The yields of **3a** were slightly decreased with SS (stainless steel) or Ni plates as the cathode (entries 9–10). Control experiments provide the electrolysis essential for this electrochemical cascade cyclization (entry 11).

Table 1. Condition optimization.

Entry	Variation from the Standard Conditions	Yield (%) [a]
1	None	93 (91 [b])
2	Without H_2SO_4	28
3	TFA instead of H_2SO_4	74
4	H_2O instead of H_2SO_4	28
5	TFE instead of H_2SO_4	10
6	THF instead of MeCN	10
7	DMSO instead of MeCN	28
8	EtOH instead of MeCN	36
9	SS plate instead of Pt plate	70
10	Ni plate instead of Pt plate	59
11	Without electrolysis	N.d.

Reaction conditions: carbon rod anode, platinum plate cathode, constant current = 15 mA, **1a** (0.3 mmol), **2** (0.9 mmol), H_2SO_4 (0.3 mmol), CH_3CN (6.0 mL), air, 3 h. [a] Yields of **3a** were determined by gas chromatography (GC) analysis by using biphenyl as the internal standard. [b] Isolated yield. N.d. = not detected.

2.2. Scope of Substrates

Next, the scope of the substrates was examined (Scheme 2). Various olefinic benzamide derivatives were compatible radical acceptors for achieving the desired transformation. Both electron-donating and electron-withdrawing substitutions on the *para*-position of the phenyl group were well tolerated, producing corresponding products in moderate to high yields (**3a** to **3g**). It is notable that substrates with a redox-sensitive functional groups smoothly completed this electrochemical reaction, for example, *N*-dimethylamino **3h**. Moreover, *ortho*-, *meta*-, and even *multi*-substituted aryl amides were successfully transformed to corresponding products in moderate yields (**3i** to **3l**). In addition, other (hetero)aryl-modified substrates also performed well in this system (**3n** to **3p**). Additionally, this electrochemical cascade cyclization was suitable for stilbene to offer the product in moderate yield (**3m**). Furthermore, a set of alkyl amides realized the desired transformation, forming target products in moderate yields (**3q** to **3w**).

2.3. Mechanistic Studies

Subsequently, radical inhibition experiments were carried out to determine the existence of radical processes (Scheme 3A). With the addition of 2 equivalents 2,2,6,6-tetramethyl-1-piperidinyloxy, the desired transformation was totally inhibited, supporting a radical process involved in this transformation. Moreover, the thiocyanate radical was trapped by 1,1-diphenylethylene under standard conditions. Then, cyclic voltammetry experiments were carried out to investigate the mechanism (Scheme 3B and Supplementary Materials). Without the acid, the oxidation peak of **1a** is not observed. In contrast, the oxidation peak potential of **1a** is detected at 2.27 V in the existence of acid. Notably, the oxidation peak potential of **2** appears at 1.48 V. With the addition of acid, two oxidation peaks of **2** are observed, promoting the secondary oxidation of thiocyanate which is similar to the halogen property {SCN^--$(SCN)_3^-$-$(SCN)_2^-$}. These CV studies disclosed ammonium thiocyanate was preferentially oxidized over **1a**.

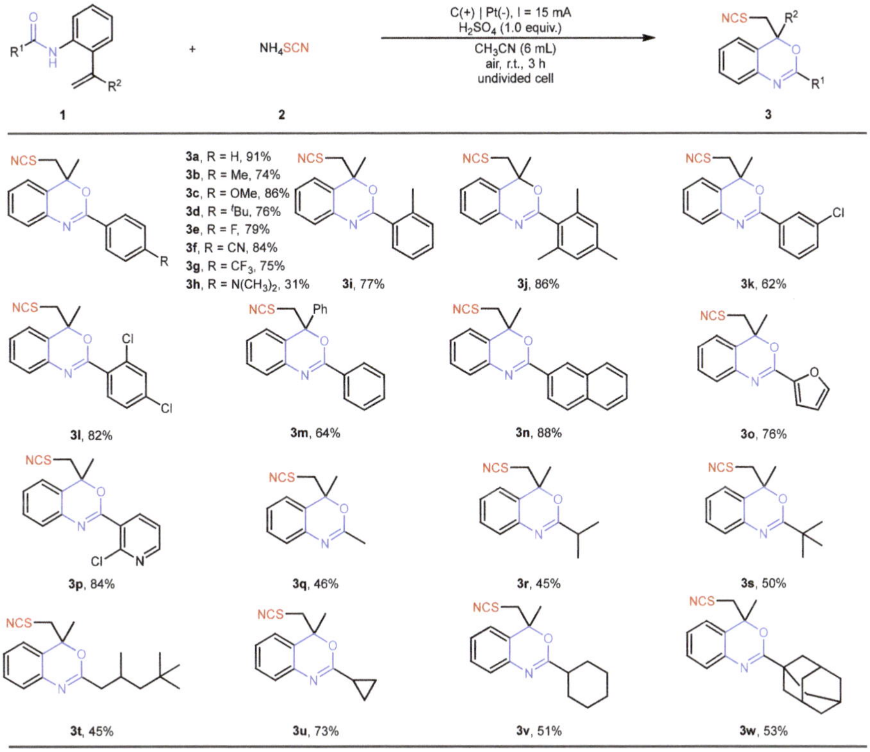

Scheme 2. Scope of substrates. Reaction conditions: carbon rod anode, platinum plate cathode, constant current = 15 mA, **1** (0.3 mmol), **2** (0.9 mmol), H$_2$SO$_4$ (0.3 mmol), CH$_3$CN (6.0 mL), air, 3 h.

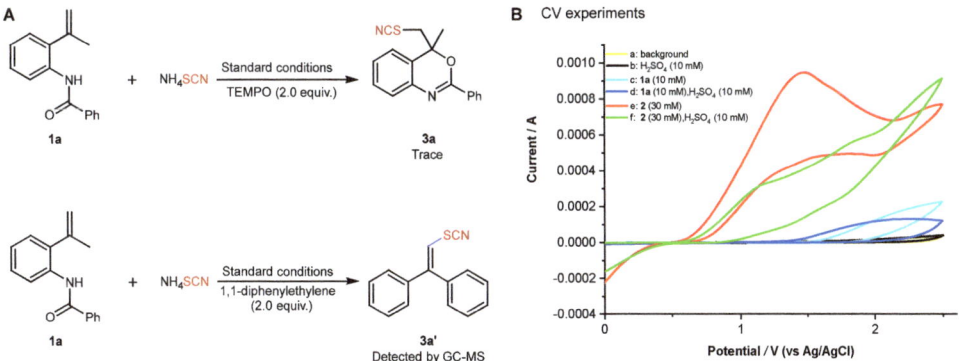

Scheme 3. Mechanistic studies. (**A**) Radical inhibition experiments. (**B**) CV experiments.

Based on the above results, a plausible mechanism was proposed (Scheme 4). In the anode, the thiocyanate anion was oxidized to form thiocyanate radical, which could react with **1a** to offer C-centered radical intermediate **I**. Then, **I** transformed to carbon cation **II** via SET in the anode. Next, the final product **3a** was generated, followed by an intramolecular nucleophilic attack and deprotonation. In the cathode, two protons were reduced to furnish hydrogen.

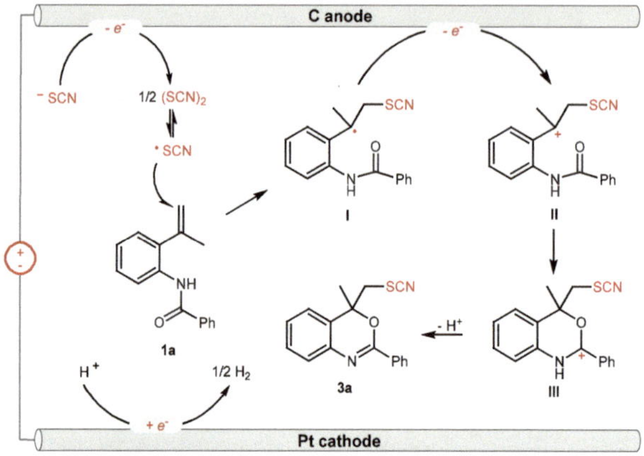

Scheme 4. Plausible mechanism.

3. Materials and Methods

General procedure for the preparation of substrates: A round-bottom flask was charged with methyltriphenylphosphonium bromide (5.36 g, 15.00 mmol) and dry THF (20.00 mL) under N_2 atmosphere, followed by the addition of potassiumtert-butoxide (1.68 g, 15.00 mmol) at 0 °C. The reaction mixture was allowed to warm to ambient temperature and stir for 0.50 h. Next, 2-aminoacetophenone (**1–1**) (1.35 g, 10.00 mmol) was added. The reaction mixture was stirred at room temperature overnight. After completion, the reaction was quenched with saturated $NaHCO_3$ solution and extracted with EtOAc (100.00 mL). The organic phase was dried over anhydrous $MgSO_4$ and concentrated under reduced pressure. The reaction mixture was purified via column chromatography to give **1–2**. To a solution of **1–2** (0.99 g, 7.40 mmol) and Et_3N (1.53 g, 11.10 mmol) in CH_2Cl_2 (15.00 mL) was added the solution of benzoylchloride (1.00 mL, 8.90 mmol) in dichloromethane (5.00 mL) dropwise at 0 °C. After completion, the reaction mixture was purified via column chromatography to give **1a**. Analogues **1a–1w** were synthesized by using similar procedures.

General procedure for electrochemical thiocyanation/cyclization cascade: In an oven-dried, undivided three-necked bottle (10 mL) equipped with a stir bar, *N*-(2-(prop-1-en-2-yl)phenyl)benzamide **1a** (0.30 mmol), ammonium thiocyanate **2** (0.90 mmol) was added to the mixture of acetonitrile (6 mL) and sulfuric acid (0.30 mmol). The bottle was equipped with a graphite rod (ϕ 6 mm, about 15 mm immersion depth in solution) as the anode and platinum plate (15 mm × 15 mm × 0.3 mm) as the cathode. The reaction mixture was stirred and electrolyzed at a constant current of 15 mA under air atmosphere at room temperature for 3 h. After completion of the reaction, as indicated by TLC and GC-MS, the pure product was obtained by flash column chromatography on silica gel.

CV experiments: Cyclic voltammetry experiments were performed in a three-electrode cell connected to a Schlenk line under air at room temperature. The working electrode was a glassy carbon electrode, the counter electrode was a platinum wire. The reference was an Ag/AgCl electrode submerged in saturated aqueous KCl solution, and 6 mL of CH_3CN containing 0.03 M H_2SO_4 was poured into the electrochemical cell in all experiments. The scan rate was 0.1 V/s, ranging from 0 V to 2.5 V. The peak potentials vs. Ag/AgCl were used.

Characterization of products: *4-methyl-2-phenyl-4-(thiocyanatomethyl)-4H-benzo[d][1,3]oxazine* (**3a**). White solid was obtained in 91% isolated yield, 79.9 mg, 0.3 mmol scale, R_f = 0.35 (petroleum ether/ethyl acetate = 10:1). ^{1}H NMR (400 MHz, $CDCl_3$) δ 8.16 (dd, *J* = 8.0, 1.7 Hz, 2H), 7.58–7.42 (m, 3H), 7.41–7.33 (m, 2H), 7.29–7.22 (m, 1H), 7.14 (d, *J* = 7.6 Hz, 1H), 3.58 (d, *J* = 13.8 Hz, 1H), 3.45 (d, *J* = 13.9 Hz, 1H), 1.91 (s, 3H). ^{13}C NMR (101 MHz, $CDCl_3$) δ 155.3, 138.6, 131.8, 131.8, 129.9, 128.3, 127.9, 127.2, 126.3, 125.9, 122.8, 112.2, 78.9, 44.4, 25.8. HRMS (ESI) *m/z*: [M + H]$^+$ Calcd for $C_{17}H_{15}N_2OS^+$ 295.0899; found 295.09245.

4-methyl-4-(thiocyanatomethyl)-2-(p-tolyl)-4H-benzo[d][1,3]oxazine (3b). Colorless oil was obtained in 74% isolated yield, 68.1 mg, 0.3 mmol scale, R_f = 0.35 (petroleum ether/ethyl acetate = 10:1). ^{1}H NMR (400 MHz, CDCl$_3$) δ 8.11–7.96 (m, 2H), 7.40–7.30 (m, 2H), 7.28–7.19 (m, 3H), 7.11 (dd, J = 7.4, 1.2 Hz, 1H), 3.54 (d, J = 13.8 Hz, 1H), 3.41 (d, J = 13.8 Hz, 1H), 2.40 (s, 3H), 1.88 (s, 3H). ^{13}C NMR (101 MHz, CDCl$_3$) δ 155.4, 142.3, 138.7, 129.8, 129.0, 128.9, 127.9, 126.9, 126.2, 125.7, 122.8, 112.2, 78.6, 44.2, 25.6, 21.5. HRMS (ESI) m/z: [M + H]$^+$ Calcd for C$_{18}$H$_{17}$N$_2$OS$^+$ 309.1056; found 309.1062.

2-(4-methoxyphenyl)-4-methyl-4-(thiocyanatomethyl)-4H-benzo[d][1,3]oxazine (3c). Colorless oil was obtained in 86% isolated yield, 83.8 mg, 0.3 mmol scale, R_f = 0.32 (petroleum ether/ethyl acetate = 10:1). ^{1}H NMR (400 MHz, CDCl$_3$) δ 8.14–8.07 (m, 2H), 7.38–7.29 (m, 2H), 7.21 (td, J = 7.3, 1.7 Hz, 1H), 7.10 (dd, J = 7.6, 1.4 Hz, 1H), 6.97–6.91 (m, 2H), 3.84 (s, 3H), 3.55 (d, J = 13.8 Hz, 1H), 3.40 (d, J = 13.8 Hz, 1H), 1.87 (s, 3H). 13C NMR (101 MHz, CDCl3) δ 162.5, 155.1, 138.8, 129.72, 129.69, 126.6, 126.1, 125.4, 124.0, 122.7, 113.6, 112.2, 78.5, 55.3, 44.1, 25.4. HRMS (ESI) m/z: [M + H]+ Calcd for C$_{18}$H$_{17}$N$_2$O$_2$S$^+$ 325.1005; found 325.1013.

2-(4-(tert-butyl)phenyl)-4-methyl-4-(thiocyanatomethyl)-4H-benzo[d][1,3]oxazine (3d). White solid was obtained in 76% isolated yield, 79.8 mg, 0.5 mmol scale, Rf = 0.39 (petroleum ether/ethyl acetate = 10:1). 1H NMR (400 MHz, CDCl3) δ 8.12–8.05 (m, 2H), 7.53–7.45 (m, 2H), 7.41–7.33 (m, 2H), 7.29–7.21 (m, 1H), 7.14 (dd, J = 7.6, 1.3 Hz, 1H), 3.59 (d, J = 13.7 Hz, 1H), 3.45 (d, J = 13.8 Hz, 1H), 1.91 (s, 3H), 1.35 (s, 9H). 13C NMR (101 MHz, CDCl3) δ 155.4, 138.8, 129.9, 129.0, 127.8, 127.0, 126.4, 125.83, 125.79, 125.4, 122.8, 112.3, 78.7, 44.3, 34.9, 31.1, 25.7. HRMS (ESI) m/z: [M + H]+ Calcd for C$_{21}$H$_{23}$N$_2$OS$^+$ 351.1525; found 351.1547.

2-(4-fluorophenyl)-4-methyl-4-(thiocyanatomethyl)-4H-benzo[d][1,3]oxazine (3e). White solid was obtained in 79% isolated yield, 73.7 mg, 0.3 mmol scale, Rf = 0.30 (petroleum ether/ethyl acetate = 10:1). 1H NMR (400 MHz, CDCl3) δ 8.22–8.12 (m, 2H), 7.40–7.30 (m, 2H), 7.24 (td, J = 7.4, 1.7 Hz, 1H), 7.17–7.08 (m, 3H), 3.56 (d, J = 13.9 Hz, 1H), 3.42 (d, J = 14.0 Hz, 1H), 1.89 (s, 3H). 13C NMR (101 MHz, CDCl3) δ 166.2, 163.7, 154.3, 138.4, 130.2, 130.1, 129.9, 127.89, 127.86, 127.2, 126.1, 125.8, 122.8, 115.5, 115.3, 112.1, 79.0, 44.3, 25.8. 19F NMR (376 MHz, CDCl3) δ −107.52. HRMS (ESI) m/z: [M + H]+ Calcd for C$_{17}$H$_{14}$FN$_2$OS$^+$ 313.0803; found 313.0805.

4-(4-methyl-4-(thiocyanatomethyl)-4H-benzo[d][1,3]oxazin-2-yl)benzonitrile (3f). White solid was obtained in 84% isolated yield, 79.7 mg, 0.3 mmol scale, R_f = 0.20 (petroleum ether/ethyl acetate = 5:1). ^{1}H NMR (400 MHz, CDCl$_3$) δ 8.33–8.23 (m, 2H), 7.77–7.69 (m, 2H), 7.45–7.34 (m, 2H), 7.34–7.25 (m, 1H), 7.14 (dd, J = 7.6, 1.4 Hz, 1H), 3.58 (d, J = 14.1 Hz, 1H), 3.46 (d, J = 14.1 Hz, 1H), 1.91 (s, 3H). ^{13}C NMR (101 MHz, CDCl$_3$) δ 153.2, 137.8, 135.8, 131.9, 130.0, 128.2, 128.0, 126.2, 126.0, 122.8, 118.3, 114.6, 111.8, 79.6, 44.4, 26.2. HRMS (ESI) m/z: [M + H]$^+$ Calcd for C$_{18}$H$_{14}$FN$_3$OS$^+$ 320.0852; found 320.0863.

4-methyl-4-(thiocyanatomethyl)-2-(4-(trifluoromethyl)phenyl)-4H-benzo[d][1,3]oxazine (3g). White solid was obtained in 75% isolated yield, 81.3 mg, 0.3 mmol scale, R_f = 0.21 (petroleum ether/ethyl acetate = 5:1). ^{1}H NMR (400 MHz, CDCl$_3$) δ 8.28 (d, J = 8.1 Hz, 2H), 7.71 (d, J = 8.3 Hz, 2H), 7.43–7.34 (m, 2H), 7.32–7.25 (m, 1H), 7.13 (dd, J = 7.6, 1.3 Hz, 1H), 3.57 (d, J = 14.0 Hz, 1H), 3.45 (d, J = 14.0 Hz, 1H), 1.91 (s, 3H). ^{13}C NMR (101 MHz, CDCl$_3$) δ 153.8, 138.1, 135.1 (d, J = 1.5 Hz), 133.0 (q, J = 32.7 Hz), 130.0, 128.2, 127.8, 126.1 (d, J = 1.8 Hz), 125.2 (q, J = 3.8 Hz),125.1, 123.8 (q, J = 273.7 Hz) 122.9, 112.0, 79.4, 44.4, 26.1. ^{19}F NMR (376 MHz, CDCl3) δ -62.76. HRMS (ESI) m/z: [M + H]+ Calcd for C$_{18}$H$_{14}$F$_3$N$_2$OS$^+$ 363.0772; found 363.0773.

4-(4-(isothiocyanatomethyl)-4-methyl-4H-benzo[d][1,3]oxazin-2-yl)-N,N-dimethylaniline (3h). White solid was obtained in 31% isolated yield, 31.3 mg, 0.3 mmol scale, Rf = 0.39 (petroleum ether/ethyl acetate = 10:1). 1H NMR (400 MHz, CDCl3) δ 8.40 (d, J = 2.0 Hz, 1H), 8.11 (dd, J = 8.4, 2.0 Hz, 1H), 7.44–7.33 (m, 2H), 7.28–7.22 (m, 3H), 7.13 (dd, J = 7.4, 1.2 Hz, 1H), 3.58 (d, J = 13.9 Hz, 1H), 3.45 (d, J = 13.9 Hz, 1H), 2.78 (s, 6H), 1.92 (s, 3H). 13C NMR (101 MHz, CDCl3) δ 154.2, 153.8, 138.4, 130.0, 129.1, 128.9, 128.0, 127.4, 126.2, 126.0, 122.9, 122.8, 120.6, 112.1, 111.1, 78.8, 44.4, 44.3, 25.9. HRMS (ESI) m/z: [M + H]+ Calcd for C$_{19}$H$_{20}$N$_3$OS$^+$ 395.0095; found 395.1007.

4-methyl-4-(thiocyanatomethyl)-2-(o-tolyl)-4H-benzo[d][1,3]oxazine (3i). Colorless oil was obtained in 77% isolated yield, 70.5 mg, 0.3 mmol scale, Rf = 0.35 (petroleum ether/ethyl acetate = 10:1). 1H NMR (400 MHz, CDCl3) δ 7.86–7.80 (m, 1H), 7.41–7.30 (m, 3H), 7.30–7.22 (m, 3H), 7.12 (dd, J = 7.7, 1.4 Hz, 1H), 3.60 (d, J = 13.7 Hz, 1H), 3.47 (d, J = 13.7 Hz, 1H), 2.65 (s, 3H), 1.88 (s, 3H). 13C NMR (101 MHz, CDCl3) δ 156.8, 138.4, 138.3, 131.6, 131.5, 130.6, 129.8, 129.5, 127.3, 125.8, 125.7, 125.4, 122.8, 112.0, 79.2, 44.4, 26.4, 21.8. HRMS (ESI) m/z: [M + H]+ Calcd for $C_{18}H_{17}N_2OS^+$ 309.1056; found 309.1071.

2-mesityl-4-methyl-4-(thiocyanatomethyl)-4H-benzo[d][1,3]oxazine (3j). White solid was obtained in 86% isolated yield, 86.3 mg, 0.3 mmol scale, Rf = 0.36 (petroleum ether/ethyl acetate = 10:1). 1H NMR (400 MHz, CDCl3) δ 7.40–7.33 (m, 1H), 7.31–7.25 (m, 2H), 7.11 (dd, J = 8.0, 1.4 Hz, 1H), 6.89 (s, 2H), 3.69 (d, J = 13.7 Hz, 1H), 3.48 (d, J = 13.6 Hz, 1H), 2.36 (s, 6H), 2.28 (s, 3H), 1.85 (s, 3H). 13C NMR (101 MHz, CDCl3) δ 157.4, 139.1, 137.8, 135.7, 130.5, 129.9, 128.3, 127.5, 125.9, 124.6, 123.0, 111.9, 79.8, 45.2, 28.4, 21.1, 19.5. HRMS (ESI) m/z: [M + H]+ Calcd for $C_{20}H_{21}N_2OS^+$ 337.1369; found 337.1380.

2-(3-chlorophenyl)-4-methyl-4-(thiocyanatomethyl)-4H-benzo[d][1,3]oxazine (3k). White solid was obtained in 62% isolated yield, 60.7 mg, 0.3 mmol scale, Rf = 0.35 (petroleum ether/ethyl acetate = 10:1). 1H NMR (400 MHz, CDCl3) δ 8.06 (t, J = 1.9 Hz, 1H), 7.99–7.94 (m, 1H), 7.43–7.37 (m, 1H), 7.34–7.24 (m, 3H), 7.22–7.16 (m, 1H), 7.05 (dd, J = 7.6, 1.4 Hz, 1H), 3.48 (d, J = 13.9 Hz, 1H), 3.36 (d, J = 13.9 Hz, 1H), 1.82 (s, 3H). 13C NMR (101 MHz, CDCl3) δ 153.9, 138.2, 134.4, 133.6, 131.7, 129.9, 129.6, 127.8, 127.6, 126.1, 126.0, 125.8, 122.8, 111.9, 79.3, 44.4, 26.0. HRMS (ESI) m/z: [M + H]+ Calcd for $C_{17}H_{14}ClN_2OS^+$ 329.0510; found 329.0519.

2-(2,4-dichlorophenyl)-4-methyl-4-(thiocyanatomethyl)-4H-benzo[d][1,3]oxazine (3l). White solid was obtained in 82% isolated yield, 88.4 mg, 0.3 mmol scale, Rf = 0.29 (petroleum ether/ethyl acetate = 10:1). 1H NMR (400 MHz, CDCl3) δ 7.68 (d, J = 8.4 Hz, 1H), 7.38 (d, J = 2.1 Hz, 1H), 7.31 (ddd, J = 8.5, 7.2, 1.4 Hz, 1H), 7.27–7.19 (m, 3H), 7.04 (dd, J = 7.6, 1.4 Hz, 1H), 3.54 (d, J = 13.9 Hz, 1H), 3.43 (d, J = 13.9 Hz, 1H), 1.82 (s, 3H). 13C NMR (101 MHz, CDCl3) δ 154.8, 137.9, 137.0, 133.9, 132.2, 130.5, 130.3, 130.0, 128.0, 127.1, 126.0, 125.3, 123.0, 111.9, 80.5, 44.6, 26.8. HRMS (ESI) m/z: [M + H]+ Calcd for $C_{17}H_{13}N_2Cl_2OS^+$ 363.0120; found 363.0129.

2,4-diphenyl-4-(thiocyanatomethyl)-4H-benzo[d][1,3]oxazine (3m). Colorless oil was obtained in 64% isolated yield, 68.4 mg, 0.3 mmol scale, R_f = 0.27 (petroleum ether/ethyl acetate = 5:1). ^{1}H NMR (400 MHz, CDCl3) δ 8.30–8.21 (m, 2H), 7.54–7.43 (m, 3H), 7.42–7.37 (m, 2H), 7.36–7.24 (m, 6H), 7.20–7.14 (m, 1H), 4.05–3.88 (m, 2H). 13C NMR (101 MHz, CDCl3) δ 155.4, 140.0, 139.5, 131.8, 131.6, 130.1, 129.0, 128.8, 128.4, 127.9, 126.9, 126.1, 125.6, 124.7, 124.0, 112.1, 82.3, 43.6. HRMS (ESI) m/z: [M + H]+ Calcd for $C_{22}H_{17}N_2OS^+$ 357.1056; found 357.1084.

4-methyl-2-(naphthalen-2-yl)-4-(thiocyanatomethyl)-4H-benzo[d][1,3]oxazine (3n). Colorless oil was obtained in 88% isolated yield, 90.3 mg, 0.3 mmol scale, Rf = 0.31 (petroleum ether/ethyl acetate = 5:1). 1H NMR (400 MHz, CDCl3) δ 8.63 (d, J = 1.7 Hz, 1H), 8.26 (dd, J = 8.7, 1.8 Hz, 1H), 7.98–7.94 (m, 1H), 7.90–7.81 (m, 2H), 7.58–7.47 (m, 2H), 7.42–7.34 (m, 2H), 7.27–7.19 (m, 1H), 7.12–7.07 (m, 1H), 3.55 (d, J = 13.9 Hz, 1H), 3.42 (d, J = 13.9 Hz, 1H), 1.91 (s, 3H). 13C NMR (101 MHz, CDCl3) δ 155.3, 138.6, 134.9, 132.6, 129.8, 129.04, 129.02, 128.6, 128.0, 127.7, 127.2, 126.5, 126.3, 125.8, 124.3, 122.8, 112.2, 78.9, 44.2, 25.7. HRMS (ESI) m/z: [M + H]+ Calcd for $C_{21}H_{17}N_2OS^+$ 345.1056; found 345.1060.

2-(furan-2-yl)-4-methyl-4-(thiocyanatomethyl)-4H-benzo[d][1,3]oxazine (3o). Yellow oil was obtained in 76% isolated yield, 64.8 mg, 0.3 mmol scale, Rf = 0.22 (petroleum ether/ethyl acetate = 5:1). 1H NMR (400 MHz, CDCl3) δ 7.62 (dd, J = 1.7, 0.8 Hz, 1H), 7.42–7.33 (m, 2H), 7.27–7.21 (m, 1H), 7.15 (dd, J = 3.5, 0.8 Hz, 1H), 7.14–7.09 (m, 1H), 6.54 (dd, J = 3.5, 1.8 Hz, 1H), 3.58 (d, J = 14.0 Hz, 1H), 3.39 (d, J = 13.9 Hz, 1H), 1.89 (s, 3H). 13C NMR (101 MHz, CDCl3) δ 148.2, 145.9, 145.7, 137.9, 129.9, 127.2, 126.2, 125.8, 122.8, 115.5, 112.0, 111.9, 78.8, 43.9, 25.5. HRMS (ESI) m/z: [M + H]+ Calcd for $C_{15}H_{13}N_2O_2S^+$ 285.0692; found 285.0721.

2-(2-chloropyridin-3-yl)-4-methyl-4-(thiocyanatomethyl)-4H-benzo[d][1,3]oxazine (3p). White solid was obtained in 84% isolated yield, 82.9 mg, 0.3 mmol scale, Rf = 0.21 (petroleum ether/ethyl acetate = 5:1). 1H NMR (400 MHz, CDCl3) δ 8.49 (dd, J = 4.8, 2.0 Hz, 1H), 8.17 (dd, J = 7.6, 2.0 Hz, 1H), 7.45–7.30 (m, 4H), 7.14 (dd, J = 7.9, 1.3 Hz, 1H), 3.65 (d, J = 14.0 Hz, 1H), 3.55 (d, J = 14.0 Hz, 1H), 1.94 (s, 3H). 13C NMR (101 MHz, CDCl3) δ 154.2, 150.8, 149.3,

140.0, 137.7, 130.0, 128.8, 128.2, 126.0, 125.2, 123.1, 122.2, 111.8, 80.8, 44.7, 27.1. HRMS (ESI) m/z: [M + H]+ Calcd for $C_{16}H_{13}N_3ClOS^+$ 330.0462; found 330.0471.

2,4-dimethyl-4-(thiocyanatomethyl)-4H-benzo[d][1,3]oxazine (3q). Colorless oil was obtained in 46% isolated yield, 32.1 mg, 0.3 mmol scale, Rf = 0.32 (petroleum ether/ethyl acetate = 5:1). 1H NMR (400 MHz, CDCl3) δ 7.32 (td, J = 7.6, 1.5 Hz, 1H), 7.25–7.13 (m, 2H), 7.03 (dd, J = 7.7, 1.4 Hz, 1H), 3.48 (d, J = 13.9 Hz, 1H), 3.27 (d, J = 14.0 Hz, 1H), 2.19 (s, 3H), 1.80 (s, 3H). 13C NMR (101 MHz, CDCl3) δ 159.0, 138.0, 129.8, 127.0, 125.3, 125.0, 122.8, 112.2, 78.7, 45.0, 26.5, 21.4. HRMS (ESI) m/z: [M + H]+ Calcd for $C_{12}H_{13}N_2OS^+$ 233.0743; found 233.0743.

2-isopropyl-4-methyl-4-(thiocyanatomethyl)-4H-benzo[d][1,3]oxazine (3r). Colorless oil was obtained in 66% isolated yield, 51.5 mg, 0.3 mmol scale, Rf = 0.33 (petroleum ether/ethyl acetate = 2:1). 1H NMR (400 MHz, CDCl3) δ 7.39–7.34 (td, J = 7.6, 1.5 Hz, 1H), 7.32–7.24 (m, 2H), 7.06 (dd, J = 7.6, 1.5 Hz, 1H), 3.63–3.42 (m, 2H), 1.86 (s, 6H), 1.83 (s, 3H). 13C NMR (101 MHz, CDCl3) δ 158.8, 137.0, 130.0, 128.2, 126.2, 125.4, 122.9, 111.8, 111.6, 80.6, 55.5, 44.2, 26.7, 26.6. HRMS (ESI) m/z: [M + H]+ Calcd for $C_{14}H_{16}N_2OS^+$ 260.1055; found 260.1056.

2-(tert-butyl)-4-methyl-4-(thiocyanatomethyl)-4H-benzo[d][1,3]oxazine (3s). Colorless oil was obtained in 50% isolated yield, 41.2 mg, 0.3 mmol scale, Rf = 0.35 (petroleum ether/ethyl acetate =10:1). 1H NMR (400 MHz, CDCl3) δ 7.32 (t, J = 7.3 Hz, 1H), 7.21 (d, J = 8.0 Hz, 2H), 7.04 (d, J = 8.1 Hz, 1H), 3.53 (d, J = 13.6 Hz, 1H), 3.42 (d, J = 13.7 Hz, 1H), 1.74 (s, 3H), 1.28 (s, 9H). 13C NMR (101 MHz, CDCl3) δ 166.3, 138.3, 129.6, 126.8, 125.6, 125.5, 122.5, 112.3, 78.0, 44.1, 37.2, 27.4, 26.0. HRMS (ESI) m/z: [M + H]+ Calcd for $C_{15}H_{18}N_2OS^+$ 275.1212; found 275.1213.

4-methyl-4-(thiocyanatomethyl)-2-(2,4,4-trimethylpentyl)-4H-benzo[d][1,3]oxazine (3t). Colorless oil was obtained in 45% isolated yield, 44.6 mg, 0.3 mmol scale, Rf = 0.35 (petroleum ether/ethyl acetate = 10:1). 1H NMR (400 MHz, CDCl3) δ 7.35–7.29 (m, 1H), 7.23–7.17 (m, 2H), 7.06–7.01 (m, 1H), 3.52 (dd, J = 13.8, 8.3 Hz, 1H), 3.34 (t, J = 14.1 Hz, 1H), 2.52–2.34 (m, 1H), 2.30–2.16 (m, 1H), 2.15–2.06 (m, 1H), 1.79 (d, J = 6.5 Hz, 3H), 1.39–1.28 (m, 1H), 1.20–1.09 (m, 1H), 1.04 (dd, J = 6.6, 3.4 Hz, 3H), 0.92 (d, J = 4.8 Hz, 9H). 13C NMR (101 MHz, CDCl3) δ 160.8 (d, J = 5.8 Hz), 138.0 (d, J = 5.3 Hz), 129.8 (d, J = 4.0 Hz), 126.8 (d, J = 3.6 Hz), 125.5 (d, J = 6.3 Hz), 125.2, 122.7 (d, J = 10.6 Hz), 112.1 (d, J = 1.8 Hz), 78.4 (d, J = 3.9 Hz), 50.5 (d, J = 25.3 Hz), 44.7 (dd, J = 39.5, 19.6 Hz), 31.0 (d, J = 2.9 Hz), 30.0, 27.5 (d, J = 10.3 Hz), 26.8, 26.5, 22.5 (d, J = 19.5 Hz). HRMS (ESI) m/z: [M + H]+ Calcd for $C_{19}H_{27}N_2OS^+$ 331.1839; found 331.1843.

2-cyclopropyl-4-methyl-4-(thiocyanatomethyl)-4H-benzo[d][1,3]oxazine (3u). Colorless oil was obtained in 73% isolated yield, 56.6 mg, 0.3 mmol scale, Rf = 0.22 (petroleum ether/ethyl acetate = 5:1). 1H NMR (400 MHz, CDCl3) δ 7.33–7.27 (td, J = 7.6, 1.4 Hz, 1H), 7.21–7.12 (m, 2H), 7.02 (dd, J = 7.9, 1.3 Hz, 1H), 3.48 (d, J = 13.8 Hz, 1H), 3.32 (d, J = 13.9 Hz, 1H), 1.80–1.67 (m, 4H), 1.16–1.04 (m, 2H), 0.97–0.85 (m, 2H). 13C NMR (101 MHz, CDCl3) δ 162.0, 138.4, 129.7, 126.3, 125.6, 124.6, 122.6, 112.1, 78.4, 44.2, 25.8, 14.4, 7.4, 6.9. HRMS (ESI) m/z: [M + H]+ Calcd for $C_{14}H_{15}N_2OS^+$ 259.0900; found 259.0906.

2-cyclohexyl-4-methyl-4-(thiocyanatomethyl)-4H-benzo[d][1,3]oxazine (3v). White solid was obtained in 51% isolated yield, 45.5 mg by 1H NMR, 0.3 mmol scale, Rf = 0.39 (petroleum ether/ethyl acetate = 5:1). 1H NMR (400 MHz, CDCl3) δ 7.35–7.29 (m, 1H), 7.23–7.17 (m, 2H), 7.07–7.02 (m, 1H), 3.51 (d, J = 13.7 Hz, 1H), 3.36 (d, J = 13.7 Hz, 1H), 2.42–2.29 (m, 1H), 2.00–1.91 (m, 2H), 1.87–1.78 (m, 2H), 1.76 (s, 3H), 1.74–1.65 (m, 1H), 1.57–1.45 (m, 2H), 1.38–1.19 (m, 3H). 13C NMR (101 MHz, CDCl3) δ 164.3, 138.2, 129.8, 126.8, 125.7, 125.2, 122.7, 112.2, 78.1, 44.5, 43.6, 26.1, 25.7, 25.7, 25.6. HRMS (ESI) m/z: [M + H]+ Calcd for $C_{17}H_{21}N_2OS^+$ 301.1369; found 301.1379.

2-(adamantan-1-yl)-4-methyl-4-(thiocyanatomethyl)-4H-benzo[d][1,3]oxazine (3w). Colorless oil was obtained in 53% isolated yield, 46.7 mg, 0.3 mmol scale, Rf = 0.44 (petroleum ether/ethyl acetate = 5:1). 1H NMR (400 MHz, CDCl3) δ 7.34–7.29 (m, 1H), 7.23–7.17 (m, 2H), 7.04 (dd, J = 7.6, 1.4 Hz, 1H), 3.51 (d, J = 13.6 Hz, 1H), 3.40 (d, J = 13.6 Hz, 1H), 2.11–2.02 (m, 3H), 1.95 (d, J = 2.9 Hz, 6H), 1.74 (d, J = 2.7 Hz, 9H). 13C NMR (101 MHz, CDCl3) δ 165.9, 138.5, 129.6, 126.7, 125.9, 125.5, 122.6, 112.4, 77.8, 44.2, 39.03, 38.98, 36.5, 28.0, 25.9. *HRMS (ESI) m/z*: [M + H]+ Calcd for $C_{21}H_{25}N_2OS^+$ 353.1682; found 3531694.

4. Conclusions

We have developed an electrochemical method to produce various benzoxazines under extra catalyst-free and oxidant-free conditions. The good functional group tolerance, excellent performance under air, and scalability demonstrated the application potential of this method. We believe this method not only provides a synthetic route towards thiocyanato-containing benzoxazines but also has a potential to inspire other electrochemical thiocyanations.

Supplementary Materials: The following supporting information can be downloaded at: https://www.mdpi.com/article/10.3390/catal13030631/s1, Figure S1: cyclic voltammetry experiments; NMR spectra [37].

Author Contributions: Conceptualization, H.Y. and A.L.; methodology, J.H. and H.W.; writing—original draft preparation, S.W.; writing—review and editing, all authors; visualization, J.H.; supervision, A.L.; project administration, A.L.; funding acquisition, H.Y. and A.L. All authors have read and agreed to the published version of the manuscript.

Funding: We are grateful for the financial support provided by the National Key R&D Program of China No. 2021YFA1500100 (A.L.), the National Natural Science Foundation of China 22031008 (A.L.), the Science Foundation of Wuhan 2020010601012192 (A.L.).

Data Availability Statement: The data underlying this study are available in the published article and its Supporting Information.

Conflicts of Interest: The authors declare no conflict of interest.

References

1. Vitaku, E.; Smith, D.T.; Njardarson, J.T. Analysis of the structural diversity, substitution patterns, and frequency of nitrogen heterocycles among U.S. FDA approved pharmaceuticals. *J. Med. Chem.* **2014**, *57*, 10257–10274. [CrossRef]
2. Hopkinson, M.N.; Richter, C.; Schedler, M.; Glorius, F. An overview of N-heterocyclic carbenes. *Nature* **2014**, *510*, 485–496. [CrossRef] [PubMed]
3. Seregin, I.V.; Gevorgyan, V. Direct Transition Metal-Catalyzed Functionalization of Heteroaromatic Compounds. *Chem. Soc. Rev.* **2007**, *36*, 1173–1193. [CrossRef] [PubMed]
4. Taylor, A.P.; Robinson, R.P.; Fobian, Y.M.; Blakemore, D.C.; Jones, L.H.; Fadeyi, O. Modern advances in heterocyclic chemistry in drug discovery. *Org. Biomol. Chem.* **2016**, *14*, 6611–6637. [CrossRef]
5. Ghosh, N.N.; Kiskan, B.; Yagci, Y. Polybenzoxazines—New high performance thermosetting resins: Synthesis and properties. *Prog. Ploym. Sci.* **2007**, *32*, 1344–1391. [CrossRef]
6. Lyu, Y.; Ishida, H. Natural-sourced benzoxazine resins, homopolymers, blends and composites: A review of their synthesis, manufacturing and applications. *Prog. Ploym. Sci.* **2019**, *99*, 101168. [CrossRef]
7. Zhang, B.; Studer, A. Recent advances in the synthesis of nitrogen heterocycles via radical cascade reactions using isonitriles as radical acceptors. *Chem. Soc. Rev.* **2015**, *44*, 3505–3521. [CrossRef]
8. Fuentes, N.; Kong, W.; Fernández-Sánchez, L.; Merino, E.; Nevado, C. Cyclization cascades via N-amidyl radicals toward highly functionalized heterocyclic scaffolds. *J. Am. Chem. Soc.* **2015**, *137*, 964–973. [CrossRef]
9. Morris, S.A.; Wang, J.; Zheng, N. The Prowess of Photogenerated Amine Radical Cations in Cascade Reactions: From Carbocycles to Heterocycles. *Acc. Chem. Res.* **2016**, *49*, 1957–1968. [CrossRef]
10. Lu, P.; Wang, Y. Strategies for heterocyclic synthesis via cascade reactions based on ketenimines. *Synlett* **2010**, *2010*, 165–173.
11. Xie, Q.; Long, H.-J.; Zhang, Q.-Y.; Tang, P.; Deng, J. Enantioselective Syntheses of 4H-3,1-Benzoxazines via Catalytic Asymmetric Chlorocyclization of o-Vinylanilides. *J. Org. Chem.* **2020**, *85*, 1882–1893. [CrossRef] [PubMed]
12. Wang, Y.-M.; Wu, J.; Hoong, C.; Rauniyar, V.; Toste, F.D. Enantioselective halocyclization using reagents tailored for chiral anion phase-transfer catalysis. *J. Am. Chem. Soc.* **2012**, *134*, 12928–12931. [CrossRef] [PubMed]
13. Wei, W.-J.; Zhan, L.; Jiang, C.-N.; Tang, H.-T.; Pan, Y.-M.; Ma, X.-L.; Mo, Z.-Y. Electrochemical oxidative cascade cyclization of olefinic amides and alcohols leading to the synthesis of alkoxylated 4H-3,1-benzoxazines and indolines. *Green Chem.* **2023**, *25*, 928–933. [CrossRef]
14. Chu, X.-Q.; Xu, X.-P.; Meng, H.; Ji, S.-J. Synthesis of functionalized benzoxazines by copper-catalyzed C(sp^3)–H bond functionalization of acetonitrile with olefinic amides. *RSC Adv.* **2015**, *5*, 67829–67832. [CrossRef]
15. Wang, J.; Sang, R.; Chong, X.; Zhao, Y.; Fan, W.; Li, Z.; Zhao, J. Copper-catalyzed radical cascade oxyalkylation of olefinic amides with simple alkanes: Highly efficient access to benzoxazines. *Chem. Commun.* **2017**, *53*, 7961–7964. [CrossRef]
16. Wu, J.; Zong, Y.; Zhao, C.; Yan, Q.; Sun, L.; Li, Y.; Zhao, J.; Ge, Y.; Li, Z. Silver or cerium-promoted free radical cascade difunctionalization of o-vinylanilides with sodium aryl- or alkylsulfinates. *Org. Biomol. Chem.* **2019**, *17*, 794–797. [CrossRef]

17. Han, B.; Ding, X.; Zhang, Y.; Gu, X.; Qi, Y.; Liang, S. Mn(OAc)$_3$-Promoted Sulfonation-Cyclization Cascade via the SO$_3^-$ Radical: The Synthesis of Heterocyclic Sulfonates. *Org. Lett.* **2022**, *24*, 8255–8260. [CrossRef]
18. Yang, H.; Duan, X.-H.; Zhao, J.-F.; Guo, L.-N. Transition-metal-free tandem radical thiocyanooxygenation of olefinic amides: A new route to SCN-containing heterocycles. *Org. Lett.* **2015**, *17*, 1998–2001. [CrossRef]
19. DiRocco, D.A.; Dykstra, K.; Krska, S.; Vachal, P.; Conway, D.V.; Tudge, M. Late-stage functionalization of biologically active heterocycles through photoredox catalysis. *Angew. Chem. Int. Ed.* **2014**, *53*, 4802–4806. [CrossRef] [PubMed]
20. Pawlowski, R.; Stanek, F.; Stodulski, M. Recent advances on metal-free, visible-light- induced catalysis for assembling nitrogen- and oxygen-based heterocyclic scaffolds. *Molecules* **2019**, *24*, 1533. [CrossRef]
21. Deng, Q.-H.; Chen, J.-R.; Wei, Q.; Zhao, Q.-Q.; Lu, L.-Q.; Xiao, W.-J. Visible-light-induced photocatalytic oxytrifluoromethylation of *N*-allylamides for the synthesis of CF$_3$-containing oxazolines and benzoxazines. *Chem. Commun.* **2015**, *51*, 3537–3540. [CrossRef] [PubMed]
22. Fu, W.; Han, X.; Zhu, M.; Xu, C.; Wang, Z.; Ji, B.; Hao, X.-Q.; Song, M.-P. Visible-light-mediated radical oxydifluoromethylation of olefinic amides for the synthesis of CF$_2$H-containing heterocycles. *Chem. Commun.* **2016**, *52*, 13413–13416. [CrossRef] [PubMed]
23. Sun, S.; Zhou, C.; Cheng, J. Synthesis of 4-cyanoethylated benzoxazines by visible-light-promoted radical oxycyanomethylation of olefinic amides with bromoacetonitrile. *Tetrahedron Lett.* **2019**, *60*, 150926. [CrossRef]
24. Francke, R.; Little, R.D. Redox catalysis in organic electrosynthesis: Basic principles and recent developments. *Chem. Soc. Rev.* **2014**, *43*, 2492–2521. [CrossRef] [PubMed]
25. Jiang, Y.; Xu, K.; Zeng, C. Use of electrochemistry in the synthesis of heterocyclic structures. *Chem. Rev.* **2018**, *118*, 4485–4540. [CrossRef]
26. Wiebe, A.; Gieshoff, T.; Möhle, S.; Rodrigo, E.; Zirbes, M.; Waldvogel, S.R. Electrifying organic synthesis. *Angew. Chem. Int. Ed.* **2018**, *57*, 5594–5619. [CrossRef] [PubMed]
27. Horn, E.J.; Rosen, B.R.; Baran, P.S. Synthetic organic electrochemistry: An enabling and innately sustainable method. *ACS Cent. Sci.* **2016**, *2*, 302–308. [CrossRef]
28. Ghosh, D.; Ghosh, S.; Hajra, A. Electrochemical Functionalization of Imidazopyridine and Indazole: An Overview. *Adv. Synth. Catal.* **2021**, *363*, 5047–5071. [CrossRef]
29. Dyga, M.; Hayrapetyan, D.; Rit, R.K.; Gooßen, L.J. Electrochemical ipso-Thiocyanation of Arylboron Compounds. *Adv. Synth. Catal.* **2019**, *361*, 3548–3553. [CrossRef]
30. Kokorekin, V.A.; Sigacheva, V.L.; Petrosyan, V.A. New data on heteroarene thiocyanation by anodic oxidation of NH$_4$SCN. The processes of electroinduced nucleophilic aromatic substitution of hydrogen. *Tetrahedron Lett.* **2014**, *55*, 4306–4309. [CrossRef]
31. Fotouhi, L.; Nikoofar, K. Electrochemical thiocyanation of nitrogen-containing aromatic and heteroaromatic compounds. *Tetrahedron Lett.* **2013**, *54*, 2903–2905. [CrossRef]
32. Qumruddeen; Yadav, A.; Kant, R.; Tripathi, C.B. Lewis Base/Brønsted Acid Cocatalysis for Thiocyanation of Amides and Thioamides. *J. Org. Chem.* **2020**, *85*, 2814–2822. [CrossRef] [PubMed]
33. He, T.-J.; Zhong, W.-Q.; Huang, J.-M. The synthesis of sulfonated 4*H*-3,1-benzoxazines via an electro-chemical radical cascade cyclization. *Chem. Commun.* **2020**, *56*, 2735–2738. [CrossRef]
34. Lu, F.; Xu, J.; Li, H.; Wang, K.; Ouyang, D.; Sun, L.; Huang, M.; Jiang, J.; Hu, J.; Alhumade, H.; et al. Electrochemical oxidative radical cascade cyclization of olefinic amides and thiophenols towards the synthesis of sulfurated benzoxazines, oxazolines and iminoisobenzofurans. *Green Chem.* **2021**, *23*, 7982–7986. [CrossRef]
35. Li, H.; Lu, F.; Xu, J.; Hu, J.; Alhumade, H.; Lu, L.; Lei, A. Electrochemical oxidative selenocyclization of olefinic amides towards the synthesis of iminoisobenzofurans. *Org. Chem. Front.* **2022**, *9*, 2786–2791. [CrossRef]
36. Qian, S.; Xu, P.; Zheng, Y.; Huang, S. Electrochemical oxythiocyanation of ortho-olefinic amides: Access to diverse thiocyanated benzoxazines. *Tetrahedron Lett.* **2023**, *116*, 154341. [CrossRef]
37. Guo, J.; Hao, Y.; Li, G.; Wang, Z.; Liu, Y.; Li, Y.; Wang, Q. Efficient synthesis of SCF3-substituted tryptanthrins by a radical tandem cyclization. *Org. Biomol. Chem.* **2020**, *18*, 1994–2001. [CrossRef] [PubMed]

Disclaimer/Publisher's Note: The statements, opinions and data contained in all publications are solely those of the individual author(s) and contributor(s) and not of MDPI and/or the editor(s). MDPI and/or the editor(s) disclaim responsibility for any injury to people or property resulting from any ideas, methods, instructions or products referred to in the content.

Article

A Second-Generation Palladacycle Architecture Bearing a N-Heterocyclic Carbene and Its Catalytic Behavior in Buchwald–Hartwig Amination Catalysis

Sylwia Ostrowska [1], Lorenzo Palio [1], Agnieszka Czapik [2], Subhrajyoti Bhandary [1], Marcin Kwit [2], Kristof Van Hecke [1] and Steven P. Nolan [1,*]

[1] Department of Chemistry, Center for Sustainable Chemistry, Ghent University, Krijgslaan 281 (S-3), 9000 Ghent, Belgium
[2] Faculty of Chemistry, Adam Mickiewicz University, Uniwersytetu Poznanskiego 8, 61-614 Poznan, Poland
* Correspondence: steven.nolan@ugent.be; Tel.: +32-9-2644458

Abstract: Palladacyclic architectures have been shown as versatile motifs in cross-coupling reactions. NHC-ligated palladacycles possessing unique electronic and steric properties have helped to stabilize the catalytically active species and provide additional control over reaction selectivity. Here, we report on a synthetic protocol leading to palladacycle complexes using a mild base and an environmentally desirable solvent, with a focus on complexes bearing backbone-substituted N-heterocyclic carbene ligands. The readily accessible complexes exhibit high catalytic activity in the Buchwald–Hartwig amination. This is achieved using low catalyst loading and mild reaction conditions in a green solvent.

Keywords: palladacycle; weak base route; NHC; green solvent; C-C and C-heteroatom bonds formation; cross-couplings

Citation: Ostrowska, S.; Palio, L.; Czapik, A.; Bhandary, S.; Kwit, M.; Van Hecke, K.; Nolan, S.P. A Second-Generation Palladacycle Architecture Bearing a N-Heterocyclic Carbene and Its Catalytic Behavior in Buchwald–Hartwig Amination Catalysis. *Catalysts* **2023**, *13*, 559. https://doi.org/10.3390/catal13030559

Academic Editors: Ewa Kowalska and Yuichi Kobayashi

Received: 14 February 2023
Revised: 7 March 2023
Accepted: 8 March 2023
Published: 10 March 2023

Copyright: © 2023 by the authors. Licensee MDPI, Basel, Switzerland. This article is an open access article distributed under the terms and conditions of the Creative Commons Attribution (CC BY) license (https://creativecommons.org/licenses/by/4.0/).

1. Introduction

The use of palladium in cross-coupling reaction catalysis has seen huge growth in the last 50 years since its discovery in the 1970s [1–4]. Various transition metal complexes are easily prepared from palladium and most of them have shown high catalytic activity compatible with most functional groups [5]. Although monoligated Pd(0) species constitute active catalytic species in cross-coupling reactions [6], such complexes are often unstable in air and difficult to synthesize [7]. The use of Pd(II) precatalysts is in most cases the solution to this problem, requiring an additional activation step from Pd(II) to Pd(0) species that then enters the catalytic cycle. The activation or reduction occurs in situ, can be promoted by reaction conditions, and usually involves a reductive elimination leading to a monoligated Pd(0)-L complex. Several types of Pd(II) precatalysts are nowadays widely used in organic synthesis, among which palladacyclic compounds occupy a prominent position [3,8–17].

Palladacycles are compounds that have at least one Pd–C bond in their molecular architectures, which are further intramolecularly stabilized by a dative bond between the metal and a built-in donor heteroatom Y (typically Y = N, P, S, or O) yielding a five- or six-membered chelate ring [8]. Since the seminal research of Cope and Siekman in 1965 on the isolation of stable cyclopalladated azobenzenes [18,19] and the subsequent pioneering contributions from Herrmann and Beller that showed the capacity of cyclopalladated ligand **1** (Figure 1) to catalyze the Heck coupling with an unprecedented catalytic activity, (due to easily formed palladium nanoparticles, in this instance), there have been significant developments in the design of new palladacyclic frameworks [20]. Various cyclopalladated ligand systems have been reported that have been successfully used in C–C and C–N cross-couplings [16,17]. However, almost all precatalysts required an exogenous additive or, in some cases, one single catalytic cycle to be activated. Another category of precatalysts that has emerged as a powerful catalytic system is the recently developed Buchwald palladacycles **2–3** (Figure 1) [21]. These are formed from Pd and a hemilabile ligand that subsequently dissociates during the initial catalytic cycle.

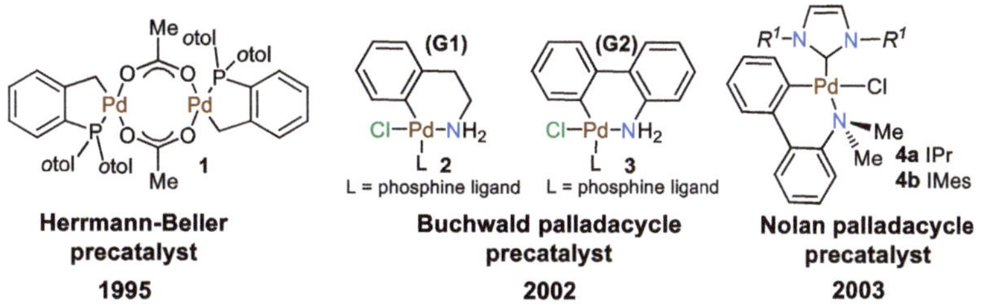

Figure 1. Herrmann–Beller precatalyst (**1**) and Buchwald precatalyst generations **G1** (**2a**), **G2** (**2b**), Solvias precatalyst (**3**) and Nolan precatalyst (**4**).

The first and second generation (**G1** and **G2**) (Figure 1) precatalysts have shown excellent activity in Suzuki–Miyaura [21,22] and Sonogashira [23] couplings, amination [24], and C–H arylation [25] reactions. For example, with the **G2** precatalyst, Suzuki–Miyaura coupling reactions of five-membered 2-heterocyclic boronic acids with (hetero)aryl halides were achieved under extremely mild reaction conditions in a short reaction time [22]. However, these precatalysts still possess some drawbacks. The **G1** displays a short lifetime in solution and cannot be activated with a weak base at room temperature, and its preparation involves the handling of unstable organometallic intermediates. For the second generation **G2**, in addition to being poorly soluble in organic solvents, it is also not stable in solution for extended periods of time, and bulkier phosphines such as BrettPhos, an important ligand in C–N bond formation, tBuXPhos, tBuBrettPhos, and RockPhos could not been incorporated to lead to well-defined and isolable complexes.

Historically, tertiary phosphines have dominated the area as supporting ligands of first-choice [26]. During the past two decades, N-heterocyclic carbenes (NHCs) have seen interest grow exponentially, as ligands in Pd-mediated cross-coupling reactions and the development of well-defined complexes have played vital roles in their now ubiquitous use [1–3,5]. The synthesis of such stable, well-defined complexes has been made possible thanks to the high Pd–NHC bond stability that prevents active catalyst decomposition [1,3,27]. In particular, considering the unique higher σ-donating and weaker π-accepting abilities compared to phosphine relatives, NHCs [3,14,27] form stronger metal–ligand bonds. This characteristic of NHCs has been deployed in the formation of new types of palladacycles, coined NHC–C,N-Pdcycles [1,28]. It is worthy of note that NHCs not only endow the Pd center with increased electronic density facilitating the oxidative addition step, but also contribute to the structural stability of active LPd(0) species, rather than the formation of Pd nanoparticles [13,15,29] throughout the catalytic cycle. Therefore, researchers have focused on this interesting field of palladacycles and intensively explored their potential applications [12,30]. Today, many examples of palladacycles exist that show high catalytic activity in Suzuki–Miyaura [31–42], Sonogashira [41–43], Heck [44–46], Buchwald–Hartwig [47–50] coupling reactions, and carbonylation reactions [51–55].

We have previously reported on the synthesis of amino-palladacycles **4a** and **4b** (Figure 1) [56] which build on the palladacyclic architecture bearing secondary phosphine ligands (**3** in Figure 1) reported by a group at Solvias [57].

The activity of the most efficient IPr-containing palladacycle **4a**, was investigated in the Buchwald–Hartwig, α-ketone arylation, reductive dehalogenation, and Suzuki–Miyaura reactions [56,58]. The reactions could be performed at low catalyst loadings (1−0.05 mol%) and under mild conditions (rt to 65 °C). Precatalyst **4a** proved to be quite versatile and displayed a wide reaction scope in numerous cross-coupling reactions. In the Buchwald–Hartwig amination, primary and secondary alkyl and aryl amines were coupled in high yields [56]. Similarly, in the α-ketone arylation, aryl and alkyl ketones reacted well [59,60].

In general, the known design strategies for NHC-C, Y-Pdcycles include two synthetic methods. The first makes use of stable precursors of N-heterocyclic carbene ligands, a simple palladium salt and a suitable organic compound as a donor source which, under the influence of a strong base present in the system, lead to the production of the palladacycle. The second method uses palladacycle dimers possessing bridging halides which are cleaved by free N-heterocyclic carbenes, giving a monomeric palladacycle (Figure 2a).

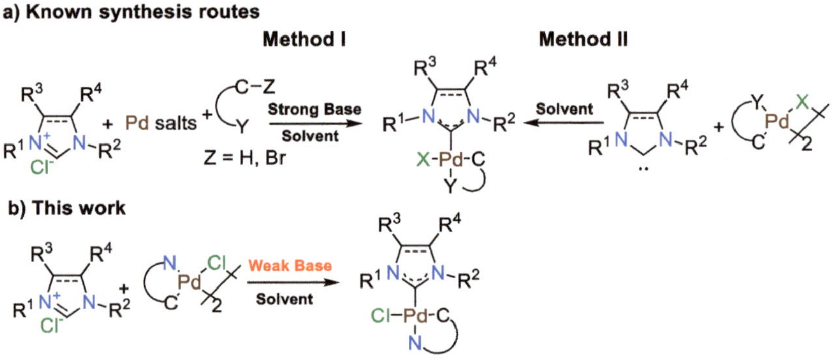

Figure 2. (**a**) Design strategies/synthetic routes to NHC–C,N-Pdcycles; (**b**) Synthesis for NHC-C,N-Pdcycles using the weak base route.

Palladacycles of the latter generations bearing N-heterocyclic carbenes (NHCs) as supporting ligands have not been studied. To our knowledge, no examples of **G2**-like precatalyst bearing NHCs as ligands are reported in the literature. We now report on such a study dealing with the synthesis of this type of palladacycle via a weak base route using environmentally safe solvents (Figure 2b). Furthermore, their role as precatalysts in the Buchwald–Hartwig aryl amination reaction has been examined.

2. Results and Discussion

2.1. Synthesis of [Pd(NHC)(NH₂)(CC)Cl] Palladacycles

Analyzing previous reports on the use of the weak base route in the synthesis of metal complexes of transition metals [60–63], we began our study by examining the reaction of 1,3-bis(2,6-diisopropylphenyl)imidazolium chloride (IPr·HCl) with the palladacycle dimer **5** using K_2CO_3 as a base and acetone as a solvent (Table 1, entry 1). Gratifyingly, the targeted [Pd(IPr)(NH₂)(CC)Cl] complex **7** was obtained at a very good yield (91%) after 2 h when the reaction was heated in acetone at 60 °C (Table 1, entry 1). Then, wanting to optimize reaction parameters for the synthesis of **7**, we conducted experiments where solvents, bases, and temperatures were modified. As shown in Table 1, after lowering the temperature to 40 °C, the reaction time required to reach full conversion had to be extended to 12 h, and the pure complex was only obtained in an 82% yield (Table 1, entry 2). Next, the effects of the base on the reaction yield was examined. When NEt₃ was used, no product was observed, even when reaction times were extended to 24 h (Table 1, entry 3). The use of sodium acetate led to product **7** at a 66% yield after 6 h.

Replacing acetone with toluene, a 52% yield of the product was obtained with reactions being conducted at 40 °C and 60 °C for 18 h (Table 1, entries 5–6). When the system was heated to 80 °C, a 77% efficiency was obtained after 6 h of reaction, while heating the system to 100 °C allowed a 90% yield after 1 h (Table 1, entry 7–8). The use of ethyl acetate as a solvent led to a 71% yield when the reaction was carried out at 40 °C, an 82% yield when the reaction was conducted at 50 °C, and a 75% yield was obtained when the reaction was performed at 60 °C (Table 1, entries 9–11). Ethanol proved to be a totally ineffective solvent, producing a <5% yield after 3 h (Table 1, entry 12); moreover, decomposition of the complex and precipitation of palladium black were observed under these conditions.

With optimal reaction conditions in hand, which coincidentally happen to be the initial conditions examined, an attempt was made to synthesize complexes bearing other NHC ligands. Palladacycles with 1,3-bis(2,6-diisopropylphenyl)imidazol-2-ylidene (IPr) and 1,3-bis(2,4,6-trimethylphenyl)imidazol-2-ylidene (IMes) ligands were synthesized in very good isolated yields of 82% and 95%, respectively, (Figure 3). In both cases, excellent yields and purities were obtained with a simple work-up procedure. Unfortunately, attempts to synthesize a complex bearing the sterically much more demanding 1,3-bis [2,6-bis(diphenylmethyl)-4-methylphenyl] imidazol-2-ylidene (IPr*) ligand proved unsuccessful. Furthermore, [Pd(IPr)(NH$_2$)(CC)Cl] and [Pd(IMes)(NH$_2$)(CC)Cl] complexes were synthesized on a gram-scale with high isolated yields (Figure 3), highlighting the scalability of this simple procedure. It should be noted that when carrying out the reaction to obtain **7** on a gram scale, acetone was used to give complete conversion after 16 h; however, the yield of the pure complex was only 46%, an issue that emerges on larger scale reactions because of the moderate solubility of both the base and the IPr·HCl salt in this solvent. Using toluene instead leads to improved solubility, as well as greatly reduced decomposition and by-product formation. In toluene, the pure product is obtained at a 74% yield.

Table 1. Selected entries for the optimization of reaction conditions leading to palladacycle 7.

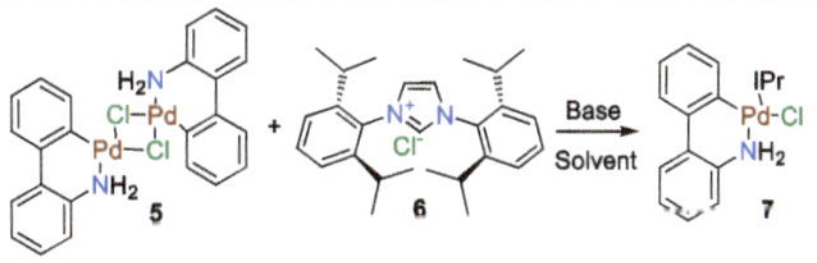

Entry	Solvent	Base	T (°C)	Time (h)	Conv. [a] (%)	Yield [a] (%)
1	Acetone	K$_2$CO$_3$	60	2	>99	91
2	Acetone	K$_2$CO$_3$	40	12	>99	82
3	Acetone	Et$_3$N	60	24	0	0
4	Acetone	NaOAc	60	6	>99	66
5	Toluene	K$_2$CO$_3$	40	18	>99	52
6	Toluene	K$_2$CO$_3$	60	18	>99	52
7	Toluene	K$_2$CO$_3$	80	6	>99	58
8	Toluene	K$_2$CO$_3$	100	1	>99	90
9	EtOAc	K$_2$CO$_3$	40	3	>99	71
10	EtOAc	K$_2$CO$_3$	50	3	>99	82
11	EtOAc	K$_2$CO$_3$	60	3	>99	75
12	EtOH	K$_2$CO$_3$	40	3	>99	3

Reaction condition: Pd dimer = 1eq; (IPr·HCl) = 2.1 eq; base = 3 eq; Pd dimer loading = 50 mg; [a] All conversions and yields were determined by ^{1}H NMR, using 1,3,5-trimethoxybenzene as internal standard.

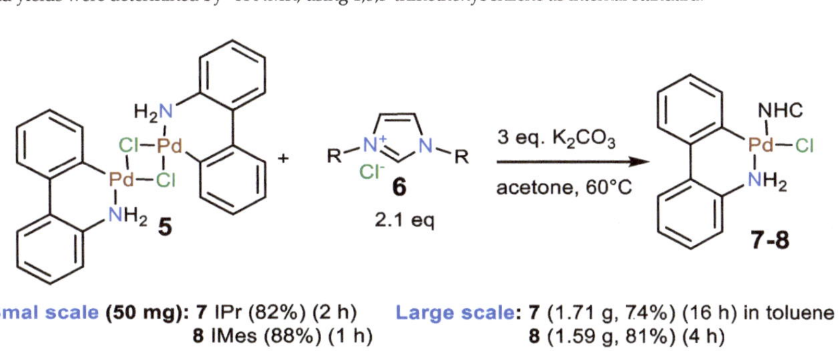

Smal scale (50 mg): 7 IPr (82%) (2 h)   Large scale: 7 (1.71 g, 74%) (16 h) in toluene
                    8 IMes (88%) (1 h)   8 (1.59 g, 81%) (4 h)

Figure 3. Scheme of the synthesis of palladacycle [Pd(NHC)(NH$_2$)(CC)Cl] complexes.

All synthesized palladium complexes were fully characterized using NMR spectroscopy and elemental analysis. In particular, the disappearance of the imidazolium proton in the ^{1}H NMR spectra and the presence of the carbene carbon signal at ca. 160 ppm in the ^{13}C NMR spectra are clearly indicative of the formation of the complexes of interest. Moreover, crystals of **7** and **8**, suitable for X-ray diffraction study on single crystals, were grown with the slow diffusion of pentane into a saturated CH$_2$Cl$_2$ solution of the complexes; the corresponding molecular structures are showed in Figure 4.

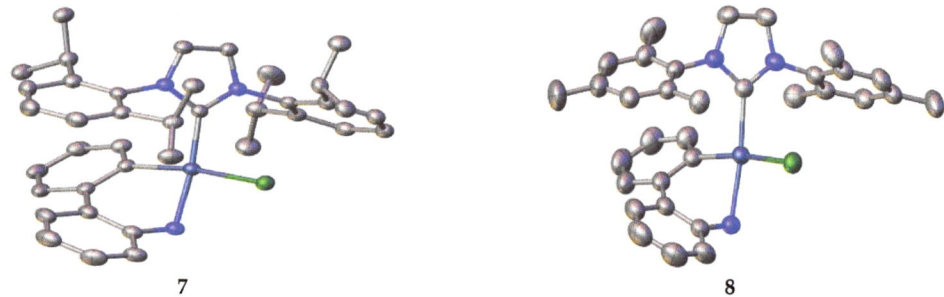

Figure 4. (**Left**) X-ray structure of palladacycle **7**; selected bond distances (Å) and angles (deg): Pd1-Cl5 2.4146(9), Pd1-N3 2.095(3), Pd1-C1 1.996(3), Pd1 C67 2.004(4), CCDC 2241786; (**Right**) X-ray structure of palladacycle **8**; selected bond distances (Å) and angles (deg): Pd1-Cl1 2.4331 (8), Pd1-C1 2.001 (3), Pd1-N3 2.109 (2), Pd1-C22 1.996 (3), C1-Pd1-N3 173.63 (12), CCDC 2225057. Ellipsoids are drawn at the 30% probability level, and hydrogen atoms have been omitted for clarity.

2.2. [Pd(NHC)(NH$_2$)(CC)Cl] Palladacycles as Catalysts in the Buchwald–Hartwig Amination Reaction

To test the catalytic activity of the synthesized complexes, we selected the Buchwald–Hartwig amination reaction as a test reaction. The coupling of 4-chloroanisole with morpholine in THF and 2-MeTHF with KO^{t}Bu, 1 mol% of **7** at 80 °C was chosen as a model reaction [64]. A very high yield (91%) of the product was achieved after 4 h in 2-MeTHF (Table 2, entries 1–2). The role and identity of the solvent was examined; moving from Me-THF to dioxane does affect the efficiency of the reaction. Indeed, the use of 1,4-dioxane allowed a 98% conversion to be obtained when the reaction was carried out at 80 °C, and 100% when the reaction was carried out at 100 °C (Table 2, entries 3–4).

Searching for a greener solvent allowed us to make use of cyclopentyl methyl ether (CPME). The use of CPME over other commonly employed ether solvents, such as THF, diethyl ether, or 1,4-dioxane, is preferred in view of its stability under acidic and basic conditions; it also does not lead to the formation of peroxides, thereby decreasing risks associated with this reaction [65]. The model reaction in this solvent in the presence of KO^{t}Bu gave an excellent conversion after 1 h when the reaction was carried out at 80 °C (Table 2, entries 5–8). Conducting this reaction at a lower temperature, i.e., 60 °C or 70 °C, led to no conversion (Table 2, entries 9–10). The ideal activation temperature appears to be 80 °C, as increasing the temperature does not affect conversion. The use of K$_2$CO$_3$ and Cs$_2$CO$_3$ as weak bases in the reaction did not lead to catalyst activation (Table 2, entries 11, 13), while the use of NaOAc gave only moderate conversion (Table 2, entry 10). The possibility of reducing the catalyst loading was also examined (Table 2, entries 14–18). The reaction proceeds using a 0.5 mol% catalyst loading. The last optimization step examined the activity of palladacycle dimer **5** and palladacycle **8**. The starting palladacycle dimer **5** proved inactive in the control experiment and complex **8** yielded an 88% conversion under optimized reaction conditions (Table 2, entries 19–20).

Table 2. Buchwald–Hartwig amination; selection of optimal conditions.

MeO–C₆H₄–Cl + HN(morpholine) →[7, solvent, base] MeO–C₆H₄–N(morpholine)

Entry	Loading Cat. [mol%]	Solvent	Base	Temp. [°C]	Time [h]	Conversion [%] [a]
1	1	THF	KO^{t}Bu	80	4	83
2	1	Me-THF	KO^{t}Bu	80	4	91
3	1	1,4-dioxane	KO^{t}Bu	80	4	98
4	1	1,4-dioxane	KO^{t}Bu	100	4	100
5	1	CPME	KO^{t}Bu	100	4	99
6	1	CPME	KO^{t}Bu	80	4	100
7	1	CPME	KO^{t}Bu	80	2	100
8	1	CPME	KO^{t}Bu	80	1	94
9	1	CPME	KO^{t}Bu	60	24	NR
10	1	CPME	KO^{t}Bu	70	24	NR
11	1	CPME	K$_2$CO$_3$	80	24	NR
12	1	CPME	NaOAc	80	2	62 (63) [b]
13	1	CPME	Cs$_2$CO$_3$	80	2	NR
14	0.5	CPME	KO^{t}Bu	80	2	100
15	0.5	CPME	KO^{t}Bu	80	1	98
16	0.2	CPME	KO^{t}Bu	80	2	8 (8) [b]
17	0.3	CPME	KO^{t}Bu	80	2	14 (15) [b]
18	0.4	CPME	KO^{t}Bu	80	2	33 (54) [b]
19 [c]	0.5	CPME	KO^{t}Bu	80	24	NR
20 [d]	0.5	CPME	KO^{t}Bu	80	2	84 (84) [b]

Reaction conditions: 4-chloroanisole (1.0 mmol); morpholine (1.2 mmol); base (1.2 mmol); cat. Pd (7); solvent (2.0 mL); N$_2$; [a] GC yield, dodecane as internal standard; [b] 24 h; [c] complex 5 was used; [d] complex 8 was used.

With the optimal reaction conditions in hand (Table 2, entry 14) Buchwald–Hartwig amination reactions were carried out between various (hetero)aryl chlorides and amines in the presence of 0.5 mol% of 7 (Figure 5).

Aryl chlorides were used as effective coupling partners in these reactions. Despite the unreactive nature of the C−Cl bond, these reactions required low catalyst loading and reaction times, ranging from 1 h to a maximum of 4 h. It was interesting to observe that varying the nature of the aryl group substituents had a minimal influence on reaction rates. Sterically hindered aryl chlorides, such as 2-methoxychlorobenzene or 2-methylchlorobenzene, have rates slightly lower than unhindered relatives. We were pleased to find that a large variety of substrates, such as heterocyclic alkylamines, dialkylamines, aryl−alkylamines, and primary amines are all efficient coupling partners. Aryl chlorides bearing electron donating and withdrawing functional groups reacted smoothly with secondary alkyl and arylalkyl amines. A series of secondary amines (aliphatic and aromatic) was also successfully coupled using this methodology. Various anilines, including sterically hindered di-ortho-substituted examples, were well-tolerated and typically produced excellent yields with these as nucleophiles. The only exception found was for the less-active and very bulky derivatives, where lower yields were obtained.

This catalytic system proved less efficient for more challenging coupling partners, such as heteroaryl chlorides and primary amines. Heteroaryl chlorides are known for possible catalyst deactivation and poor solubility that cause significant difficulties in C−N coupling. The use of a long-chain halide turned out to be equally ineffective under these conditions, even after extending the reaction time to 24 h. The comparison of the catalytic activity of known palladacycles in the Buchwald–Hartwig reaction is presented in Table S1.

The mechanism of the Buchwald–Hartwig amination reaction was first described more than 20 years ago [66–68]. It is generally assumed to proceed through steps similar to those known for palladium-catalyzed C–C coupling reactions. These steps include oxidative addition of the aryl halide to the Pd (0) groupings, addition of an amine to the oxidative addition complex, deprotonation of the amine, and then reductive elimination.

Based on our previous studies [56] and those of the Indolese group [57,69], we propose a mechanism (Figure 6) in which the activation of the palladacyclic catalyst is initiated by the attack of the alkoxide on the palladium, resulting in the formation of palladium alkoxide. Subsequently, this electron-rich palladium type (0), stabilized by the presence of the NHC ligand, enters the catalytic cycle where oxidative addition of aryl halides takes place. The reaction proceeds in the classical manner.

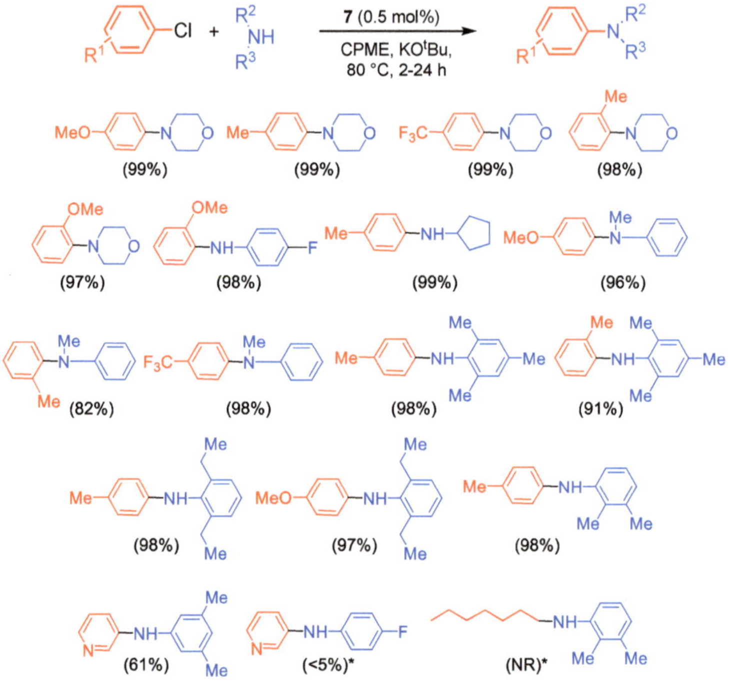

Figure 5. Scope of the Buchwald–Hartwig amination catalyzed by **7**. Reaction conditions: aryl chlorides (1.0 mmol); amine (1.2 mmol); base (1.2 mmol); cat. Pd (**7**); solvent (2.0 mL); N$_2$; Isolated yield; * GC yield, 24 h.

The palladacycle activation step proposed in the mechanism is supported by experiments. Specifically, we performed a reaction in which 1 eq. palladacycle **7** was treated with 1.1 eq. KOtBu in CPME by heating this mixture at 80 °C for 1 h. After completion of the reaction, the ^{1}H NMR spectrum (CDCl$_3$) revealed the formation of a new complex at a 28% yield. Moreover, decomposition compounds of the palladium complex that are difficult to identify were also observed. Due to the overlap of signals relevant to the interpretation of the tert-butyl group, we were unable to determine unequivocally whether the new complex we obtained is the alkoxide form of palladium. We therefore repeated this reaction in deuterated benzene as a solvent to facilitate the spectroscopic analysis.

The ^{1}H NMR spectrum after 5 min of heating revealed the disappearance of a signal from the CH$_3$ groups in KO^{t}Bu (1.04 ppm) and the formation of a new signal from the CH$_3$ groups in -O^{t}Bu (1.05 ppm), shifted slightly towards a higher field. All other signals characteristic of the starting material had also shifted. For example, the signal for C$\underline{H}$(CH$_3$)$_2$ was shifted from 1.43 ppm to 1.48 ppm. We additionally performed a reaction in which 1 eq. palladacycle **7** was reacted with 1 eq. 1-chloro-4-methoxybenzene and 1.1 eq. KO^{t}Bu in CPME by heating this mixture at 80 °C for 1 hr. After completion of the reaction, we observed a signal at 11.2 min on the gas chromatogram which suggested the presence of an aminobiphenyl fragment of the palladacycle released under these conditions. This retention time was confirmed by comparison with an authentic sample. These experiments strongly support the "activation" of the Pd(II) precatalyst by an alkoxide base as the key event leading to an in-cycle catalytically active species.

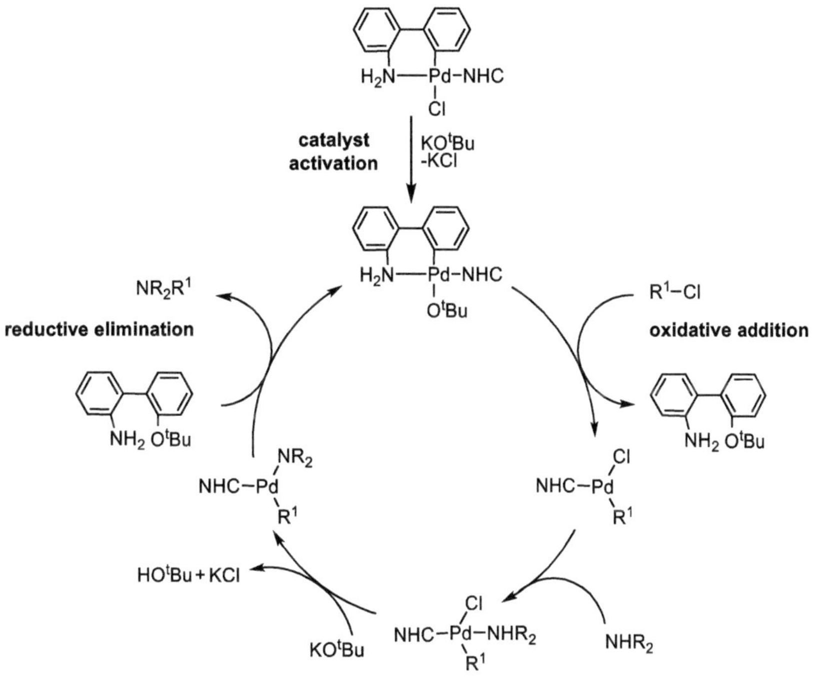

Figure 6. Mechanistic proposal.

3. Materials and Methods

3.1. Materials

All reactions were performed under N_2 unless otherwise mentioned. All solvents and other reagents were purchased from commercial sources (ChemLab (Zedelgem, Belgium), Sigma-Aldrich (Overijse, Belgium), Umicore (Brussels, Belgium)), and used as received without further purification unless otherwise stated. ^{1}H NMR and ^{13}C-{^{1}H} NMR spectra were recorded on Bruker Avance-400 or 300 (Billerica, United States) instruments at 298 K. Chemical shifts (ppm) in ^{1}H and ^{13}C NMR spectra are referenced to the residual solvent peak (CDCl$_3$: δ H = 7.26 ppm, δ C = 77.2 ppm). Coupling constants (*J*) are given in hertz. Abbreviations used in the designation of the signals are: s = singlet, d = doublet, t = triplet, m = multiplet, q = quartet. All GC analyses were performed on an Agilent 7890 A Gas Chromatograph (Santa Clara, United States) with an FID detector using J&W HP-5 column (30 m, 0.32 mm). Elemental analyses were recorded on elementar Analysensysteme GmbH—vario EL III Element Analyzer, (Langenselbold, Germany).

3.2. Methods

3.2.1. General Procedure for Synthesis of [Pd(NHC)(NH$_2$)(CC)Cl] Complexes (Small Scale)

A 4 mL scintillation vial equipped with a septum cap and a stirring bar was charged with Di-μ-chlorobis[2'-(amino-N)[1,1'-biphenyl]-2-yl-C]dipalladium(II) (1 eq., 50 or 100 mg scale), NHC·HCl (2.1 eq.), K$_2$CO$_3$ (3 eq.), and acetone (1 mL). The reaction mixture was stirred at 60 °C for 1–2 h. The solvent was removed under vacuum and purification of the product was carried out with filtration through a Millipore membrane filter with ethyl acetate (3 mL). Evaporation of the solvent, washing with pentane (3 × 3 mL), and drying under high vacuum afforded the products as microcrystalline powders.

3.2.2. General Procedure for Synthesis of [Pd(NHC)(NH$_2$)(CC)Cl] Complexes (Larger Scale)

A 15 mL scintillation vial equipped with a septum cap and a stirring bar was charged with Di-μ-chlorobis[2'-(amino-N)[1,1'-biphenyl]-2-yl-C]dipalladium(II) (1.00 g, 1.51 mmol), IPr·HCl (1.35 g, 3.17 mmol), K$_2$CO$_3$ (0.626 g, 4.53 mmol), and acetone (3 mL). The reaction mixture was stirred at 60 °C for 16 h. The solvent was removed under vacuum and purification of the product was carried out with filtration through a Millipore membrane filter with ethyl acetate (6 mL). Evaporation of the solvent, washing with pentane (5 × 5 mL), and drying under high vacuum afforded the product **7** as microcrystalline yellow powder (1.71 g, 74%).

A 15 mL scintillation vial equipped with a septum cap and a stirring bar was charged with Di-μ-chlorobis[2'-(amino-N)[1,1'-biphenyl]-2-yl-C]dipalladium(II) (1.00 g, 1.51 mmol), IMes·HCl (1.08 g, 3.17 mmol), K$_2$CO$_3$ (0.626 g, 4.53 mmol), and acetone or toluene (3 mL). The reaction mixture was stirred at 60 °C for 4 h. The solvent was removed under vacuum and purification of the product was carried out with filtration through a Millipore membrane filter with ethyl acetate (6 mL). Evaporation of the solvent, washing with pentane (5 × 5 mL), and drying under high vacuum afforded the product **8** as microcrystalline yellow powder (1.59 g, 81%).

3.2.3. Procedures for the Catalytic Tests Buchwald–Hartwig Reaction

The appropriate amount of palladium complex (0.2–1 mol%), base (1.2 mmol), and a stirring bar were charged into a 4 mL scintillation vial. If the base was sensitive to air and moisture, it was weighed inside a glovebox. Under an argon atmosphere, 4-chloroanisole (1.0 mmol), morpholine (1.2 mmol), and an appropriate degassed solvent (2.0 mL) were added to the vial at room temperature. The reaction mixture was stirred at the indicated temperature for the indicated time. The course of the reaction was monitored with gas chromatography using dodecane as an internal standard.

3.2.4. General Procedure for the Buchwald–Hartwig Reaction

[Pd(IPr)(NH$_2$)(CC)Cl] (0.5 mol%), aryl chloride (1.0 mmol), amine (1.2 mmol) (if the substrate was solid), and a stirring bar were charged into a 4 mL scintillation vial. The vial was charged with KO^{t}Bu (1.2 mmol) under inert atmosphere (in the glovebox as a standard procedure), the vial was closed with a cap, and the vial was taken outside of the glovebox. Under argon atmosphere, aryl chloride (1.0 mmol), amine (1.2 mmol) (if the substrate was liquid), and degassed dry cyclopentyl methyl ether (2.0 mL) were added at room temperature and the reaction mixture was stirred at 80 °C for the indicated time. After the indicated time, the crude mixture was purified with filtration through silica gel and the product was isolated by the removal of volatiles under reduced pressure.

4. Conclusions

In summary, a new class of catalysts with a potentially broad spectrum of activity in cross-coupling chemistry has been synthesized and fully characterized. The catalysts consist of a palladacycle scaffold stabilized by the presence of a highly donating, sterically demanding NHC ligand. The catalysts are well-defined, air stable, and very active in the cross-coupling of aryl chlorides with amines. Their synthesis is simple and is achieved by mixing NHC·HCl with a palladacycle and K$_2$CO$_3$ in acetone at 60 °C in air. The proposed mechanism of activation is based on the generation of aryl-alkoxy palladium species. The palladium (0) species formed upon elimination of the ether is stabilized by coordination to an electron-rich, sterically demanding NHC.

Supplementary Materials: The following supporting information can be downloaded at: https://www.mdpi.com/article/10.3390/catal13030559/s1, screening tables, preparation of palladacycles, analytical data palladacycles, biaryls, amino aryls, ^{1}H and ^{13}C NMR spectra, elemental analysis, more detailed materials and methods; Figure S1: Molecular structure of compound **7** and atoms numbering scheme. The hydrogen atoms omitted for clarity. Displacement ellipsoids shown at the 30% probability level; Figure S2: Molecular structure of compound **8** and atoms numbering scheme. The hydrogen atoms omitted for clarity. Displacement ellipsoids shown at the 30% probability level; Table S1: Buchwald-Hartwig amination catalyzed by NHC-Pd$_{cycles}$; Table S2: Selected crystal data and structure refinement details (**7**); Table S3: Selected geometric parameters (Å, °) (**7**); Table S4: Selected crystal data and structure refinement details (**8**); Table S5: Selected geometric parameters (Å, °) (**8**) [14,50,56,64,65,70–86].

Author Contributions: Conceptualization, S.P.N. and S.O.; methodology, S.O. and L.P.; investigation, S.O. and L.P.; XRD analysis, A.C., S.B., M.K. and K.V.H.; writing—original draft preparation, S.O.; writing—review and editing, S.O. and S.P.N.; supervision, S.P.N.; funding acquisition, S.P.N. All authors have read and agreed to the published version of the manuscript.

Funding: S.O. is grateful for the generous support by the Polish National Agency for Academic Exchange (Bekker Fellowship). S.P., K.V.H. and S.B. acknowledge The Research Foundation—Flanders (FWO) for a research grant (G0A6823N and 1275221N). We are grateful to the SBO (D2M to SPN) and the BOF (starting and senior grants to SPN) as well as the iBOF C3 project for financial support. Umicore is gratefully acknowledged for gift of materials.

Data Availability Statement: All experimental data is contained in the article and Supplementary Materials.

Acknowledgments: The FWO, BOF and the Polish National Academy for Academic Exchange are acknowledged for funding.

Conflicts of Interest: The authors declare no conflict of interest.

References

1. Marion, N.; Nolan, S.P. Well-Defined N-Heterocyclic Carbenes-Palladium(II) Precatalysts for Cross-Coupling Reactions. *Acc. Chem. Res.* **2008**, *41*, 1440–1449. [CrossRef]
2. Ostrowska, S.; Scattolin, T.; Nolan, S.P. N-Heterocyclic Carbene Complexes Enabling the α-Arylation of Carbonyl Compounds. *Chem. Commun.* **2021**, *57*, 4354–4375. [CrossRef]
3. Nolan, S.P. *N-Heterocyclic Carbenes: Effective Tools for Organometallic Synthesis*, 1st ed.; Wiley-VCH: Hoboken, NJ, USA, 2014; ISBN 9783527334902.
4. Maluenda, I.; Navarro, O. Recent Developments in the Suzuki-Miyaura Reaction: 2010–2014. *Molecules* **2015**, *20*, 7528–7557. [CrossRef]
5. Fortman, G.C.; Nolan, S.P. N-Heterocyclic Carbene (NHC) Ligands and Palladium in Homogeneous Cross-Coupling Catalysis: A Perfect Union. *Chem. Soc. Rev.* **2011**, *40*, 5151–5169. [CrossRef] [PubMed]
6. Christmann, U.; Vilar, R. Monoligated Palladium Species as Catalysts in Cross-Coupling Reactions. *Angew. Chemie Int. Ed.* **2005**, *44*, 366–374. [CrossRef] [PubMed]
7. Biscoe, M.R.; Fors, B.P.; Buchwald, S.L. A New Class of Easily Activated Palladium Precatalysts for Facile C-N Cross-Coupling Reactions and the Low Temperature Oxidative Addition of Aryl Chlorides. *J. Am. Chem. Soc.* **2008**, *130*, 6686–6687. [CrossRef] [PubMed]
8. Dupont, J.; Pfeffer, M. *Palladacycles*; WILEY-VCH GmbH & Co. KGaA: Weinheim, Germany, 2008; ISBN 9783527319527.
9. Vila, J.M.; Pereira, M.T.; Lucio-Martínez, F.; Reigosa, F. Palladacycles as Efficient Precatalysts for Suzuki-Miyaura Cross-Coupling Reactions. *Palladacycles* **2019**, *1*, 1–20. [CrossRef]
10. Albrecht, M. Cyclometalation Using D-Block Transition Metals: Fundamental Aspects and Recent Trends. *Chem. Rev.* **2010**, *110*, 576–623. [CrossRef] [PubMed]
11. Ghedini, M.; Aiello, I.; Crispini, A.; Golemme, A.; La Deda, M.; Pucci, D. Azobenzenes and Heteroaromatic Nitrogen Cyclopalladated Complexes for Advanced Applications. *Coord. Chem. Rev.* **2006**, *250*, 1373–1390. [CrossRef]
12. Dupont, J.; Consorti, C.S.; Spencer, J. The Potential of Palladacycles: More than Just Precatalysts. *Chem. Rev.* **2005**, *105*, 2527–2571. [CrossRef]
13. Alonso, D.A.; Nájera, C. Oxime-Derived Palladacycles as Source of Palladium Nanoparticles. *Chem. Soc. Rev.* **2010**, *39*, 2891–2902. [CrossRef]
14. Wang, K.; Fan, R.; Wei, X.; Fang, W. Palladacyclic N-Heterocyclic Carbene Precatalysts for Transition Metal Catalysis. *Green Synth. Catal.* **2022**, *3*, 327–338. [CrossRef]
15. Nájera, C. Oxime-Derived Palladacycles: Applications in Catalysis. *ChemCatChem* **2016**, *8*, 1865–1881. [CrossRef]
16. Li, H.; Johansson Seechurn, C.C.C.; Colacot, T.J. Development of Preformed Pd Catalysts for Cross-Coupling Reactions, beyond the 2010 Nobel Prize. *ACS Catal.* **2012**, *2*, 1147–1164. [CrossRef]
17. Beletskaya, I.P.; Cheprakov, A.V. Palladacycles in Catalysis—A Critical Survey. *J. Organomet. Chem.* **2004**, *689*, 4055–4082. [CrossRef]

18. Cope, A.C.; Siekman, R.W. Formation of Covalent Bonds from Platinum or Palladium to Carbon by Direct Substitution. *J. Am. Chem. Soc.* **1965**, *87*, 3272–3273. [CrossRef]
19. Cope, A.; Friedrichlb, E. N,N-Dimethylbenzylamines. *J. Am. Chem. Soc.* **1968**, *90*, 909–913. [CrossRef]
20. Herrmann, W.A.; Brossmer, C.; Öfele, K.; Reisinger, C.-P.; Priermeier, T.; Beller, M.; Fischer, H. Palladacycles as Structurally Defined Catalysts for the Heck Olefination of Chloro- and Bromoarenes. *Angew. Chemie Int. Ed. Eng.* **1995**, *34*, 1844–1848. [CrossRef]
21. Bruneau, A.; Roche, M.; Alami, M.; Messaoudi, S. 2-Aminobiphenyl Palladacycles: The "Most Powerful" Precatalysts in C-C and C-Heteroatom Cross-Couplings. *ACS Catal.* **2015**, *5*, 1386–1396. [CrossRef]
22. Molander, G.A.; Barcellos, T.; Traister, K.M. Pd-Catalyzed Cross-Coupling of Potassium Alkenyltrifluoroborates with 2-Chloroacetates and 2-Chloroacetamides. *Org. Lett.* **2013**, *15*, 3342–3345. [CrossRef] [PubMed]
23. Shu, W.; Buchwald, S.L. Use of Precatalysts Greatly Facilitate Palladium-Catalyzed Alkynylations in Batch and Continuous-Flow Conditions. *Chem. Sci.* **2011**, *2*, 2321–2325. [CrossRef]
24. Maiti, D.; Fors, B.P.; Henderson, J.L.; Nakamura, Y.; Buchwald, S.L. Palladium-Catalyzed Coupling of Functionalized Primary and Secondary Amines with Aryl and Heteroaryl Halides: Two Ligands Suffice in Most Cases. *Chem. Sci.* **2011**, *2*, 57–68. [CrossRef] [PubMed]
25. Gorelsky, S.I.; Lapointe, D.; Fagnou, K. Analysis of the Palladium-Catalyzed (Aromatic)C-H Bond Metalation- Deprotonation Mechanism Spanning the Entire Spectrum of Arenes. *J. Org. Chem.* **2012**, *77*, 658–668. [CrossRef]
26. Martin, R.; Buchwald, S.L. Palladium-Catalyzed Suzuki-Miyaura Cross-Coupling Reactions Employing Dialkylbiaryl Phosphine Ligands. *Acc. Chem. Res.* **2008**, *41*, 1461–1473. [CrossRef] [PubMed]
27. Hopkinson, M.N.; Richter, C.; Schedler, M.; Glorius, F. An Overview of N-Heterocyclic Carbenes. *Nature* **2014**, *510*, 485–496. [CrossRef]
28. Herrmann, W.A.; Schwarz, J.; Gardiner, M.G. High-Yield Syntheses of Sterically Demanding Bis(N-Heterocyclic Carbene) Complexes of Palladium. *Organometallics* **1999**, *18*, 4082–4089. [CrossRef]
29. Phan, N.T.S.; Van Der Sluys, M.; Jones, C.W. On the Nature of the Active Species in Palladium Catalyzed Mizoroki-Heck and Suzuki-Miyaura Couplings—Homogeneous or Heterogeneous Catalysis, a Critical Review. *Adv. Synth. Catal.* **2006**, *348*, 609–679. [CrossRef]
30. Kapdi, A.R.; Fairlamb, I.J.S. Anti-Cancer Palladium Complexes: A Focus on PdX2L2, Palladacycles and Related Complexes. *Chem. Soc. Rev.* **2014**, *43*, 4751–4777. [CrossRef]
31. Emin Günay, M.; Gümüşada, R.; Özdemir, N.; Dinçer, M.; Çetinkaya, B. Synthesis, X-Ray Structures, and Catalytic Activities of (K2-C,N)-Palladacycles Bearing Imidazol-2-Ylidenes. *J. Organomet. Chem.* **2009**, *694*, 2343–2349. [CrossRef]
32. Kilinçarslan, R.; Günay, M.E.; Firinci, R.; Denizalti, S.; Çetinkaya, B. New Palladium(II)-N-Heterocyclic Carbene Complexes Containing Benzimidazole-2-Ylidene Ligand Derived from Menthol: Synthesis, Characterization and Catalytic Activities. *Appl. Organomet. Chem.* **2016**, *30*, 268–272. [CrossRef]
33. Micksch, M.; Tenne, M.; Strassner, T. Cyclometalated 2-Phenylimidazole Palladium Carbene Complexes in the Catalytic Suzuki-Miyaura Cross-Coupling Reaction. *Organometallics* **2014**, *33*, 3966–3976. [CrossRef]
34. Babahan, İ.; Fırıncı, R.; Özdemir, N.; Emin Günay, M. Synthesis, Characterization and Catalytic Activity of N-Heterocyclic Carbene Ligated Schiff Base Palladacycles. *Inorg. Chim. Acta* **2021**, *522*, 120360. [CrossRef]
35. Schroeter, F.; Soellner, J.; Strassner, T. Cyclometalated Palladium NHC Complexes Bearing PEG Chains for Suzuki-Miyaura Cross-Coupling in Water. *Organometallics* **2018**, *37*, 4267–4275. [CrossRef]
36. Deng, Q.; Shen, Y.; Zhu, H.; Tu, T. A Magnetic Nanoparticle-Supported N-Heterocyclic Carbene-Palladacycle: An Efficient and Recyclable Solid Molecular Catalyst for Suzuki-Miyaura Cross-Coupling of 9-Chloroacridine. *Chem. Commun.* **2017**, *53*, 13063–13066. [CrossRef]
37. Deng, Q.; Zheng, Q.; Zuo, B.; Tu, T. Robust NHC-Palladacycles-Catalyzed Suzuki−Miyaura Cross-Coupling of Amides via C-N Activation. *Green Synth. Catal.* **2020**, *1*, 75–78. [CrossRef]
38. Zuo, B.; Lu, Z.; Wu, X.; Fang, W.; Li, W.; Huang, M.; Yang, D.Y.; Deng, Q.; Tu, T. A N-Heterocyclic Carbene-Palladacycle with Constrained Aliphatic Linker: Synthesis, Characterization and Its Catalytic Application towards Suzuki-Miyaura Cross-Coupling. *Asian, J. Org. Chem.* **2021**, *10*, 3233–3236. [CrossRef]
39. Xu, C.; Li, H.M.; Xiao, Z.Q.; Wang, Z.Q.; Tang, S.F.; Ji, B.M.; Hao, X.Q.; Song, M.P. Cyclometalated Pd(Ii) and Ir(Iii) 2-(4-Bromophenyl)Pyridine Complexes with N-Heterocyclic Carbenes (NHCs) and Acetylacetonate (Acac): Synthesis, Structures, Luminescent Properties and Application in One-Pot Oxidation/Suzuki Coupling of Aryl Chlorides Containing Hydroxymethyl. *Dalt. Trans.* **2014**, *43*, 10235–10247. [CrossRef]
40. Serrano, J.L.; Perez, J.; Garcia, L.; Sanchez, G.; Garcia, J.; Lozano, P.; Zende, V.; Kapdi, A. N-Heterocyclic-Carbene Complexes Readily Prepared from Di-μ-Hydroxopalladacycles Catalyze the Suzuki Arylation of 9-Bromophenanthrene. *Organometallics* **2015**, *34*, 522–533. [CrossRef]
41. Xu, C.; Lou, X.H.; Wang, Z.Q.; Fu, W.J. N-Heterocyclic Carbene Adducts of Cyclopalladated Ferrocenylchloropyrimidine: Synthesis, Structural Characterization and Application in the Sonogashira Reaction. *Transit. Met. Chem.* **2012**, *37*, 519–523. [CrossRef]
42. Li, H.M.; Xu, C.; Duan, L.M.; Lou, X.H.; Wang, Z.Q.; Li, Z.; Fan, Y.T. N-Heterocyclic Carbene Adducts of Cyclopalladated Ferrocenylpyridine Containing Chloride or Iodide Anions: Synthesis, Crystal Structures and Application in the Coupling of Terminal Alkynes with Arylboronic Acids. *Transit. Met. Chem.* **2013**, *38*, 313–318. [CrossRef]
43. Doucet, H.; Hierso, J.C. Palladium-Based Catalytic Systems for the Synthesis of Conjugated Enynes by Sonogashira Reactions and Related Alkynylations. *Angew. Chemie Int. Ed.* **2007**, *46*, 834–871. [CrossRef]
44. Kantchev, E.A.B.; Peh, G.R.; Zhang, C.; Ying, J.Y. Practical Heck-Mizoroki Coupling Protocol for Challenging Substrates Mediated by an N-Heterocyclic Carbene-Ligated Palladacycle. *Org. Lett.* **2008**, *10*, 3949–3952. [CrossRef]

45. Green, K.A.; Maragh, P.T.; Abdur-Rashid, K.; Lough, A.J.; Dasgupta, T.P. Benzimidazol-2-Ylidene Ligated Palladacyclic Complexes of N,N-Dimethylbenzylamine—Synthesis and Application to C-C Coupling Reactions. *Inorg. Chim. Acta* **2016**, *449*, 38–43. [CrossRef]
46. Ren, G.; Cui, X.; Yang, E.; Yang, F.; Wu, Y. Study on the Heck Reaction Promoted by Carbene Adduct of Cyclopalladated Ferrocenylimine and the Related Reaction Mechanism. *Tetrahedron* **2010**, *66*, 4022–4028. [CrossRef]
47. Dorel, R.; Grugel, C.P.; Haydl, A.M. The Buchwald–Hartwig Amination After 25 Years. *Angew. Chemie Int. Ed.* **2019**, *58*, 17118–17129. [CrossRef] [PubMed]
48. Marion, N.; Nolan, S.P. N-Heterocyclic Carbenes in Gold Catalysis. *Chem. Soc. Rev.* **2008**, *37*, 1776–1782. [CrossRef] [PubMed]
49. Kantchev, E.A.B.; Ying, J.Y. Practical One-Pot, Three-Component Synthesis of N-Heterocyclic Carbene (NHC) Ligated Palladacycles Derived from n,n-Dimethylbenzylamine. *Organometallics* **2009**, *28*, 289–299. [CrossRef]
50. Zuo, B.; Shao, H.; Qu, E.; Ma, Y.; Li, W.; Huang, M.; Deng, Q. An Alkoxy Modified N-Heterocyclic Carbene-Palladacycle: Synthesis, Characterization and Application towards Buchwald-Hartwig and Suzuki-Miyaura Coupling Reactions. *ChemistrySelect* **2021**, *6*, 10121–10126. [CrossRef]
51. Wang, Y.; Yang, X.; Zhang, C.; Yu, J.; Liu, J.; Xia, C. Phosphine-Free, Efficient Double Carbonylation of Aryl Iodides with Amines Catalyzed by Water-Insoluble and Water-Soluble N-Heterocyclic Carbene-Amine Palladium Complexes. *Adv. Synth. Catal.* **2014**, *356*, 2539–2546. [CrossRef]
52. Zhang, C.; Liu, J.; Xia, C. Aryl-Palladium-NHC Complex: Efficient Phosphine-Free Catalyst Precursors for the Carbonylation of Aryl Iodides with Amines or Alkynes. *Org. Biomol. Chem.* **2014**, *12*, 9702–9706. [CrossRef]
53. Zhang, C.; Liu, J.; Xia, C. Palladium-N-Heterocyclic Carbene (NHC)-Catalyzed Synthesis of 2-Ynamides via Oxidative Aminocarbonylation of Alkynes with Amines. *Catal. Sci. Technol.* **2015**, *5*, 4750–4754. [CrossRef]
54. Hu, Y.; Guo, S. Palladacycles Bearing COOH-/Ester-Functionalized N-Heterocyclic Carbenes: Divergent Syntheses and Catalytic Applications. *Appl. Organomet. Chem.* **2019**, *33*, e4703. [CrossRef]
55. Niggli, N.E.; Baudoin, O. Design of Chiral NHC-Carboxylates as Potential Ligands for Pd-Catalyzed Enantioselective C−H Activation. *Helv. Chim. Acta* **2021**, *104*, e2100015. [CrossRef]
56. Viciu, M.S.; Kelly, R.A.; Stevens, E.D.; Naud, F.; Studer, M.; Nolan, S.P. Synthesis, Characterization, and Catalytic Activity of N-Heterocyclic Carbene (NHC) Palladacycle Complexes. *Org. Lett.* **2003**, *5*, 1479–1482. [CrossRef]
57. Schnyder, A.; Indolese, A.F.; Studer, M.; Blaser, H.-U. A New Genertion of Air-Stable, Highly Active Pd Comnplexes for C-C and C-N Coupling Reactions with Aryl Chlorides. *Angew. Chem. Int. Engl. Ed.* **2002**, *41*, 3668–3671. [CrossRef]
58. Navarro, O.; Marion, N.; Oonishi, Y.; Kelly, R.A.; Nolan, S.P. Suzuki-Miyaura, α-Ketone Arylation and Dehalogenation Reactions Catalyzed by a Versatile N-Heterocyclic Carbene-Palladacycle Complex. *J. Org. Chem.* **2006**, *71*, 685–692. [CrossRef]
59. Navarro, O.; Kelly, R.A.; Nolan, S.P. A General Method for the Suzuki-Miyaura Cross-Coupling of Sterically Hindered Aryl Chlorides: Synthesis of Di- and Tri-Ortho-Substituted Biaryls in 2-Propanol at Room Temperature. *J. Am. Chem. Soc.* **2003**, *125*, 16194–16195. [CrossRef]
60. Martynova, E.A.; Tzouras, N.V.; Pisanò, G.; Cazin, C.S.J.; Nolan, S.P. The "Weak Base Route" Leading to Transition Metal-N-Heterocyclic Carbene Complexes. *Chem. Commun.* **2021**, *57*, 3836–3856. [CrossRef]
61. Ma, X.; Guillet, S.G.; Peng, M.; Van Hecke, K.; Nolan, S.P. A Simple Synthesis of [RuCl2(NHC)(p-Cymene)] Complexes and Their Use in Olefin Oxidation Catalysis. *Dalt. Trans.* **2021**, *50*, 3959–3965. [CrossRef] [PubMed]
62. Ma, X.; Guillet, S.G.; Liu, Y.; Cazin, C.S.J.; Nolan, S.P. Simple Synthesis of [Ru(CO3)(NHC)(p-Cymene)] Complexes and Their Use in Transfer Hydrogenation Catalysis. *Dalt. Trans.* **2021**, *50*, 13012–13019. [CrossRef]
63. Liu, Y.; Scattolin, T.; Gobbo, A.; Beliš, M.; Van Hecke, K.; Nolan, S.P.; Cazin, C.S.J. A Simple Synthetic Route to Well-Defined [Pd(NHC)Cl(1-TBu-Indenyl)] Pre-Catalysts for Cross-Coupling Reactions. *Eur. J. Inorg. Chem.* **2022**, *2022*, e202100840. [CrossRef]
64. Liu, Y.; Voloshkin, V.A.; Scattolin, T.; Peng, M.; Van Hecke, K.; Nolan, S.P.; Cazin, C.S.J. Versatile and Highly Efficient Trans[Pd(NHC)Cl2(DMS/THT)] Precatalysts for C−N and C−C Coupling Reactions in Green Solvents. *Eur. J. Org. Chem.* **2022**, *2022*, e202200309. [CrossRef]
65. Zinser, C.M.; Warren, K.G.; Nahra, F.; Al-Majid, A.; Barakat, A.; Islam, M.S.; Nolan, S.P.; Cazin, C.S.J. Palladate Precatalysts for the Formation of C-N and C-C Bonds. *Organometallics* **2019**, *38*, 2812–2817. [CrossRef]
66. Surry, D.S.; Buchwald, S.L. Biaryl Phosphane Ligands in Palladium-Catalyzed Amination. *Angew. Chem. Int. Ed.* **2008**, *47*, 6338–6361. [CrossRef] [PubMed]
67. Shekhar, S.; Ryberg, P.; Hartwig, J.F. Reevaluation of the Mechanism of the Amination of Aryl Halides Catalyzed by BINAP-Ligated Palladium Complexes. *J. Am. Chem. Soc.* **2006**, *128*, 3584–3591. [CrossRef] [PubMed]
68. Muci, A.R.; Buchwald, S.L. Practical Palladium Catalysts for C-N and C-O Bond Formation. *Top. Curr. Chem.* **2002**, *219*, 131–209. [CrossRef]
69. Titcomb, R.L.; Caddick, S.; Geoffrey, F.; Cloke, N.; Wilson, D.J.; McKerrecher, D. Unexpected reactivity of two-coordinate palladium–carbene complexes; synthetic and catalytic implications. *Chem. Commun.* **2001**, *15*, 1388–1389. [CrossRef]
70. Meiries, S.; Le Duc, G.; Chartoire, A.; Collado, A.; Speck, K.; Arachchige, K.S.A.; Slawin, A.M.Z.; Nolan, S.P. Large yet Flexible N-Heterocyclic Carbene Ligands for Palladium Catalysis. *Chem.-A Eur. J.* **2013**, *19*, 17358–17368. [CrossRef] [PubMed]
71. Li, J.; Wang, Z.X. Nickel-Catalyzed Amination of Aryl 2-Pyridyl Ethers via Cleavage of the Carbon–Oxygen Bond. *Org. Lett.* **2017**, *19*, 3723–3726. [CrossRef]
72. Yang, J.; Li, P.; Zhang, Y.; Wang, L. Dinuclear N-heterocyclic carbene palladium(II) complexes as efficient catalysts for the Buchwald–Hartwig amination. *J. Organomet. Chem.* **2014**, *766*, 73–78. [CrossRef]

73. Krinsky, J.L.; Martínez, A.; Godard, C.; Castillón, S.; Claver, C. Modular Synthesis of Functionalisable Alkoxy-Tethered N-Heterocyclic Carbene Ligands and an Active Catalyst for Buchwald–Hartwig Aminations. *Adv. Synth. Catal.* **2014**, *356*, 460–474. [CrossRef]
74. Zhang, L.; Wang, W.; Fan, R. One-Pot Synthesis of Diarylamines from Two Aromatic Amines via Oxidative Dearomatization–Imino Exchange–Reductive Aromatization. *Org. Lett.* **2013**, *15*, 2018–2021. [CrossRef] [PubMed]
75. Chen, K.; Kang, Q.K.; Li, Y.; Wu, W.Q.; Zhu, H.; Shi, H. Catalytic Amination of Phenols with Amines. *J. Am. Chem. Soc.* **2022**, *144*, 1144–1151. [CrossRef] [PubMed]
76. Zhang, Z.; Wu, X.; Han, J.; Wu, W.; Wang, L. Direct arylation of tertiary amines via aryne intermediates using diaryliodonium salts. *Tetrahedron Lett.* **2018**, *59*, 1737–1741. [CrossRef]
77. Zhang, Z.M.; Xu, Y.T.; Shao, L.X. Synthesis of N-heterocyclic carbene-Pd(II)-5-phenyloxazole complexes and initial studies of their catalytic activity toward the Buchwald-Hartwig amination of aryl chlorides. *J. Organomet. Chem.* **2021**, *940*, 121683. [CrossRef]
78. Cai, L.; Qian, X.; Song, W.; Liu, T.; Tao, X.; Li, W.; Xie, X. Effects of solvent and base on the palladium-catalyzed amination: PdCl2(Ph3P)2/Ph3P-catalyzed selective arylation of primary anilines with aryl bromides. *Tetrahedron* **2014**, *70*, 4754–4759. [CrossRef]
79. Broggi, J.; Clavier, H.; Nolan, S.P. N-Heterocyclic Carbenes (NHCs) Containing N-C-Palladacycle Complexes: Synthesis and Reactivity in Aryl Amination Reactions. *Organometallics* **2008**, *27*, 5525–5531. [CrossRef]
80. Liu, F.; Hu, Y.-Y.; Li, D.; Zhou, Q.; Lu, J.-M. N-Heterocyclic carbene-palladacyclic complexes: Synthesis, characterization and their applications in the C-N coupling and α-arylation of ketones using aryl chlorides. *Tetrahedron* **2018**, *74*, 5683–5690. [CrossRef]
81. Deng, Q.; Zhang, Y.; Zhu, H.; Tu, T. Robust Acenaphthoimidazolylidene Palladacycles: Highly Efficient Catalysts for the Amination of N-Heteroaryl Chlorides. *Chem. Asian J.* **2017**, *12*, 2364–2368. [CrossRef]
82. *CrysAlis PRO 1.171.42.75a*, Rigaku Oxford Diffraction; Rigaku Corporation (and its Global Subsidiaries): Tokyo, Japan, 2022.
83. Sheldrick, G.M. SHELXT-Integrated Space-Group and Crystal-Structure Determination. *Acta Crystallogr.* **2015**, *A71*, 3–8. [CrossRef]
84. Sheldrick, G.M. Crystal Structure Refinement with SHELXL. *Acta Crystallogr.* **2015**, *C71*, 3–8. [CrossRef]
85. Dolomanov, O.V.; Bourhis, L.J.; Gildea, R.J.; Howard, J.A.K.; Puschmann, H. OLEX2: A Complete Structure Solution, Refinement and Analysis Program. *J. Appl. Cryst.* **2009**, *42*, 339–341. [CrossRef]
86. Macrae, C.F.; Edgington, P.R.; McCabe, P.; Pidcock, E.; Shields, G.P.; Taylor, R.; Towler, M.; van de Streek, J. Mercury: Visualization and Analysis of Crystal Structures. *J. Appl. Cryst.* **2006**, *39*, 453–457. [CrossRef]

Disclaimer/Publisher's Note: The statements, opinions and data contained in all publications are solely those of the individual author(s) and contributor(s) and not of MDPI and/or the editor(s). MDPI and/or the editor(s) disclaim responsibility for any injury to people or property resulting from any ideas, methods, instructions or products referred to in the content.

Article

De Novo Synthesis of Polysubstituted 3-Hydroxypyridines Via "Anti-Wacker"-Type Cyclization [†]

Kazuya Ito [1], Takayuki Doi [1] and Hirokazu Tsukamoto [1,2,*]

[1] Graduate School of Pharmaceutical Sciences, Tohoku University, 6-3 Aza-Aoba, Aramaki, Aoba-ku, Sendai 980-8578, Japan
[2] Department of Pharmaceutical Sciences, Yokohama University of Pharmacy, 601 Matano-cho, Totsuka-ku, Yokohama 245-0066, Japan
* Correspondence: hirokazu.tsukamoto@hamayaku.ac.jp
[†] This paper is dedicated to late Professor Jiro Tsuji.

Abstract: We report an efficient method to prepare polysubstituted 3-hydroxypyridines from amino acids, propargyl alcohols, and arylboronic acids. The process involves Pd(0)-catalyzed *anti*-selective arylative cyclizations of *N*-propargyl-*N*-tosyl-aminoaldehydes with arylboronic acids ("anti-Wacker"-type cyclization), oxidation of the resulting 5-substituted-3-hydroxy-1,2,3,6-tetrahydropyridines to 3-oxo derivatives, and elimination of *p*-toluenesulfinic acid. This method provides diverse polysubstituted 3-hydroxypyridines, whose hydroxy group can be further substituted by a cross-coupling reaction via a triflate.

Keywords: "anti-Wacker"-type cyclization; 3-hydroxypyridine; arylboronic acids; palladium; amino acids; propargyl alcohols

Citation: Ito, K.; Doi, T.; Tsukamoto, H. De Novo Synthesis of Polysubstituted 3-Hydroxypyridines Via "Anti-Wacker"-Type Cyclization. *Catalysts* **2023**, *13*, 319. https://doi.org/10.3390/catal13020319

Academic Editors: Ewa Kowalska and Yuichi Kobayashi

Received: 29 December 2022
Revised: 20 January 2023
Accepted: 24 January 2023
Published: 1 February 2023

Copyright: © 2023 by the authors. Licensee MDPI, Basel, Switzerland. This article is an open access article distributed under the terms and conditions of the Creative Commons Attribution (CC BY) license (https:// creativecommons.org/licenses/by/ 4.0/).

1. Introduction

Pyridines are important motifs found in natural products, pharmaceutical molecules, and agricultural chemicals [1–7]. Therefore, a wide variety of methods for the synthesis of pyridine and its derivatives have been developed; the proposed strategies rely on the modification of a pre-existing aromatic core [8–26] or the implementation of de novo synthetic technologies [27–43]. However, it is still difficult to introduce multiple substituents into the pyridine skeleton in a perfectly regioselective manner. For example, 3-hydroxypyridines [27–31] have been identified as not only bioactive compounds [1–7] but also useful intermediates for transformation into more functionalized pyridines [30]. In contrast to 2- and 4-hydroxypyridines, which readily tautomerize to the corresponding pyridones, 3-hydroxypyridines cannot form keto tautomers and are transformed into 3-substituted pyridines through a cross-coupling reaction of their triflates under palladium catalysis [44–49]. Donohoe [50] and Yanagisawa [51] independently reported de novo syntheses of 3-hydroxypyridines by the ring-closing metathesis of *N*-allyl-*N*-(2-oxobut-3-en-1-yl)amino derivative **1** (Scheme 1, top equation). However, the synthesis of each of the 3-hydroxypyridines **3** requires the preparation of the respective precursors from building blocks that are difficult to obtain, including polysubstituted allylic alcohols and alkenylmetal species. Alkylative cyclization to convert a single precursor into multiple cyclized products with a wide variety of substituents could be more desirable for the diversity-oriented synthesis of 3-hydroxypyridines. Herein, we describe a practical and regioselective synthesis of 3-hydroxypyridines by the Pd(0)-catalyzed *anti*-selective arylative-, alkylative, or alkynylative cyclizations ("anti-Wacker"-type cyclization [52–54]) of alkynals **4**, which can be easily prepared from available amino acid derivatives and propargyl alcohols (Scheme 1, bottom equation). Polysubstituted 3-hydroxypyridines are obtained with a simple two-step sequence: oxidation of cyclization products **5** and subsequent desulfinative aromatization developed by Boger [50,51,55]. The "anti-Wacker"-type cyclization proceeds through the newly proposed "anti-Wacker"-type oxidative addition of alkynyl electrophiles that do not form oxapalladacycles, transmetalation with organometallic reagents, and reductive elimination [52,53]. Both substituents at the alkyne terminus and phosphine ligands affect the regioselectivity of the cyclization reaction, with a combination of terminal alkynes and triphenylphosphine and that of aryl- and 1-alkynyl substituents and tricyclohexylphosphine favoring the formation of endocyclic products over exocyclic products. However, the effect of substituents at the α-positions of the alkyne and carbonyl on the diastereoselectivity and how many substituents are allowed are poorly understood [54]. N-Tosyl-tethered aldehyde **4** with an α-substituent of the carbonyl group can be readily prepared from amino acids. The tosyl-protecting group promotes not only the N-propargylation step of the substrate preparation but also the cyclization step by the electron-withdrawing inductive effect and the Thorpe–Ingold effect, which is eliminated after oxidation of the resulting allylic alcohols **5** to afford 3-hydroxypyridines **3**. Through the synthesis of multisubstituted 3-hydroxypyidines, we explored the scope and limitations of the "anti-Wacker"-type cyclization and established the structures of six-membered endocyclic products.

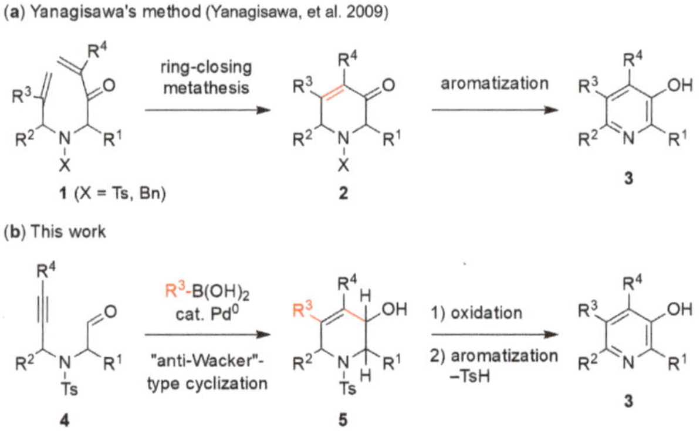

Scheme 1. Synthetic methods for 3-hydroxypyridines. (**a**) Yanagisawa's method [51], (**b**) this work.

2. Results and Discussion

The preparation of alkyne-aldehydes **4a–o** for the "anti-Wacker"-type cyclization was commenced with the N-alkylation of N-tosyl amino acid methyl esters via the S_N2 reaction with propargyl bromides or the Mitsunobu reaction with propargyl alcohols (Scheme 2). The terminal alkyne in methyl esters **6a·b** was also able to be substituted with aryl, 1-alkenyl, and 1-alkynyl groups by the Sonogashira or Cadiot–Chodkiewicz coupling reactions [56,57]. The ester intermediates were subsequently reduced with DIBAL to give aldehydes **4a–o**.

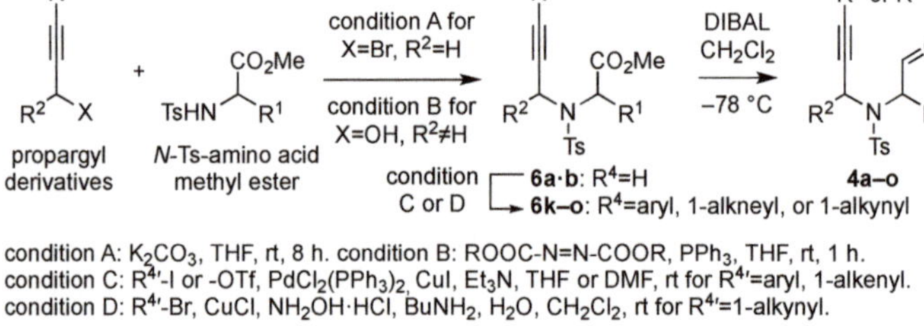

condition A: K_2CO_3, THF, rt, 8 h. condition B: ROOC-N=N-COOR, PPh_3, THF, rt, 1 h.
condition C: $R^{4'}$-I or -OTf, $PdCl_2(PPh_3)_2$, CuI, Et_3N, THF or DMF, rt for $R^{4'}$=aryl, 1-alkenyl.
condition D: $R^{4'}$-Br, CuCl, $NH_2OH·HCl$, $BuNH_2$, H_2O, CH_2Cl_2, rt for $R^{4'}$=1-alkynyl.

Scheme 2. Preparation of alkyne-aldehydes **4a–o**.

The terminal alkyne **4a** ($R^1=R^2=R^4=H$) derived from glycine underwent the Pd(PPh$_3$)$_4$-catalyzed "anti-Wacker"-type cyclization with arylboronic acids **7A–C** upon heating at 80 °C in methanol to afford 5-substituted-3-hydroxy-1,2,3,6-tetrahydropyridines **5aA–C** in good to moderate yields (Scheme 3). In addition to the aryl group, the alkyl and alkynyl groups were also effectively introduced into products **5aD** and **5aE** using the triethylborane **7D** and alkynylcopper species generated in situ from phenylacetylene **7E** along with a catalytic amount of copper iodide, respectively [52,53]. The α-substituted aldehydes **4b–e** derived from alanine, leucine, phenylalanine, and valine also participated in the arylative cyclization with **7A** to furnish **5b–eA** with *cis*-disubstituents as the predominant products in high yields [54]. The stereochemical outcome observed herein provides useful information about the transition states. The observed *cis*-diastereoselectivity would result from the steric effect of the substituent at the pseudoequatorial position of the twist boat transition state shown in Figure 1, where there is maximum overlap between the π-orbital of the incoming alkyne and the π*-orbital of the carbonyl [58,59]. The bulky isopropyl group in **4e** would increase the gauche interaction with the *N*-Ts group and be partially oriented in the pseudoaxial position, leading to lower *cis*-diastereoselectivity.

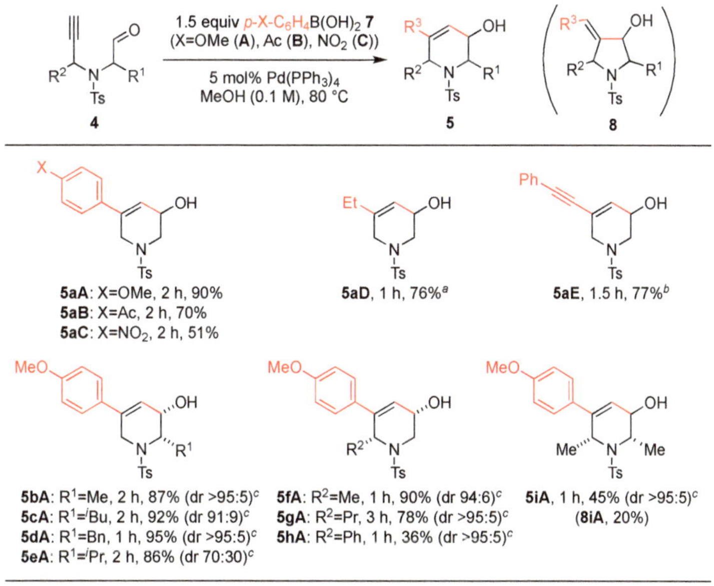

a reaction with 1.5 equiv Et$_3$B instead of arylboronic acid. *b* reaction with 2.0 equiv PhCCH and 6 mol% CuI instead of arylboronic acid. *c* determined by ^{1}H-NMR analysis.

Scheme 3. Tetrahydropyridines prepared by Pd/PPh$_3$-catalyzed cyclizations of terminal alkyne-aldehydes.

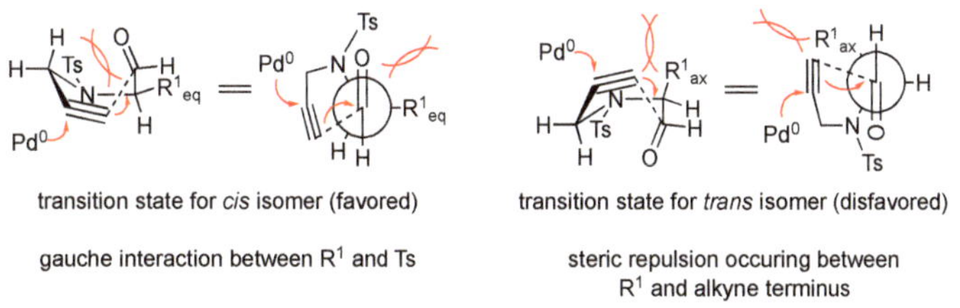

Figure 1. Diastereoselective cyclizations of **4b–e** with a substituent at the α-position to the carbonyl group.

On the other hand, a substituent (R^2) at the α-position of the alkyne functionalities in glycine-derived terminal alkyne-aldehydes **4f–h** dramatically affected the yield of products **5f–hA**, with the sterically demanding phenyl group resulting in much lower yields (Scheme 3). Interestingly, *cis*-diastereoselectivity was consistently high, regardless of the steric bulkiness of the substituents. The nucleophilic attack of a Pd(0) species would be hindered more significantly by the propargyl substituent at the pseudoequatorial position than by the substituent at the pseudoaxial position (Figure 2). The favored transition state with the substituent at the pseudoaxial position leads to *cis*-disubstituted products. Surprisingly, the introduction of two *cis*-oriented substituents at both the α-positions of the aldehyde and alkyne moieties led to the formation of not only the endocyclic product **5iA** but also the exocyclic product **8iA**. To the best of our knowledge, this is the only example of the formation of both endocyclic and exocyclic products during the arylative cyclization of terminal alkyne-aldehydes under Pd(PPh$_3$)$_4$ catalysis.

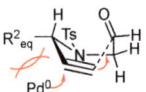 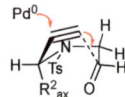

transition state for *trans* isomer (disfavored)   transition state for *cis* isomer (favored)

large steric repulsion between R^2 and Pd^0 no steric repulsion between R^2 and Pd^0

Figure 2. Diastereoselective cyclizations of **4f–h** with a substituent at the propargyl position.

The arylative cyclizations of internal alkyne-aldehydes **4j–o** with *p*-methoxyphenylboronic acid **7A** under the catalysis of the strongly σ-donating tricyclohexylphosphine-ligated palladium also provided 4,5-disubstituted-3-hydroxy-1,2,3,6-tetrahydropyridines **5j–oA**, along with **8j–oA**, in good to moderate yields (Scheme 4). For the predominant endocyclic closure that affords the tetrahydropyridines, the alkyl, aryl, 1-alkynyl, or polysubstituted 1-alkenyl groups at the alkyne terminus were necessary [53]. The arylative cyclization of alkyl-substituted alkyne-aldehyde **4j** was relatively slow and gave an inseparable mixture of **5jA** and **8jA**. The former tetrahydropyridine **5jA** can be alternatively prepared with a two-step sequence: arylative cyclization of conjugated diyne-aldehyde **4m**, followed by chemoselective hydrogenation of the internal alkyne **5mA** in the presence of a *tetra*-substituted alkene [53]. Further substitution at the α-position of the carbonyl group in **4o** preserves the high yield of product **5oA** with two *cis*-oriented substituents, which would also result from the similar transition state shown in Figure 1.

The 5-substituted 3-hydroxy-1,2,3,6-tetrahydropyridines were transformed into the corresponding 3-hydroxypyridines through the following two steps (Scheme 5). The Dess–Martin oxidation of the hydroxy group in **5** afforded enone **2**, although that of the acid-sensitive **5eA** required the addition of sodium bicarbonate to prevent acid-mediated dehydration. Subsequent elimination of the *p*-toluenesulfinic acid moiety in **2** was achieved using 1,8-diazabicyclo [5.4.0]undec-7-ene (DBU) to furnish the desired multiply substituted 3-hydroxypyridines **3** in good yields. For reasons unknown, the eliminated product was not formed in the case of **2mA** with the 1-alkynyl group at the C4 position.

Finally, the hydroxyl group at the C3 position of **3cA** was substituted with an aryl group via triflate **9cA**. After a brief screening of the Suzuki–Miyaura cross-coupling reaction, we found that the use of lithium chloride [60] successfully transformed triflate **9cA** into 2-substituted 3,5-diarylpyridine **10** in excellent yield (Scheme 6).

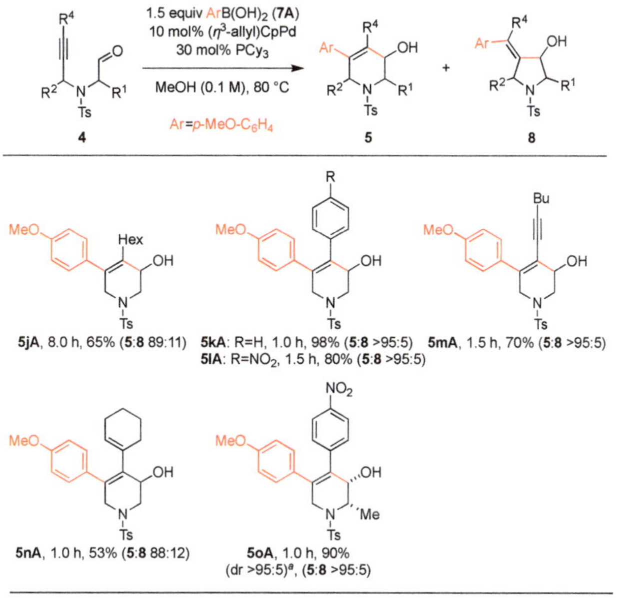

[a] determined by ^{1}H-NMR analysis.

Scheme 4. Tetrahydropyridines prepared by Pd/PCy$_3$-catalyzed arylative cyclizations of internal alkyne-aldehydes.

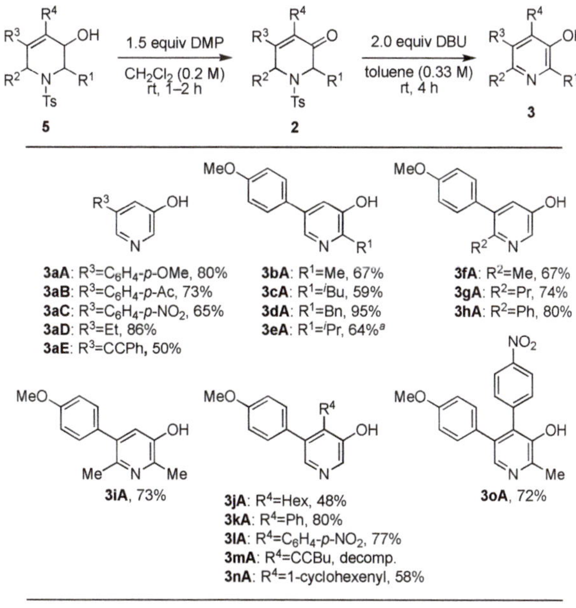

Reactions were performed on a 0.014–0.30 mmol scale. Isolated yields of **3** in 2 steps from **5** were given. [a] $NaHCO_3$ (2 equiv) was added in the oxidation step.

Scheme 5. Conversion of 3-hydroxytetrahydropyridines **5** to 3-hydroxypyridines **3**.

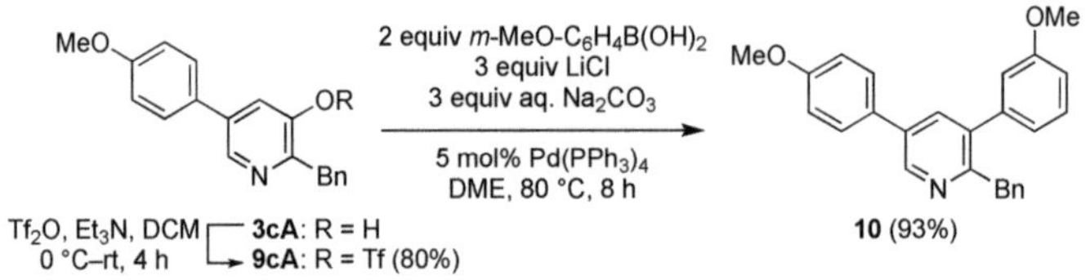

Scheme 6. Suzuki–Miyaura cross-coupling of triflate **9cA** derived from **3cA**.

3. Materials and Methods

3.1. General Techniques

All commercially available reagents and anhydrous solvents including tetrahydrofuran (THF), dichloromethane (DCM), and 1,2-dimethoxyethane (DME) were purchased and used without further purification. Anhydrous methanol (MeOH), *N,N*-dimethylformamide (DMF), and toluene were obtained using distillation from magnesium, calcium hydride, and sodium, respectively. All reactions were monitored using thin-layer chromatography (TLC) performed using 0.25 mm silica gel glass plates (60 F_{254}) using UV light and ethanolic *p*-anisaldehyde-sulfuric acid, ethanolic molybdatophosphoric acid, aqueous cerium sulfate-hexaammonium heptamolybdate-sulfuric acid, or aqueous potassium permanganate-potassium carbonate-sodium hydroxide solutions as visualizing agents. Flash column chromatography was carried out with silica gel (spherical, neutral, 100–210 μm grade). Preparative thin-layer chromatography was performed using 0.75 mm Wakogel® B-5F PLC plates. Yields refer to chromatographically and spectroscopically homogenous materials. Melting points were measured with a melting point apparatus and were uncorrected. Only the strongest and/or structurally important absorptions of infrared (IR) spectra are reported in reciprocal centimeters (cm^{-1}). The ^{1}H-NMR spectra (400 MHz or 600 MHz) and ^{13}C{^{1}H}NMR spectra (100 MHz or 151 MHz) were recorded in the indicated solvent. Chemical shifts (δ) are reported in delta (δ) units, parts per million (ppm). Chemical shifts for the ^{1}H-NMR spectra are given relative to signals for internal tetramethylsilane (0 ppm) or residual nondeuterated solvents, i.e., chloroform (7.26 ppm). Chemical shifts for the ^{13}C-NMR spectra are given relative to the signal for chloroform-*d* (77.0 ppm). Multiplicities are reported as the following abbreviations: s (singlet), d (doublet), t (triplet), q (quartet), m (multiplet), and br (broad). Coupling constants (*J*) are represented in hertz (Hz). The ^{1}H and ^{13}C-NMR chemical shifts were assigned using a combination of COSY, NOESY, HMQC, and HMBC. Low- and high-resolution mass spectra were measured using TOF-MS with EI, FAB, or ESI probes.

3.2. Materials

Ynals **4a**, **4b**, **4d**, **4j**, **4k**, **4m**, and **4n** were prepared according to the literature procedure [61–63]. Ynals **4c** and **4e** were prepared from *N*-tosyl amino acid methyl ester [64] and propargyl bromide. Ynals **4f**, **4g**, and **4i** were prepared from *N*-tosyl amino acid methyl ester and propargyl alcohols. Ynal **4h** was prepared from *N*-benzylidene-*p*-toluenesulfonamide [65], ethynylmagnesium bromide, and methyl bromoacetate. Ynals **4l** and **4o** were prepared through Sonogashira reaction of terminal alkyne **6a·b** with 1-iodo-4-nitrobenzene. The details of procedures for the preparation of ynals are described in Supplementary Materials.

3.3. Methods

3.3.1. General Procedure for the Pd(PPh$_3$)$_4$-Catalyzed Arylative Cyclizations of Terminal Alkyne-Aldehyde **4A–i** with Arylboronic Acid **7A–C**

To a test tube containing **4a–i** (1 equiv), arylboronic acid **7A–C** (1.5 equiv), and Pd(PPh$_3$)$_4$ (5 mol%) was added anhydrous MeOH (0.1 M) under argon. The resulting mixture was sealed with a screw cap and agitated at 80 °C for the time described in Scheme 3. The reaction mixture was cooled down to room temperature and then treated with polymer-supported diethanolamine (PL-DEAM™, 1.72 mmol/g, 3 equiv, X g) and

THF (10 × X mL) to remove an excess of **7A–C**. The mixture was agitated at room temperature for 2 h. The mixture was filtered, and the resin was thoroughly rinsed with CHCl$_3$. The filtrate was concentrated in vacuo and the residue was purified with preparative TLC or silica gel column chromatography to give endocyclic products **5(a–i)(A–C)** in the yield described in Scheme 3.

Procedure for 5-(4-methoxyphenyl)-1-tosyl-1,2,3,6-tetrahydropyridin-3-ol (**5aA**)

Method: **5aA** (16.6 mg, 90%) was obtained from **4a** (12.9 mg, 0.0513 mmol), **7A** (11.4 mg), and Pd(PPh$_3$)$_4$ (2.9 mg) and isolated with silica gel column chromatography eluting with 15% EtOAc/hexane. Spectra data of **5aA** were in agreement with those reported in the literature [52].

Procedure for 1-(4-(5-hydroxy-1-tosyl-1,2,5,6-tetrahydropyridin-3-yl)phenyl)ethan-1-one (**5aB**)

Method: **5aB** (12.3 mg, 70%) was obtained from **4a** (11.9 mg, 0.0474 mmol), **7B** (12.3 mg), and Pd(PPh$_3$)$_4$ (2.9 mg) and isolated with preparative TLC eluting with 20% EtOAc/toluene. Spectra data of **5aB** were in agreement with those reported in the literature [53].

Procedure for 5-(4-nitrophenyl)-1-tosyl-1,2,3,6-tetrahydropyridin-3-ol (**5aC**)

Method: **5aC** (28.2 mg, 51%) was obtained from **4a** (37.7 mg, 0.150 mmol), **7C** (37.6 mg), and Pd(PPh$_3$)$_4$ (8.6 mg) and isolated with preparative TLC eluting with 20% EtOAc/toluene.

Pale-brown oil. IR (neat): 3620–3200, 1681, 1604, 1344, 1271, 1167, 1094, 819, 755, 660 cm^{-1}; ^{1}H-NMR (400 MHz, CDCl$_3$): δ 7.19 (d, J = 8.8 Hz, 2H), 7.73 (d, J = 8.0 Hz, 2H), 7.49 (d, J = 8.8 Hz, 2H), 7.37 (d, J = 8.0 Hz, 2H), 6.35 (s, 1H), 4.45 (ddd, J = 4.8, 4.0, 5.6 Hz, 1H), 4.09 (d, J = 16.0 Hz, 1H), 3.76 (d, J = 16.0 Hz, 1H), 3.37 (dd, J = 12.0, 4.8 Hz, 1H), 3.24 (dd, J = 12.0, 4.0 Hz, 1H), 2.60 (d, J = 5.6 Hz, 1H), 2.44 (s, 3H); ^{13}C-NMR (151 MHz, CDCl$_3$): δ 147.7, 144.3, 143.7, 135.0, 132.9, 130.0, 128.1, 127.7, 126.3, 124.0, 63.7, 49.7, 46.2, 21.6. LRMS (EI) m/z (relative intensity) 374 ([M]$^+$, 2), 356 (3), 184 (100), 155 (61). HRMS (EI, [M]$^+$): m/z calcd for C$_{18}$H$_{18}$N$_2$O$_5$S, 374.0936; found, 374.0956.

Procedure for (2S,3S)-5-(4-methoxyphenyl)-2-methyl-1-tosyl-1,2,3,6-tetrahydropyridin-3-ol (**5bA**)

Method: **5bA** (33.1 mg, 87%, dr >95:<5) was obtained from **4b** (27.1 mg, 0.102 mmol), **7A** (22.8 mg), and Pd(PPh$_3$)$_4$ (5.8 mg) and isolated with preparative TLC eluting with 10% EtOAc/toluene.

Colorless oil. Rf 0.40 (50% EtOAc/hexane). [α]$_D^{23}$ −5.8 (c 0.60, CHCl$_3$). IR (neat): 3497, 1608, 1335, 1515, 1160, 1030, 816, 752, 659 cm^{-1}; ^{1}H-NMR (400 MHz, CDCl$_3$): δ 7.73 (d, J = 8.0 Hz, 2H) 7.32–7.28 (m, 4H), 6.87 (d, J = 8.8 Hz, 2H), 5.82 (s, 1H), 4.49 (m, 1H), 4.47 (d, J = 16.0 Hz, 1H), 4.34 (m, 1H), 3.81 (s, 3H), 3.74 (d, J = 16.0 Hz, 1H), 2.42 (s, 3H), 1.88 (br-s, 1H), 0.91 (d, J = 6.8 Hz, 3H); ^{13}C-NMR (100 MHz, CDCl$_3$): δ 159.9, 143.5, 137.0, 133.5, 129.8, 129.5, 127.0, 126.3, 123.2, 114.0, 67.0, 55.3, 50.8, 41.5, 21.5, 9.4. HRMS (ESI, [M + Na]$^+$) m/z calcd for C$_{20}$H$_{23}$NNaO$_4$S 396.1240, found 396.1242.

Procedure for 2-isobutyl-5-(4-methoxyphenyl)-1-tosyl-1,2,3,6-tetrahydropyridin-3-ol (**5cA**)

Method: **5cA** (19.1 mg, 92%, dr 91:9) was obtained from **4c** (15.4 mg, 0.0501 mmol), **7A** (11.5 mg), and Pd(PPh$_3$)$_4$ (3.0 mg) and isolated with preparative TLC eluting with 10% EtOAc/toluene.

For (2S,3S)-**5cA** as a major diastereomer: Colorless oil. Rf 0.38 (10% EtOAc/toluene). [α]$_D^{22}$ −131 (c 0.52, CHCl$_3$). IR (neat): 3505, 2955, 1608, 1514, 1331, 1158, 817, 745, 660 cm^{-1}; ^{1}H-NMR (400 MHz, CDCl$_3$): δ 7.69 (d, J = 8.4 Hz, 2H), 7.28 (d, J = 8.4 Hz, 2H), 7.24 (d, J = 8.4 Hz, 2H), 6.88 (d, J = 8.4 Hz, 2H), 5.76 (s, 1H), 4.52 (d, J = 18.0 Hz, 1H), 4.33–4.20 (m, 2H), 3.83 (d, J = 18.0 Hz, 1H), 3.82 (s, 3H), 2.34 (s, 3H), 1.80–1.64 (m, 2H), 1.36 (m, 2H), 0.94 (d, J = 6.4 Hz, 3H), 0.91 (d, J = 6.8 Hz, 3H); ^{13}C-NMR (100 MHz, CDCl$_3$): δ 159.6, 143.3, 138.0, 129.7, 129.3, 126.8, 126.5, 126.2, 123.4, 114.0, 65.7, 55.3, 52.7, 41.5, 33.0, 24.3, 23.9, 21.5 (one signal missing due to an overlap). HRMS (ESI, [M + Na]$^+$) m/z calcd for $C_{23}H_{29}NNaO_4S$ 438.1710, found 438.1707.

For (2S,3R)-**5cA** as a minor diastereomer: Colorless oil. Rf 0.42 (10% EtOAc/toluene). IR (neat): 3600–3200, 2926, 2869, 1607, 1515, 1335, 1247, 1158, 1093, 1031, 827, 754, 655 cm^{-1}; ^{1}H-NMR (600 MHz, CDCl$_3$): δ 7.79 (d, J = 8.2 Hz, 2H), 7.33–7.28 (m, 4H), 6.89 (d, J = 8.8 Hz, 2H), 6.12 (d, J = 6.1 Hz, 1H), 4.51 (d, J = 17.5 Hz, 1H), 4.19 (t, J = 7.2 Hz, 1H), 3.99 (dd, J = 10.5, 6.1 Hz, 1H), 3.82 (s, 3H), 3.76 (d, J = 17.5 Hz, 1H), 2.43 (s, 3H), 1.96 (d, J = 10.5, 1H), 1.61–1.51 (m, 1H), 1.20–1.14 (m, 2H), 0.88 (d, J = 6.5 Hz, 3H), 0.83 (d, J = 6.5 Hz, 3H); ^{13}C-NMR (151 MHz, CDCl$_3$): δ 160.0, 143.6, 137.2, 136.4, 129.74, 129.70, 127.4, 126.5, 120.6, 114.1, 66.7, 56.8, 55.4, 41.7, 37.6, 25.1, 22.7, 22.6, 21.5. HRMS (ESI, [M + Na]$^+$) m/z calcd for $C_{23}H_{29}NNaO_4S$ 438.1710, found 438.1707.

Procedure for (2R*, 3R*)-2-benzyl-5-(4-methoxyphenyl)-1-tosyl-1,2,3,6-tetrahydro-pyridin-3-ol (**5dA**)

Method: **5dA** (42.5 mg, 95%, dr >95:<5) was obtained from **4d** (33.0 mg, 0.100 mmol), **7A** (22.8 mg), and Pd(PPh$_3$)$_4$ (5.8 mg) and isolated with preparative TLC eluting with 50% EtOAc/hexane.

Pale-yellow oil. Rf 0.50 (50% EtOAc/hexane). IR (neat): 3492, 1607, 1514, 1248, 1157, 1096, 752, 660 cm^{-1}; ^{1}H-NMR (400 MHz, CDCl$_3$): δ 7.34 (d, J = 6.8 Hz, 2H), 7.27–7.13 (m, 7H), 7.08 (d, J = 8.4 Hz, 2H), 6.90 (d, J = 8.8 Hz, 2H), 5.94 (s, 1H), 4.64 (dd, J = 5.4, 6.0 Hz, 1H), 4.57 (ddd, J = 4.8, 5.4, 9.6 Hz, 1H), 4.47 (d, J = 18.0 Hz, 1H), 3.86 (d, J = 18.0 Hz, 1H), 3.11 (dd, J = 4.8, 14.2 Hz, 1H), 2.56 (dd, J = 9.6, 14.2 Hz, 1H), 2.36 (s, 3H), 1.77 (d, J = 6.0 Hz, 1H); ^{13}C-NMR (100 MHz, CDCl$_3$): δ 159.6, 143.0, 138.6, 137.1, 133.4, 129.5, 129.4, 129.1, 128.4, 126.9, 126.3, 126.2, 123.6, 114.0, 66.7, 56.4, 55.3, 41.6, 31.2, 21.4. HRMS (ESI, [M + Na]$^+$) m/z calcd for $C_{26}H_{27}NNaO_4S$ 472.1553, found 472.1548.

Procedure for 2-isopropyl-5-(4-methoxyphenyl)-1-tosyl-1,2,3,6-tetrahydropyridin-3-ol (**5eA**)

Method: **5eA** (33.8 mg, 86%, dr 70: 30) was obtained from **4e** (15.4 mg, 0.525 mmol), **7A** (11.5 mg), and Pd(PPh$_3$)$_4$ (3.0 mg) and isolated with preparative TLC eluting with 15% EtOAc/toluene.

For (2S,3S)-**5eA** as a major diastereomer: Colorless oil. Rf 0.39 (15% EtOAc/toluene). [α]$_D^{23}$ −82 (c 0.58, CHCl$_3$). IR (neat): 3509, 2962, 1608, 1515, 1464, 1333, 1251, 1159, 1090, 1046, 816, 758, 663 cm^{-1}; ^{1}H-NMR (400 MHz, CDCl$_3$): δ 7.69 (d, J = 8.4 Hz, 2H), 7.31–7.20 (m, 4H), 6.87 (d, J = 6.8 Hz, 2H), 5.88 (s, 1H), 4.44 (d, J = 18.0 Hz, 1H), 4.35 (m, 1H), 3.98–3.90 (m, 2H), 3.80 (s, 3H), 2.39 (s, 3H), 2.00 (m, 2H), 1.14 (d, J = 6.8 Hz, 3H), 0.94 (d, J = 6.8 Hz, 3H); ^{13}C-NMR (100 MHz, CDCl$_3$): δ 159.7, 143.2, 138.0, 133.0, 129.7, 129.4, 126.7, 126.2, 124.2, 114.0, 67.3, 60.2, 55.3, 43.5, 27.0, 21.5, 20.9 (one signal missing due to an overlap). HRMS (ESI, [M + Na]$^+$) m/z calcd for $C_{15}H_{19}NNaO_4S$ 424.1553, found 424.1551.

For (2S,3R)-**5eA** as a minor diastereomer: Colorless oil. Rf 0.44 (15% EtOAc/toluene). [α]$^{23}_D$ −122 (c 2.25 in CHCl$_3$). IR (neat): 3600–3260, 2964, 1607, 1515, 1457, 1326, 1250, 1156, 1093, 1033, 826, 760, 657 cm^{-1}; ^{1}H-NMR (400 MHz, CDCl$_3$): δ 7.81 (d, J = 8.4 Hz, 2H), 7.31–7.20 (m, 4H), 6.87 (d, J = 8.8 Hz, 2H), 6.11 (d, J = 5.5 Hz, 1H), 4.46 (d, J = 18.3 Hz, 1H), 4.32–4.20 (m, 1H), 3.89–3.73 (m, 5H), 2.40 (s, 3H), 1.84 (d, J = 9.3 Hz, 1H), 1.76–1.61 (m, 1H), 1.01 (d, J = 6.3 Hz, 3H), 0.94 (d, J = 6.6 Hz, 3H); ^{13}C-NMR (100 MHz, CDCl$_3$): δ 159.9, 143.4, 137.7, 136.3, 129.6, 129.5, 127.3, 126.3, 120.5, 114.0, 64.7, 64.5, 55.3, 41.9, 27.5, 21.5, 20.8, 20.3. HRMS (ESI, [M + Na]$^+$) m/z calcd for $C_{15}H_{19}NNaO_4S$ 424.1553, found 424.1551.

Procedure for 5-(4-methoxyphenyl)-6-methyl-1-tosyl-1,2,3,6-tetrahydropyridin-3-ol (**5fA**)

Method: **5fA** (17.0 mg, 90%, dr 94: 6) was obtained from **4f** (13.3 mg, 0.0507 mmol), **7A** (11.4 mg), and Pd(PPh$_3$)$_4$ (2.9 mg) and isolated with preparative TLC eluting with 10% EtOAc/toluene (developed six times).

For (3R*,6S*)-**5fA** as a major diastereomer: Colorless oil. Rf 0.37 (17% EtOAc/toluene). IR (neat): 3492, 1607, 1514, 1248, 1157, 252, 660 cm^{-1}; ^{1}H-NMR (400 MHz, CDCl$_3$): δ 7.73 (d, J = 7.2 Hz, 2H) 7.27–2.22 (m, 4H), 6.87 (d, J = 8.8 Hz, 2H), 6.29 (s, 1H), 4.96 (q, J = 6.8 Hz, 1H), 4.18 (dd, J = 6.4, 10.0 Hz, 1H), 4.07 (dd, J = 6.4, 13.6 Hz, 1H), 3.81 (s, 3H), 2.90 (dd,

J = 10.0 Hz, 13.6 Hz, 1H), 2.39 (s, 3H), 2.15 (br-s, 1H), 1.16 (d, J = 6.8 Hz, 3H); ^{13}C-NMR (100 MHz, CDCl$_3$): δ 159.5, 143.4, 142.0, 137.9, 130.4, 129.7, 127.5, 126.7, 125.2, 114.0, 63.1, 55.3, 50.6, 43.5, 21.4, 18.3. HRMS (ESI, [M + Na]$^+$) m/z calcd for C$_{20}$H$_{23}$NNaO$_4$S 396.1240, found 396.1239.

For (3R*,6R*)-**5fA** as a minor diastereomer: Colorless oil. Rf 0.38 (17% EtOAc/toluene). IR (neat): 3600–3160 (br), 2979, 2934, 2838, 1607, 1513, 1335, 1247, 1155, 1122, 1088, 1013, 815, 741, 654 cm^{-1}; ^{1}H-NMR (600 MHz, CDCl$_3$): δ 7.81 (d, J = 8.6 Hz, 2H) 7.33–7.24 (m, 4H), 6.89 (d, J = 8.8 Hz, 2H), 5.93 (d, J = 4.1 Hz, 1H), 5.05 (q, J = 6.9 Hz, 1H), 4.18–4.12 (m, 1H), 3.90 (d, J = 14.4 Hz, 1H), 3.82 (s, 3H), 3.33 (d, J = 14.4 Hz, 1H), 2.42 (s, 3H), 1.95 (d, J = 10.3 Hz, 1H), 1.06 (d, J = 6.9 Hz, 3H); ^{13}C-NMR (151 MHz, CDCl$_3$): δ 159.9, 144.1, 143.5, 137.6, 130.5, 129.8, 127.7, 127.3, 122.2, 114.1, 63.6, 55.3, 50.5, 45.1, 21.5, 16.6. HRMS (ESI, [M + Na]$^+$) m/z calcd for C$_{20}$H$_{23}$NNaO$_4$S 396.1240, found 396.1238.

Procedure for (3R*,6S*)-5-(4-methoxyphenyl)-6-propyl-1-tosyl-1,2,3,6-tetrahydropyridin-3-ol (**5gA**)

Method: **5gA** (21.6 mg, 78%, dr >95:<5) was obtained from **4g** (20.2 mg, 0.0689 mmol), **7A** (15.7 mg), and Pd(PPh$_3$)$_4$ (4.0 mg) and isolated with preparative TLC eluting with 10% EtOAc/toluene.

Pale-yellow oil. Rf 0.30 (33% EtOAc/hexane). IR (neat): 3494, 2959, 2934, 1606, 1513, 1336, 1248, 825, 761, 661 cm^{-1}; ^{1}H-NMR (400 MHz, CDCl$_3$): δ 7.72 (d, J = 8.0 Hz, 2H) 7.25 (d, J = 8.0 Hz, 2H), 7.21 (d, J = 8.0 Hz, 2H), 6.89 (d, J = 8.0 Hz, 2H), 5.55 (s, 1H), 4.82 (t, J = 10.0 Hz, 1H), 4.08 (dd, J = 6.8 Hz, 14.0 Hz, 1H), 3.90 (dd, J = 6.8, 10.0 Hz, 1H), 3.83 (s, 3H), 2.96 (dd, J = 10.0 Hz, 14.0 Hz, 1H), 2.41 (s, 3H), 1.77 (br-s, 1H), 1.61–1,30 (m, 4H), 0.84 (t, J = 7.2 Hz, 3H); ^{13}C-NMR (100 MHz, CDCl$_3$): δ 159.5, 143.4, 141.7, 138.1, 130.8, 129.6, 127.3, 126.8, 124.8, 114.1, 62.0, 55.3, 43.8, 43.8, 34.6, 21.5, 19.9, 13.6. HRMS (ESI, [M + Na]$^+$) m/z calcd for C$_{22}$H$_{27}$NNaO$_4$S 424.1553, found 424.1550.

Procedure for (3R*,6S*)-5-(4-methoxyphenyl)-6-phenyl-1-tosyl-1,2,3,6-tetrahydropyridin-3-ol (**5hA**)

Method: **5hA** (2.9 mg, 36%, dr >95:<5) was obtained from **4h** (16.4 mg, 0.0501 mmol), **7A** (11.4 mg), and Pd(PPh$_3$)$_4$ (2.9 mg) and isolated with preparative TLC eluting with 15% EtOAc/toluene.

Pale-yellow oil. Rf 0.40 (50% EtOAc/hexane). IR (neat): 3491, 1606, 1513, 1335, 1250, 1160, 1034, 815, 744, 704, 661 cm^{-1}; ^{1}H-NMR (400 MHz, CDCl$_3$): δ 7.63 (d, J = 8.4 Hz, 2H) 7.39 (d, J = 6.8 Hz, 2H), 7.30–7.15 (m, 7H), 6.77 (d, J = 8.8 Hz, 2H), 6.05 (s, 1H), 6.02 (s, 1H), 4.18 (dd, J = 7.6, 10.3 Hz, 1H), 3.88 (dd, J = 7.6 Hz, 14.1 Hz, 1H), 3.74 (s, 3H), 2.84 (dd, J = 10.3, 14.1 Hz, 1H), 2.37 (s, 3H), 1.82 (br-s, 1H); ^{13}C-NMR (100 MHz, CDCl$_3$): δ 159.4, 143.4, 137.8, 137.68, 137.66, 130.0, 129.6, 129.0, 128.5, 128.0, 127.2, 127.0, 126.5, 113.9, 62.8, 57.5, 55.2, 43.6, 21.5. HRMS (ESI, [M + Na]$^+$) m/z calcd for C$_{25}$H$_{25}$NNaO$_4$S 458.1397, found 458.1398.

Procedure for (2S,6R)-5-(4-methoxyphenyl)-2,6-dimethyl-1-tosyl-1,2,3,6-tetrahydropyridin-3-ol (**5iA**) and (2S,5R)-4-((E)-4–methoxybenzylidene)-2,5-dimethyl-1-tosylpyrrolidin-3-ol (**8iA**)

Method: **5iA** (98 mg, 45%, dr >95:<5) and **8iA** (44 mg, 20%, dr >95:<5) were obtained from **4i** (158 mg, 0.566 mmol), **7A** (129 mg), and Pd(PPh$_3$)$_4$ (49.0 mg) and isolated with preparative TLC eluting with 40% EtOAc/hexane.

For **5iA**: Pale-yellow oil. Rf 0.30 (40% EtOAc/hexane). [α]$_D^{21}$ –163 (c 0.55, CHCl$_3$). IR (neat): 3492, 1607, 1514, 1248, 1157, 752, 660 cm^{-1}; ^{1}H-NMR (400 MHz, CDCl$_3$): δ 7.74 (d, J = 8.0 Hz, 2H) 7.32–7.20 (m, 4H), 6.88 (d, J = 8.8 Hz, 2H), 5.56 (s, 1H), 4.97 (q, J = 7.2 Hz, 1H), 4.26 (m, 1H), 4.13 (m, 1H), 3.82 (s, 3H), 2.40 (s, 3H), 1.78 (br-s, 1H), 1.31 (d, J = 7.2 Hz, 3H), 1.23 (d, J = 7.2 Hz, 3H); ^{13}C-NMR (100 MHz, CDCl$_3$): δ 159.4, 143.3, 142.0, 138.4, 130.9, 129.8, 127.9, 126.8, 123.8, 113.9, 65.8, 55.3, 49.8, 49.7, 22.1, 21.5, 14.8. HRMS (ESI, [M + Na]$^+$) m/z calcd for C$_{21}$H$_{25}$NNaO$_4$S 410.1397, found 410.1396.

For **8iA**: Pale-yellow oil. Rf 0.33 (40% EtOAc/hexane). IR (neat): 3491, 1606, 1513, 1250, 1160, 744, 661 cm^{-1}; ^{1}H-NMR (600 MHz, CDCl$_3$): δ 7.63 (d, J = 8.0 Hz, 2H), 7.19 (d, J = 8.0 Hz, 2H), 7.12 (d, J = 8.8 Hz, 2H), 6.89 (d, J = 8.8 Hz, 2H), 6.35 (s, 1H), 4.77 (q, J = 6.9 Hz, 1H), 4.22 (dd, J = 5.5, 6.5 Hz, 1H), 3.84 (s, 3H), 3.66 (dq, J = 5.5, 6.5 Hz, 1H), 2.37

(s, 3H), 1.65 (d, J = 5.5 Hz, 1H), 1.56 (d, J = 6.5 Hz, 3H), 1.38 (d, J = 6.9 Hz, 3H); ^{13}C-NMR (151 MHz, CDCl$_3$): δ 159.1, 143.3, 141.3, 135.3, 129.9, 129.6, 128.0, 127.2, 125.1, 113.9, 76.9, 58.2, 56.9, 55.3, 23.5, 21.5, 16.5. HRMS (ESI, [M + Na]$^+$) m/z calcd for C$_{21}$H$_{25}$NNaO$_4$S 410.1397, found 410.1396.

3.3.2. Procedure for 5-ethyl-1-tosyl-1,2,3,6-tetrahydropyridin-3-ol (**5aD**)

To a test tube containing **4a** (50.3 mg, 0.200 mmol) and Pd(PPh$_3$)$_4$ (11.6 mg, 5 mol%) were added anhydrous MeOH (2.0 mL) and 1.0 M Et$_3$B solution in THF (0.30 mL, 1.5 equiv) under argon. The resulting mixture was sealed with a screw cap and agitated at 80 °C for 1 h. The reaction mixture was cooled down to room temperature and then concentrated in vacuo. The residue was purified with preparative TLC eluting with 20% EtOAc/toluene to give **5aD** (42.8 mg, 76%) as a colorless oil. Spectra data of **5aD** were in agreement with those reported in the literature [52].

3.3.3. Procedure for 5-(phenylethynyl)-1-tosyl-1,2,3,6-tetrahydropyridin-3-ol (**5aE**)

To a test tube containing **4a** (25.1 mg, 0.100 mmol), CuI (1.2 mg, 6 mol%), PhCCH (22 µL, 2.0 equiv), and Pd(PPh$_3$)$_4$ (5.8 mg, 5 mol%) was added anhydrous MeOH (1.0 mL) under argon. The resulting mixture was sealed with a screw cap and agitated at 80 °C for 1.5 h. The reaction mixture was cooled down to room temperature and then concentrated in vacuo. The residue was purified with preparative TLC eluting with 20% EtOAc/toluene to give **5aE** (27.3 mg, 77%) as a pale-yellow oil. Spectra data of **5aE** were in agreement with those reported in the literature [53].

3.3.4. General Procedure for the Pd/PCy$_3$-Catalyzed Arylative Cyclizations of Internal Alkyne-Aldehyde **4j–o** with **7A**

To a test tube containing **4j–o** (1 equiv), p-methoxyphenylboronic acid (**7A**, 1.5 equiv), (η^3-allyl)CpPd (10 mol%), and PCy$_3$ (30 mol%) was added anhydrous MeOH (0.10 M) under argon. The resulting mixture was sealed with a screw cap and agitated at 80 °C for the time described in Scheme 4. The reaction mixture was cooled down to room temperature and then treated with PL-DEAM^{TM} (1.72 mmol/g, 2 equiv, X g) and THF (10 × X mL) to remove an excess of **7A**. The mixture was agitated at room temperature for 2 h. The mixture was filtered, and the resin was thoroughly rinsed with CHCl$_3$. The filtrate was concentrated in vacuo and the residue was purified with preparative TLC to give **5(j–o)A** along with a small amount of **8(j–o)A** in the yield described in Scheme 4.

Procedure for 4-hexyl-5-(4-methoxyphenyl)-1-tosyl-1,2,3,6-tetrahydropyridin-3-ol (**5jA**) and (E)-4-(1-(4-methoxyphenyl)heptyli–dene)-1-tosylpyrrolidin-3-ol (**8jA**)

Method: **5jA** (28.4 mg, 65%) and **8jA** (3.6 mg, 8%) were obtained from **4j** (33.5 mg, 0.0999 mmol), **7A** (23.0 mg), (η^3-allyl)CpPd (1.1 mg), and PCy$_3$ (4.2 mg) and isolated with preparative TLC eluting with 15% EtOAc/toluene (developed four times). Spectra data of **5jA** and **8jA** were in agreement with those reported in the literature [53].

Procedure for 5-(4-methoxyphenyl)-4-phenyl-1-tosyl-1,2,3,6-tetrahydropyridin-3-ol (**5kA**)

Method: **5kA** (32.5 mg, 98%) was obtained from **4k** (24.6 mg, 0.0751 mmol), **7A** (17.1 mg), (η^3-allyl)CpPd (1.5 mg), and PCy$_3$ (5.7 mg) and isolated with preparative TLC eluting with 20% EtOAc/toluene. Spectra data of **5kA** were in agreement with those reported in the literature [53].

Procedure for 5-(4-methoxyphenyl)-4-(4-nitrophenyl)-1-tosyl-1,2,3,6-tetrahydropyridin-3-ol (**5lA**)

Method: **5lA** (15.8 mg, 80%) was obtained from **4l** (15.3 mg, 0.0411 mmol), **7A** (11.4 mg), (η^3-allyl)CpPd (1.0 mg), and PCy$_3$ (3.8 mg) and isolated with preparative TLC eluting with 20% EtOAc/toluene.

Pale-yellow oil. Rf 0.40 (50% EtOAc/hexane). IR (neat): 3600–3160 (br), 2925, 1598, 1514, 1449, 1344, 1250, 1166, 1092, 1032, 760, 661 cm^{-1}; ^{1}H NMR (400 MHz, CDCl$_3$): δ 8.00 (d, J = 8.8 Hz, 2H) 7.73 (d, J = 8.6 Hz, 2H), 7.37 (d, J = 8.8 Hz, 2H), 7.22 (d, J = 8.6 Hz, 2H),

6.88 (d, J = 8.8 Hz, 2H), 6.70 (d, J = 8.8 Hz, 2H), 4.59–4.51 (m, 1H), 4.39 (d, J = 17.1 Hz, 1H), 3.97 (dd, J = 12.1, 2.4 Hz, 1H), 3.74 (s, 3H), 3.37 (d, J = 17.1 Hz, 1H), 2.92 (dd, J = 12.1, 2.7 Hz, 1H), 2.45 (s, 3H), 2.46–2.36 (m, 1H); ^{13}C-NMR (100 MHz, CDCl$_3$): δ 159.4, 146.48, 146.45, 144.3, 136.8, 133.7, 132.4, 130.2, 130.1, 130.0, 129.1, 127.9, 123.3, 114.0, 67.1, 55.2, 50.8, 49.6, 21.5. HRMS (ESI, [M + Na]$^+$) m/z calcd for $C_{25}H_{24}N_2NaO_6S$ 503.1247, found 503.1246.

Procedure for 4-(hex-1-ynyl)-5-(4-methoxyphenyl)-1-(toluene-4-sulfonyl)-1,2,3,6-tetrahydropyridin-3-ol (**5mA**)

Method: **5mA** (12.0 mg, 70%) was obtained from **4m** (14.0 mg, 0.0422 mmol), **7A** (9.6 mg), (η^3-allyl)CpPd (0.8 mg), and PCy$_3$ (3.2 mg) and isolated with preparative TLC eluting with 25% EtOAc/toluene. Spectra data of **5mA** were in agreement with those reported in the literature [53].

Procedure for 4-(cyclohex-1-en-1-yl)-5-(4-methoxyphenyl)-1-tosyl-1,2,3,6-tetrahydropyridin-3-ol (**5nA**) and (*E*)-4-(cyclohex-1-en-1-yl(4-methoxyphenyl)methylene)-1-tosylpyrrolidin-3-ol (**8nA**)

Method: **5nA** (13.0 mg, 53%) and **8nA** (1.6 mg, 7%) were obtained from **4n** (18.5 mg, 0.0558 mmol), **7A** (12.7 mg), (η^3-allyl)CpPd (1.2 mg), and PCy$_3$ (4.7 mg) and isolated with preparative TLC eluting with 40% EtOAc/hexane. Spectra data of **5nA** and **8nA** were in agreement with those reported in the literature [53].

Procedure for (2*S*,3*S*)-5-(4-methoxyphenyl)-2-methyl-4-(4-nitrophenyl)-1-tosyl-1,2,3,6-tetrahydropyridin-3-ol (**5oA**)

Method: **5oA** (21.0 mg, 90%, dr >95:<5) was obtained from **4o** (18.2 mg, 0.0471 mmol), **7A** (11.4 mg), (η^3-allyl)CpPd (1.0 mg), and PCy$_3$ (3.8 mg) and isolated with preparative TLC eluting with 20% EtOAc/toluene.

Pale-yellow oil. Rf 0.40 (50% EtOAc/hexane). [α]$_D^{22}$ −29 (*c* 0.22, CHCl$_3$). IR (neat): 2932, 1607, 1596, 1512, 1344, 1248, 1160, 1031, 757, 662 cm^{-1}; ^{1}H-NMR (400 MHz, CDCl$_3$): δ 7.97 (d, J = 8.8 Hz, 2H) 7.74 (d, J = 8.4 Hz, 2H), 7.34 (d, J = 7.6 Hz, 2H), 7.10 (d, J = 8.4 Hz, 2H), 6.82 (d, J = 8.8 Hz, 2H), 6.67 (d, J = 7.6 Hz, 2H), 4.79 (m, 1H), 4.46 (m, 1H), 4.28 (d, J = 18.0 Hz, 1H), 3.89 (d, J = 18.0 Hz, 1H), 3.72 (s, 3H), 2.45 (3H, s), 1.09 (d, J = 6.8 Hz, 3H); ^{13}C-NMR (100 MHz, CDCl$_3$): δ 158.9, 146.3, 144.9, 143.6, 143.6, 136.8, 134.5, 133.7, 130.9, 129.9, 129.5, 127.1, 122.9, 113.8, 68.1, 55.1, 51.0, 45.2, 21.5, 9.8. HRMS (ESI, [M + Na]$^+$) m/z calcd for $C_{26}H_{26}N_2NaO_6S$ 517.1404, found 517.1401.

3.3.5. General Procedure for the Transformations of Tetrahydropyridine **5** into 3-Hydroxypyridine **3**

To a solution of tetrahydropyridine **5** (1 equiv) in anhydrous DCM (0.2 M) was added Dess–Martin periodinane (1.5 equiv) at room temperature. In the oxidation of **5eA**, sodium bicarbonate (2 equiv) was added prior to Dess–Martin periodinane to prevent acid-mediated dehydration. After being stirred at the same temperature for 1 h, the reaction mixture was diluted with Et$_2$O and treated with saturated aqueous sodium thiosulfate and saturated aqueous NaHCO$_3$. The resulting mixture was stirred for 1 h and then extracted with Et$_2$O. The organic layer was washed with brine, dried over MgSO$_4$, and concentrated in vacuo to give enone, which was used for the next step without further purification.

To a solution of the crude enone (1 equiv) in anhydrous toluene (0.33 M) was added DBU (2.0 equiv) at room temperature under argon. After being stirred at the same temperature for 4 h, the reaction mixture was concentrated in vacuo. The residue was purified by preparative TLC eluting with 10% MeOH/CHCl$_3$ to give 3-hydroxypyridine **3**.

Procedure for 5-(4-methoxyphenyl)pyridin-3-ol (**3aA**)

Method: **3aA** (2.3 mg, 80%) was obtained from **5aA** (5.0 mg, 0.0139 mmol), DMPI (8.3 mg), and DBU (4.2 μL).

Pale-yellow oil. Rf 0.61 (10% MeOH/CHCl$_3$). IR (neat): 2929, 2853, 1609, 1583, 1518, 1440, 1290, 1251, 1221, 1180, 1149, 1031, 828, 755 cm^{-1}; ^{1}H-NMR (400 MHz, CDCl$_3$:CD$_3$OD = 3:1):

δ 8.24 (s, 1H), 8.06 (s, 1H), 7.51 (d, J = 8.8 Hz, 2H), 7.37 (s, 1H), 7.01 (d, J = 8.8 Hz, 2H), 3.87 (s, 3H); ^{13}C-NMR (100 MHz, CDCl$_3$:CD$_3$OD = 3:1): δ 159.6, 153.9, 138.0, 135.1, 129.6, 128.0, 120.9, 114.3, 109.2, 55.1. HRMS (ESI, [M + H]$^+$) m/z calcd for C$_{12}$H$_{12}$NO$_2$ 202.0863, found 202.0862.

Procedure for 1-(4-(5-hydroxypyridin-3-yl)phenyl)ethan-1-one (**3aB**)

Method: **3aB** (10.2 mg, 73%) was obtained from **5aB** (25.0 mg, 0.0673 mmol), DMPI (40.0 mg), and DBU (20.0 µL).

Pale-yellow oil. Rf 0.55 (10% MeOH/CHCl$_3$). IR (neat): 2925, 1684, 1604, 1267, 1162, 755, 668 cm^{-1}; ^{1}H-NMR (400 MHz, CDCl$_3$:CD$_3$OD = 3:1): δ 8.33 (s, 1H), 8.17 (s, 1H), 8.05 (d, J = 8.4 Hz, 2H), 7.67 (d, J = 8.4 Hz, 2H), 7.42 (s, 1H), 2.66 (s, 3H); ^{13}C-NMR (100 MHz, CDCl$_3$:CD$_3$OD = 3:1): δ 198.4, 154.0, 142.1, 138.5, 136.9, 136.5, 136.3, 128.9, 127.2, 121.5, 26.4. HRMS (ESI, [M + H]$^+$) m/z calcd for C$_{13}$H$_{12}$NO$_2$ 214.0863, found 214.0858.

Procedure for 5-(4-nitrophenyl)pyridin-3-ol (**3aC**)

Method: **3aC** (6.5 mg, 65%) was obtained from **5aC** (17.3 mg, 0.0462 mmol), DMPI (27.5 mg), and DBU (14.0 µL).

Pale-yellow oil. Rf 0.53 (10% MeOH/CHCl$_3$). IR (neat): 2923, 1598, 1521, 1345, 1159, 795 cm^{-1}; ^{1}H-NMR (400 MHz, CDCl$_3$:CD$_3$OD = 3:1): δ 8.34 (s, 1H), 8.33 (d, J = 8.0 Hz, 2H), 8.21 (s, 1H), 7.74 (d, J = 8.0 Hz, 2H), 7.42 (s, 1H); ^{13}C-NMR (100 MHz, CDCl$_3$:CD$_3$OD = 3:1): δ 154.1, 147.5, 143.9, 138.4, 137.6, 135.4, 127.8, 124.1, 121.5. HRMS (ESI, [M + H]$^+$) m/z calcd for C$_{11}$H$_9$N$_2$O$_3$ 217.0608, found 217.0607.

Procedure for 5-ethylpyridin-3-ol (**3aD**)

Method: **3aD** (15.2 mg, 86%) was obtained from **5aD** (40.5 mg, 0.144 mmol), DMPI (91.6 mg), and DBU (43.0 µL).

Pale-yellow oil. Rf 0.55 (10% MeOH/CHCl$_3$). IR (neat): 2968, 1585, 1438, 1225, 756, 707 cm^{-1}; ^{1}H-NMR (400 MHz, CDCl$_3$): δ 8.10 (s, 1H), 7.94 (s, 1H), 7.16 (s, 1H), 2.63 (q, J = 7.8 Hz, 2H), 1.24 (t, J = 7.8 Hz, 3H); ^{13}C-NMR (100 MHz, CDCl$_3$): δ 155.2, 141.6, 138.8, 133.6, 124.6, 25.9, 15.0. LRMS (EI) m/z (relative intensity) 123 ([M]$^+$, 100), 108 (70), 95 (12). HRMS (EI, [M]$^+$): m/z calcd for C$_7$H$_9$NO, 123.0684; found, 123.0684.

Procedure for 5-(phenylethynyl)pyridin-3-ol (**3aE**)

Method: **3aE** (4.2 mg, 50%) was obtained from **5aE** (14.5 mg, 0.0410 mmol), DMPI (24.2 mg), and DBU (12.8 µL).

Pale-yellow oil. Rf 0.57 (10% MeOH/CHCl$_3$). IR (neat): 2924, 2644, 2568, 2216, 1579, 1425, 1325, 1248, 1150, 1124, 1022, 868, 754, 688 cm^{-1}; ^{1}H-NMR (600 MHz, CDCl$_3$:CD$_3$OD = 3:1): δ 8.19 (s, 1H), 8.08 (s, 1H), 7.57–7.48 (m, 2H), 7.44–7.35 (m, 3H), 7.33 (s, 1H); ^{13}C-NMR (151 MHz, CDCl$_3$:CD$_3$OD = 3:1): δ 153.2, 142.1, 136.5, 131.2, 128.4, 128.0, 124.8, 122.0, 120.8, 91.9, 85.5. LRMS (EI) m/z (relative intensity) 195 ([M]$^+$, 100), 139 (25), 69 (11). HRMS (EI, [M]$^+$): m/z calcd for C$_{13}$H$_9$NO, 195.0684; found, 195.0700.

Procedure for 5-(4-methoxyphenyl)-2-methylpyridin-3-ol (**3bA**)

Method: **3bA** (8.7 mg, 67%) was obtained from **5bA** (22.6 mg, 0.0605 mmol), DMPI (35.9 mg), and DBU (18.1 µL).

Pale-yellow oil. Rf 0.50 (10% MeOH/CHCl$_3$). IR (neat): 2922, 1604, 1515, 1444, 1287, 1220, 1163, 773 cm^{-1}; ^{1}H-NMR (400 MHz, CDCl$_3$:CD$_3$OD = 3:1): δ 8.03 (s, 1H), 7.43 (d, J = 7.2 Hz, 2H), 7.24 (s, 1H), 6.93 (d, J = 7.2 Hz, 2H), 3.80 (s, 3H), 2.40 (s, 3H); ^{13}C-NMR (100 MHz, CDCl$_3$:CD$_3$OD = 3:1): δ 159.3, 144.5, 136.7, 135.3, 129.9, 127.7, 126.7, 119.7, 114.2, 55.1, 17.1. HRMS (ESI, [M + H]$^+$) m/z calcd for C$_{13}$H$_{14}$NO$_2$ 216.1019, found 216.1015.

Procedure for 2-isobutyl-5-(4-methoxyphenyl)pyridin-3-ol (**3cA**)

Method: **3cA** (7.8 mg, 59%) was obtained from **5cA** (21.5 mg, 0.0517 mmol), DMPI (30.7 mg), and DBU (15.5 µL).

Pale-yellow oil. Rf 0.55 (10% MeOH/CHCl$_3$). IR (neat): 2955, 1608, 1608, 1521, 1393, 1252, 1165, 1033, 830, 772 cm^{-1}; ^{1}H-NMR (400 MHz, CDCl$_3$:CD$_3$OD = 3:1): δ 8.14 (s, 1H), 7.50 (d, J = 8.8 Hz, 2H), 7.30 (s, 1H), 6.99 (d, J = 8.8 Hz, 2H), 3.86 (s, 3H), 2.71 (d, J = 7.8 Hz, 2H), 2.15 (t–sept, J = 7.8, 6.8 Hz, 1H), 0.96 (d, J = 6.8 Hz, 6H); ^{13}C-NMR (100 MHz, CDCl$_3$:CD$_3$OD = 3:1): δ 159.3, 151.7, 147.7, 136.6, 135.0, 129.9, 127.7, 119.9, 114.1, 55.0, 40.4, 27.9, 22.1. HRMS (ESI, [M + H]$^+$) m/z calcd for C$_{16}$H$_{20}$NO$_2$ 258.1489, found 258.1487.

Procedure for 2-benzyl-5-(4-methoxyphenyl)pyridin-3-ol (**3dA**)

Method: **3dA** (82.2 mg, 95%) was obtained from **5dA** (133 mg, 0.296 mmol), DMPI (176 mg), and DBU (88.4 μL).

Pale-yellow oil. Rf 0.55 (10% MeOH/CHCl$_3$). IR (neat): 1600, 1522, 1433, 1392, 1257, 1176, 1027, 827 cm^{-1}; ^{1}H-NMR (400 MHz, CDCl$_3$:CD$_3$OD = 3:1): δ 8.16 (d, J = 2.0 Hz, 1H), 7.47 (d, J = 8.4 Hz, 2H), 7.34 (d, J = 7.2 Hz, 2H), 7.30 (d, J = 2.0 Hz, 1H), 7.25 (dd, J = 7.2, 7.2 Hz, 2H), 7.15 (t, J = 7.2 Hz, 1H), 6.97 (d, J = 8.4 Hz, 2H), 4.19 (s, 2H), 3.85 (s, 3H); ^{13}C-NMR (100 MHz, CDCl$_3$:CD$_3$OD = 3:1): δ 159.3, 151.5, 146.7, 139.3, 137.1, 135.7, 129.8, 128.6, 128.0, 127.8, 125.7, 120.4, 114.2, 55.0, 37.7. LRMS (EI) m/z (relative intensity) 291 ([M]$^+$, 100), 274 (12). HRMS (EI, [M]$^+$): m/z calcd for C$_{19}$H$_{17}$NO$_2$, 291.1259; found, 291.1248.

Procedure for 2-isopropyl-5-(4-methoxyphenyl)pyridin-3-ol (**3eA**)

Method: **3eA** (6.8 mg, 64%) was obtained from **5eA** (16.3 mg, 0.0406 mmol), DMPI (26.3 mg), NaHCO$_3$ (6.9 mg), and DBU (12.4 μL).

Pale-yellow oil. Rf 0.55 (10% MeOH/CHCl$_3$). IR (neat): 2969, 2932, 1610, 1518, 1290, 1251, 1229, 1176, 1033, 830, 756 cm^{-1}; ^{1}H-NMR (400 MHz, CDCl$_3$:CD$_3$OD = 3:1): δ 8.33 (s, 1H), 7.43 (d, J = 8.8 Hz, 2H), 7.26 (s, 1H), 6.95 (d, J = 8.8 Hz, 2H), 3.83 (s, 3H), 3.47 (sept, J = 7.2 Hz, 1H), 1.35 (d, J = 7.2 Hz, 6H); ^{13}C-NMR (100 MHz, CDCl$_3$:CD$_3$OD = 3:1): δ 159.6, 152.6, 149.8, 138.5, 135.1, 130.0, 128.0, 120.5, 114.4, 53.3, 29.1, 21.1. HRMS (ESI, [M + H]$^+$) m/z calcd for C$_{15}$H$_{18}$NO$_2$ 244.1332, found 244.1330.

Procedure for 5-(4-methoxyphenyl)-6-methylpyridin-3-ol (**3fA**)

Method: **3fA** (9.3 mg, 67%) was obtained from **5fA** (24.2 mg, 0.0648 mmol), DMPI (38.5 mg), and DBU (19.4 μL).

Pale-yellow oil. Rf 0.55 (10% MeOH/CHCl$_3$). IR (neat): 2931, 1610, 1515, 1453, 1290, 1248, 1176, 1031, 834, 771, 707 cm^{-1}; ^{1}H-NMR (400 MHz, CDCl$_3$:CD$_3$OD = 3:1): δ 7.97 (s, 1H), 7.24 (d, J = 8.0 Hz, 2H), 7.08 (s, 1H), 6.98 (d, J = 8.0 Hz, 2H), 3.87 (s, 3H), 2.38 (s, 3H); ^{13}C-NMR (100 MHz, CDCl$_3$:CD$_3$OD = 3:1): δ 158.8, 151.7, 145.9, 137.5, 134.4, 131.7, 129.8, 124.7, 113.6, 55.0, 21.2. HRMS (ESI, [M + H]$^+$) m/z calcd for C$_{13}$H$_{14}$NO$_2$ 216.1019, found 216.1015.

Procedure for 5-(4-methoxyphenyl)-6-propylpyridin-3-ol (**3gA**)

Method: **3gA** (7.2 mg, 74%) was obtained from **5gA** (16.0 mg, 0.0398 mmol), DMPI (23.7 mg), and DBU (12.0 μL).

Pale-yellow oil. Rf 0.55 (10% MeOH/CHCl$_3$). IR (neat): 2960, 2931, 1610, 1516, 1452, 1288, 1248, 1175, 1032, 835, 755, 705 cm^{-1}; ^{1}H-NMR (400 MHz, CDCl$_3$:CD$_3$OD = 3:1): δ 8.27 (d, J = 2.4 Hz, 1H), 7.23 (d, J = 8.8 Hz, 2H), 7.17 (d, J = 2.4 Hz, 1H), 6.96 (d, J = 8.8 Hz, 2H), 3.68 (s, 3H), 2.71 (t, J = 7.8 Hz, 2H), 1.57 (tq, J = 7.8, 7.2 Hz, 2H), 0.81 (t, J = 7.2 Hz, 3H); ^{13}C-NMR (100 MHz, CDCl$_3$:CD$_3$OD = 3:1): δ 159.0, 152.6, 150.6, 138.4, 134.5, 131.8, 130.1, 126.9, 113.8, 55.3, 35.6, 23.7, 14.0. HRMS (ESI, [M + Na]$^+$) m/z calcd for C$_{15}$H$_{17}$NNaO$_2$ 266.1152, found 266.1151.

Procedure for 5-(4-methoxyphenyl)-6-phenylpyridin-3-ol (**3hA**)

Method: **3hA** (6.6 mg, 80%) was obtained from **5hA** (12.8 mg, 0.0294 mmol), DMPI (17.5 mg), and DBU (8.8 μL).

Pale-yellow oil. Rf 0.55 (10% MeOH/CHCl$_3$). IR (neat): 2917, 1610, 1514, 1447, 1290, 1249, 1177, 1030, 833, 752, 702 cm^{-1}; ^{1}H-NMR (400 MHz, CDCl$_3$:CD$_3$OD = 3:1): δ 8.14 (s,

1H), 7.44 (s, 1H), 7.25–7.18 (m, 5H), 7.06 (d, J = 6.8 Hz, 2H), 6.79 (d, J = 6.8 Hz, 2H), 3.78 (s, 3H); ^{13}C-NMR (100 MHz, CDCl$_3$:CD$_3$OD = 3:1): δ 158.8, 151.5, 141.3, 137.2, 135.7, 135.4, 133.6, 130.9, 130.3, 129.4, 128.5, 127.9, 113.4, 54.8. HRMS (ESI, [M + H]$^+$) m/z calcd for C$_{18}$H$_{16}$NO$_2$ 278.1176, found 278.1170.

Procedure for 5-(4-methoxyphenyl)-2,6-dimethylpyridin-3-ol (**3iA**)

Method: **3iA** (23.8 mg, 73%) was obtained from **5iA** (55.2 mg, 0.142 mmol), DMPI (90.7 mg), and DBU (44.5 µL).

Pale-yellow oil. Rf 0.55 (10% MeOH/CHCl$_3$). IR (neat): 2924, 1516, 1289, 1249, 1161, 1033, 840, 812, 719, 668 cm^{-1}; ^{1}H-NMR (400 MHz, CDCl$_3$:CD$_3$OD = 3:1): δ 7.36 (s, 1H), 7.22 (d, J = 8.8 Hz, 2H), 6.96 (d, J = 8.8 Hz, 2H), 3.86 (s, 3H), 2.46 (s, 3H), 2.38 (s, 3H); ^{13}C-NMR (100 MHz, CDCl$_3$:CD$_3$OD = 3:1): δ 158.7, 149.2, 144.8, 144.0, 135.0, 132.0, 129.9, 123.7, 113.6, 55.1, 21.1, 17.6. HRMS (ESI, [M + H]$^+$) m/z calcd for C$_{14}$H$_{16}$NO$_2$ 230.1176, found 230.1171.

Procedure for 4-hexyl-5-(4-methoxyphenyl)pyridin-3-ol (**3jA**)

Method: **3jA** (9.3 mg, 48%) was obtained from **5jA** (30.8 mg, 0.0694 mmol), DMPI (41.2 mg), and DBU (20.7 µL).

Pale-yellow oil. Rf 0.50 (10% MeOH/CHCl$_3$). IR (neat): 2955, 1611, 1517, 1501, 1425, 1289, 1244, 1176, 1036, 831 cm^{-1}; ^{1}H-NMR (400 MHz, CDCl$_3$): δ 8.33 (s, 1H), 7.96 (s, 1H), 7.24 (d, J = 8.0 Hz, 2H), 6.97 (d, J = 8.0 Hz, 2H), 3.87 (s, 3H), 2.64 (t, J = 2.8 Hz, 2H), 1.60–1.48 (m, 2H), 1.30–1.10 (m, 6H), 0.81 (t, J = 6.8 Hz, 3H); ^{13}C-NMR (100 MHz, CDCl$_3$): δ 159.1, 153.8, 140.1, 139.0, 138.9, 134.1, 130.4, 130.1, 113.7, 55.3, 31.4, 29.4, 28.8, 26.8, 22.5, 14.0. HRMS (ESI, [M + H]$^+$) m/z calcd for C$_{18}$H$_{24}$NO$_2$ 286.1802, found 286.1801.

Procedure for 5-(4-methoxyphenyl)-4-phenylpyridin-3-ol (**3kA**)

Method: **3kA** (22.3 mg, 80%) was obtained from **5kA** (43.6 mg, 0.100 mmol), DMPI (59.4 mg), and DBU (29.9 µL).

Pale-yellow oil. Rf 0.55 (10% MeOH/CHCl$_3$). IR (neat): 2933, 1609, 1425, 1290, 1249, 1178, 1033, 831, 750, 699 cm^{-1}; ^{1}H-NMR (400 MHz, CDCl$_3$): δ 8.43 (s, 1H), 8.16 (s, 1H), 7.35–7.25 (m, 3H), 7.20 (d, J = 6.8 Hz, 2H), 6.99 (d, J = 8.0 Hz, 2H), 6.73 (d, J = 8.0 Hz, 2H), 3.75 (s, 3H); ^{13}C-NMR (100 MHz, CDCl$_3$): δ 158.8, 151.5, 141.3, 137.2, 135.7, 135.4, 133.6, 130.9, 130.3, 129.4, 128.5, 127.9, 113.5, 55.1. HRMS (ESI, [M + H]$^+$) m/z calcd for C$_{18}$H$_{16}$NO$_2$ 278.1176, found 278.1173.

Procedure for 5-(4-methoxyphenyl)-4-(4-nitrophenyl)pyridin-3-ol (**3lA**)

Method: **3lA** (25.9 mg, 77%) was obtained from **5lA** (50.2 mg, 0.104 mmol), DMPI (62.1 mg), and DBU (33.1 µL).

Pale-yellow oil. Rf 0.51 (10% MeOH/CHCl$_3$). IR (neat): 2933, 1515, 1247, 1176, 1110, 1033, 830, 753 cm^{-1}; ^{1}H-NMR (600 MHz, CDCl$_3$:CD$_3$OD = 1:1): δ 8.21 (s, 1H), 8.16–8.09 (m, 3H), 7.39 (d, J = 8.6 Hz, 2H), 6.99 (d, J = 8.6 Hz, 2H), 6.79 (d, J = 8.6 Hz, 2H), 3.78 (s, 3H); ^{13}C-NMR (151 MHz, CDCl$_3$:CD$_3$OD = 1:1): δ 158.7, 146.5, 141.4, 140.7, 135.5, 132.6, 131.3, 130.4, 128.4, 122.4, 113.3, 54.5. (two signals missing due to an overlap). HRMS (ESI, [M + H]$^+$) m/z calcd for C$_{18}$H$_{15}$N$_2$O$_4$ 323.1026, found 323.1025.

Procedure for 4-(cyclohex-1-en-1-yl)-5-(4-methoxyphenyl)pyridin-3-ol (**3nA**)

Method: **3nA** (26.4 mg, 58%) was obtained from **5nA** (71.2 mg, 0.162 mmol), DMPI (96.2 mg), and DBU (49.0 µL).

Pale-yellow oil. Rf 0.55 (10% MeOH/CHCl$_3$). IR (neat): 2931, 1610, 1511, 1452, 1246, 1170, 1032, 832, 758, 664 cm^{-1}; ^{1}H-NMR (600 MHz, CDCl$_3$): δ 8.30 (s, 1H), 8.12 (s, 1H), 7.34 (d, J = 8.9 Hz, 2H), 6.93 (d, J = 8.9 Hz, 2H), 5.98–5.94 (m, 1H), 3.85 (s, 3H), 2.24–2.19 (m, 2H), 2.73–2.69 (m, 2H), 2.61–2.54 (m, 2H), 2.52–2.44 (m, 2H)' ^{13}C-NMR (151 MHz, CDCl$_3$): δ 159.3, 148.7, 142.1, 136.2, 135.4, 135.0, 132.6, 131.5, 130.0, 129.9, 113.7, 55.3, 28.1, 25.4, 22.5, 21.6. HRMS (ESI, [M + H]$^+$) m/z calcd for C$_{18}$H$_{20}$NO$_2$ 282.1489, found 282.1486.

Procedure for 5-(4-methoxyphenyl)-2-methyl-4-(4-nitrophenyl)pyridin-3-ol (**3oA**)

Method: **3oA** (9.3 mg, 72%) was obtained from **5oA** (20.5 mg, 0.0415 mmol), DMPI (24.6 mg), and DBU (12.4 µL).

Pale-yellow oil. Rf 0.55 (10% MeOH/CHCl$_3$). IR (neat): 2923, 1513, 1343, 1241, 1219, 1176, 1128, 1106, 1033, 829, 755 cm^{-1}; ^{1}H-NMR (400 MHz, CDCl$_3$): δ 8.14 (d, J = 8.8 Hz, 2H), 8.03 (s, 1H), 7.35 (d, J = 8.8 Hz, 2H), 6.93 (d, J = 8.8 Hz, 2H), 6.75 (d, J = 8.8 Hz, 2H), 3.77 (s, 3H), 2.56 (s, 3H); ^{13}C-NMR (100 MHz, CDCl$_3$): δ 158.7, 146.8, 145.8, 141.9, 134.9, 131.47, 131.45, 130.59, 130.56, 128.8, 123.02, 122.99, 113.6, 54.9, 18.5. HRMS (ESI, [M + H]$^+$) m/z calcd for C$_{19}$H$_{17}$N$_2$O$_4$ 337.1183, found 337.1179.

3.3.6. Procedure for 2-benzyl-3-(3-methoxyphenyl)-5-(4-methoxyphenyl)pyridine (**10**)

To a solution of **3cA** (40.0 mg, 0.137 mmol) and Et$_3$N (38.2 µL, 0.274 mmol) in anhydrous DCM (1.0 mL) was added Tf$_2$O (38.2 µL, 0.164 mmol) at 0 °C under argon. After being stirred at the same temperature for 4 h, the reaction mixture was treated with saturated aqueous NaHCO$_3$. The resulting mixture was extracted with EtOAc, washed with brine, dried over MgSO$_4$, and concentrated in vacuo. The residue was purified with preparative TLC eluting with 10% EtOAc/toluene to give triflate (46.6 mg, 80%).

To a test tube containing the above triflate (8.7 mg, 0.021 mmol), m–methoxyphenylboronic acid (6.2 mg, 2 equiv), Pd(PPh$_3$)$_4$ (1.2 mg, 5 mol%), and LiCl (2.6 mg, 3 equiv) in DME (0.3 mL) was added 2.0 M aqueous Na$_2$CO$_3$ (31 µL) under argon. The resulting mixture was sealed with a screw cap and stirred at 80 °C for 8 h. The reaction mixture was cooled down to room temperature, diluted with EtOAc, washed with water and brine, dried over MgSO$_4$, and concentrated in vacuo. The residue was purified with preparative TLC eluting with 20% EtOAc/toluene to give **10** (7.2 mg, 93%) as a brown solid.

Rf 0.70 (20% EtOAc/toluene). IR (neat): 1609, 1516, 1455, 1440, 1288, 1248, 1179, 1148, 1035, 830, 701 cm^{-1}; ^{1}H-NMR (400 MHz, CDCl$_3$): δ 8.80 (s, 1H), 7.71 (s, 1H), 7.54 (d, J = 8.8 Hz, 2H), 7.32 (t, J = 8.4 Hz, 1H), 7.24–7.10 (m, 3H), 7.06 (d, J = 8.4 Hz, 2H), 6.99 (d, J = 8.8 Hz, 2H), 6.93 (d, J = 8.0 Hz, 1H), 6.87 (d, J = 8.0 Hz, 1H), 6.75 (s, 1H), 4.18 (s, 2H), 3.85 (s, 3H), 3.72 (s, 3H); ^{13}C-NMR (100 MHz, CDCl$_3$): δ 159.7, 159.4, 155.8, 146.3, 140.9, 140.1, 137.3, 135.6, 133.9, 130.0, 129.8, 129.4, 128.8, 128.2, 128.1, 125.9, 121.5, 114.5, 113.6, 55.3, 55.2, 41.3. HRMS (ESI, [M + H]$^+$) m/z calcd for C$_{26}$H$_{24}$NO$_2$ 382.1802, found 382.1796.

4. Conclusions

In summary, we have developed a new synthetic method for polysubstituted 3-hydroxypyridines. The starting alkynals, which were readily prepared from N-tosyl amino acid esters and propargyl alcohols, were effectively converted to a wide range of 3-hydroxy-1,2,3,6-tetrahydropyridines with various organometallic reagents in the "anti-Wacker"-type cyclization. The 5-Monosubstituted 3-hydroxypyridnes, 2,5-, 4,5-, and 5,6-disubstituted 3-hydroxypyridnes, and 2,4,5- and 2,5,6-trisubstituted 3-hydroxypyridnes were obtained by the oxidation and elimination of toluenesufinic acid. This approach enables the introduction of substituents into 3-hydroxypyridines one by one in a highly regioselective manner. The hydroxy group at the C3 position can be further substituted with cross-coupling reactions via the corresponding triflate.

Supplementary Materials: The following supporting information can be downloaded at: https://www.mdpi.com/article/10.3390/catal13020319/s1, preparation of substrates, analytical data, ^{1}H and ^{13}C NMR spectra, and more detailed materials and methods. References [52,53,61–65] are cited in Supplementary Materials.

Author Contributions: Conceptualization, H.T.; investigation, K.I. and H.T.; writing—original draft preparation, K.I.; writing—review and editing, H.T. and T.D.; supervision, T.D.; funding acquisition, H.T. and T.D. All authors have read and agreed to the published version of the manuscript.

Funding: This research was partly funded by the Banyu Pharmaceutical Co. Ltd. Award in Synthetic Organic Chemistry, The Research Foundation for Pharmaceutical Sciences, SUNTRY FOUNDATION for LIFE SCIENCES, Platform Project for Supporting Drug Discovery and Life Science Research (Basis for Supporting Innovative Drug Discovery and Life Science Research (BINDS)) from AMED under Grant Number JP19am0101095 and JP19am0101100, and JSPS KAKENHI Grant Numbers JP2459004 and JP15K07849.

Data Availability Statement: All experimental data is contained in the article and Supplementary Material.

Conflicts of Interest: The authors declare no conflict of interest.

References

1. Schmidt, A. Biologically Active Mesomeric Betaines and Alkaloids, Derived from 3-Hydroxypyridine, Pyridin-N-oxide, Nicotinic Acid and Picolinic Acid: Three Types of Conjugation and Their Consequences. *Curr. Org. Chem.* **2004**, *8*, 653–670. [CrossRef]
2. Baumann, M.; Baxendale, I.R. An overview of the synthetic routes to the best selling drugs containing 6-membered heterocycles. *Beilstein J. Org. Chem.* **2013**, *9*, 2265–2319. [CrossRef] [PubMed]
3. de Ruiter, G.; Lahav, M.; van der Boom, M.E. Pyridine Coordination Chemistry for Molecular Assemblies on Surfaces. *Acc. Chem. Res.* **2014**, *47*, 3407–3416. [CrossRef] [PubMed]
4. Vitaku, E.; Smith, D.T.; Njardarson, J.T. Analysis of the Structural Diversity, Substitution Patterns, and Frequency of Nitrogen Heterocycles among U.S. FDA Approved Pharmaceuticals. *J. Med. Chem.* **2014**, *57*, 10257–10274. [CrossRef] [PubMed]
5. Altaf, A.A.; Shahzad, A.; Gul, Z.; Rasool, N.; Badshah, A.; Lal, B.; Khan, E. A Review on the Medicinal Importance of Pyridine Derivatives. *J. Drug Des. Med. Chem.* **2015**, *1*, 1–11.
6. Guan, A.-Y.; Liu, C.-L.; Sun, X.-F.; Xie, Y.; Wang, M.-A. Discovery of pyridine-based agrochemicals by using Intermediate Derivatization Methods. *Bioorg. Med. Chem.* **2016**, *24*, 342–353. [CrossRef] [PubMed]
7. Prachayasittikul, S.; Pingaew, R.; Worachartcheewan, A.; Sinthupoom, N.; Prachayasittikul, V.; Ruchirawat, S.; Prachayasittikul, V. Roles of Pyridine and Pyrimidine Derivatives as Privileged Scaffolds in Anticancer Agents. *Mini-Rev. Med. Chem.* **2017**, *17*, 869–901. [CrossRef] [PubMed]
8. Andersson, H.; Almqvist, F.; Olsson, R. Synthesis of 2-Substituted Pyridines via a Regiospecific Alkylation, Alkynylation, and Arylation of Pyridine N-Oxides. *Org. Lett.* **2007**, *9*, 1335–1337. [CrossRef]
9. Do, H.-Q.; Kashif Khan, R.M.; Daugulis, O. A General Method for Copper-Catalyzed Arylation of Arene C-H Bonds. *J. Am. Chem. Soc.* **2008**, *130*, 15185–15192. [CrossRef]
10. Li, M.; Hua, R. Gold(I)-catalyzed direct C–H arylation of pyrazine and pyridine with aryl bromides. *Tetrahedron Lett.* **2009**, *50*, 1478–1481. [CrossRef]
11. Deng, J.Z.; Paone, D.V.; Ginnetti, A.T.; Kurihara, H.; Dreher, S.D.; Weissman, S.A.; Stauffer, S.R.; Burgey, C.S. Copper-Facilitated Suzuki Reactions: Application to 2-Heterocyclic Boronates. *Org. Lett.* **2009**, *11*, 345–347. [CrossRef] [PubMed]
12. Gøgsig, T.M.; Lindhardt, A.T.; Skrydstrup, T. Heteroaromatic Sulfonates and Phosphates as Electrophiles in Iron-Catalyzed Cross-Couplings. *Org. Lett.* **2009**, *11*, 4886–4888. [CrossRef] [PubMed]
13. Wasa, M.; Worrell, B.T.; Yu, J.-Q. Pd^0/PR_3-Catalyzed Arylation of Nicotinic and Isonicotinic Acid Derivatives. *Angew. Chem. Int. Ed.* **2010**, *49*, 1275–1277. [CrossRef] [PubMed]
14. Seiple, I.B.; Su, S.; Rodriguez, R.A.; Gianatassio, R.; Fujiwara, Y.; Sobel, A.L.; Baran, P.S. Direct C-H Arylation of Electron-Deficient Heterocycles with Arylboronic Acids. *J. Am. Chem. Soc.* **2010**, *132*, 13194–13196. [CrossRef]
15. Berman, A.M.; Bergman, R.G.; Ellman, J.A. Rh(I)-Catalyzed Direct Arylation of Azines. *J. Org. Chem.* **2010**, *75*, 7863–7868. [CrossRef]
16. Luzung, M.R.; Patel, J.S.; Yin, J. A Mild Negishi Cross-Coupling of 2-Heterocyclic Organozinc Reagents and Aryl Chlorides. *J. Org. Chem.* **2010**, *75*, 8330–8332. [CrossRef]
17. Fujiwara, Y.; Dixon, J.A.; O'Hara, F.; Funder, E.D.; Dixon, D.D.; Rodriguez, R.A.; Baxter, R.D.; Herlé, B.; Sach, N.; Collins, M.R.; et al. Practical and innate carbon–hydrogen functionalization of heterocycles. *Nature* **2012**, *492*, 95–100. [CrossRef]
18. Dai, F.; Gui, Q.; Liu, J.; Yang, Z.; Chen, X.; Guo, R.; Tan, Z. Pd-catalyzed C3-selective arylation of pyridines with phenyl tosylates. *Chem. Commun.* **2013**, *49*, 4634–4636. [CrossRef]
19. Liu, B.; Huang, Y.; Lan, J.; Song, F.; You, J. Pd-catalyzed oxidative C–H/C–H cross-coupling of pyridines with heteroarenes. *Chem. Sci.* **2013**, *4*, 2163–2167. [CrossRef]
20. Sakashita, S.; Takizawa, M.; Sugai, J.; Ito, H.; Yamamoto, Y. Tetrabutylammonium 2-Pyridyltriolborate Salts for Suzuki–Miyaura Cross-Coupling Reactions with Aryl Chlorides. *Org. Lett.* **2013**, *15*, 4308–4311. [CrossRef]
21. Colombe, J.R.; Bernhardt, S.; Stathakis, C.; Buchwald, S.L.; Knochel, P. Synthesis of Solid 2-Pyridylzinc Reagents and Their Application in Negishi Reactions. *Org. Lett.* **2013**, *15*, 5754–5757. [CrossRef]
22. Larionov, O.V.; Stephens, D.; Mfuh, A.; Chavez, G. Direct, Catalytic, and Regioselective Synthesis of 2-Alkyl-, Aryl-, and Alkenyl-Substituted N-Heterocycles from N-Oxides. *Org. Lett.* **2014**, *16*, 864–867. [CrossRef]
23. Gao, G.-L.; Xia, W.; Jain, P.; Yu, J.-Q. Pd(II)-Catalyzed C3-Selective Arylation of Pyridine with (Hetero)arenes. *Org. Lett.* **2016**, *18*, 744–747. [CrossRef]
24. Zeng, Y.; Zhang, C.; Yin, C.; Sun, M.; Fu, H.; Zheng, X.; Yuan, M.; Li, R.; Chen, H. Direct C–H Functionalization of Pyridine via a Transient Activator Strategy: Synthesis of 2,6-Diarylpyridines. *Org. Lett.* **2017**, *19*, 1970–1973. [CrossRef]

25. Bull, J.A.; Mousseau, J.J.; Pelletier, G.; Charette, A.B. Synthesis of Pyridine and Dihydropyridine Derivatives by Regio- and Stereoselective Addition to N-Activated Pyridines. *Chem. Rev.* **2012**, *112*, 2642–2713. [CrossRef]
26. Pomaranski, P.; Czarnocki, Z. Arylpyridines: A Review from Selective Synthesis to Atropisomerism. *Synthesis* **2019**, *51*, 587–611.
27. Lu, J.-Y.; Keith, J.A.; Shen, W.-Z.; Schürmann, M.; Preut, H.; Jacob, T.; Arndt, H.-D. Regioselective De Novo Synthesis of Cyanohydroxypyridines with a Concerted Cycloaddition Mechanism. *J. Am. Chem. Soc.* **2008**, *130*, 13219–13221. [CrossRef]
28. Sabot, C.; Oueis, E.; Brune, X.; Renard, P.-Y. Synthesis of polysubstituted 3-hydroxypyridines via the revisited hetero-Diels–Alder reaction of 5-alkoxyoxazoles with dienophiles. *Chem. Commun.* **2012**, *48*, 768–770. [CrossRef]
29. Ishida, N.; Yuhki, T.; Murakami, M. Synthesis of Enantiopure Dehydropiperidinones from α-Amino Acids and Alkynes via Azetidin-3-ones. *Org. Lett.* **2012**, *14*, 3898–3901. [CrossRef]
30. Barday, M.; Ho, K.Y.T.; Halsall, C.T.; Aïssa, C. Regioselective Synthesis of 3-Hydroxy-4,5-alkyl-Substituted Pyridines Using 1,3-Enynes as Alkynes Surrogates. *Org. Lett.* **2017**, *19*, 178–181.
31. Erhardt, H.; Kunz, K.A.; Kirsch, S.F. Thermolysis of Geminal Diazides: Reagent-Free Synthesis of 3-Hydroxypyridines. *Org. Lett.* **2017**, *19*, 178–181. [CrossRef] [PubMed]
32. Donohoe, T.J.; Basutto, J.A.; Bower, J.F.; Rathi, A. Heteroaromatic Synthesis via Olefin Cross-Metathesis: Entry to Polysubstituted Pyridines. *Org. Lett.* **2011**, *13*, 1036–1039. [CrossRef]
33. Donohoe, T.J.; Bower, J.F.; Baker, D.B.; Basutto, J.A.; Chan, L.M.K.; Gallagher, P. Synthesis of 2,4,6-trisubstituted pyridines via an olefin cross-metathesis/Heck–cyclisation–elimination sequence. *Chem. Commun.* **2011**, *47*, 10611–10613. [CrossRef]
34. Chen, M.Z.; Micalizio, G.C. Three-Component Coupling Sequence for the Regiospecific Synthesis of Substituted Pyridines. *J. Am. Chem. Soc.* **2012**, *134*, 1352–1356. [CrossRef] [PubMed]
35. Henry, G.D. De novo synthesis of substituted pyridines. *Tetrahedron* **2004**, *60*, 6043–6061. [CrossRef]
36. Heller, B.; Hapke, M. The fascinating construction of pyridine ring systems by transition metal-catalysed [2 + 2 + 2] cycloaddition reactions. *Chem. Soc. Rev.* **2007**, *36*, 1085–1094. [CrossRef] [PubMed]
37. Groenendaal, B.; Ruijter, E.; Orru, R.V.A. 1-Azadienes in cycloaddition and multicomponent reactions towards N-heterocycles. *Chem. Commun.* **2008**, 5474–5489. [CrossRef]
38. Hill, M.D. Recent Strategies for the Synthesis of Pyridine Derivatives. *Chem. Eur. J.* **2010**, *16*, 12052–12062. [CrossRef] [PubMed]
39. Donohoe, T.J.; Bower, J.F.; Chan, L.K.M. Olefin cross-metathesis for the synthesis of heteroaromatic compounds. *Org. Biomol. Chem.* **2012**, *10*, 1322–1328. [CrossRef]
40. Allais, C.; Grassot, J.-M.; Rodriguez, J.; Constantieux, T. Metal-Free Multicomponent Syntheses of Pyridines. *Chem. Rev.* **2014**, *114*, 10829–10868. [CrossRef]
41. Wang, Q.; Wan, C.; Gu, Y.; Zhang, J.; Gao, L.; Wang, Z. A metal-free decarboxylative cyclization from natural a-amino acids to construct pyridine derivatives. *Green Chem.* **2011**, *13*, 578–581. [CrossRef]
42. Xiang, J.-C.; Wang, M.; Cheng, Y.; Wu, A.-X. Molecular Iodine-Mediated Chemoselective Synthesis of Multisubstituted Pyridines through Catabolism and Reconstruction Behavior of Natural Amino Acids. *Org. Lett.* **2016**, *18*, 24–27. [CrossRef] [PubMed]
43. Xiang, J.-C.; Cheng, Y.; Wang, Z.-X.; Ma, J.-T.; Wang, M.; Tang, B.-C.; Wu, Y.-D.; Wu, A.-X. Oxidative Trimerization of Amino Acids: Selective Synthesis of 2,3,5-Trisubstituted Pyridines. *Org. Lett.* **2017**, *19*, 2997–3000. [CrossRef]
44. Tilley, J.W.; Zawoiski, S. A Convenient Palladium-Catalyzed Coupling Approach to 2,5-Disubstituted Pyridines. *J. Org. Chem.* **1988**, *53*, 386–390. [CrossRef]
45. Vyvyan, J.R.; Dell, J.A.; Ligon, T.J.; Motanic, K.K.; Wall, H.S. Suzuki–Miyaura Cross-Coupling of 3-Pyridyl Triflates with Alk-1-enyl-2-pinacol Boronates. *Synthesis* **2010**, 3637–3644. [CrossRef]
46. Bera, M.K.; Hommes, P.; Reissig, H.-U. In Search of Oligo(2-thienyl)-Substituted Pyridine Derivatives: A Modular Approach to Di-, Tri- and Tetra(2-thienyl)pyridines. *Chem. Eur. J.* **2011**, *17*, 11383–11843. [CrossRef]
47. Doebelin, C.; Wagner, P.; Bihel, F.; Humbert, N.; Kenfack, C.A.; Mely, Y.; Bourguignon, J.-J.; Schmitt, M. Fully Regiocontrolled Polyarylation of Pyridine. *J. Org. Chem.* **2014**, *79*, 908–918. [CrossRef]
48. Zhang, E.; Tang, J.; Li, S.; Wu, P.; Moses, J.E.; Sharpless, K.B. Chemoselective Synthesis of Polysubstituted Pyridines from Heteroaryl Fluorosulfates. *Chem. Eur. J.* **2016**, *22*, 5692–5697. [CrossRef]
49. Asako, T.; Hayashi, W.; Amaike, K.; Suzuki, S.; Itami, K.; Muto, K.; Yamaguchi, J. Synthesis of multiply arylated pyridines. *Tetrahedron* **2017**, *73*, 3669–3676.
50. Donohoe, T.J.; Fishlock, L.P.; Basutto, J.A.; Bower, J.F.; Procopiou, P.A.; Thompson, A.L. Synthesis of substituted pyridines and pyridazines via ring closing metathesis. *Chem. Commun.* **2009**, 3008–3010. [CrossRef]
51. Yoshida, K.; Kawagoe, F.; Hayashi, K.; Horiuchi, S.; Imamoto, T.; Yanagisawa, A. Synthesis of 3-Hydroxypyridines Using Ruthenium-Catalyzed Ring-Closing Olefin Metathesis. *Org. Lett.* **2009**, *11*, 515–518. [CrossRef]
52. Tsukamoto, H.; Ueno, T.; Kondo, Y. Palladium(0)-Catalyzed Alkylative Cyclization of Alkynals and Alkynones: Remarkable *trans*-Addition of Organoboronic Reagents. *J. Am. Chem. Soc.* **2006**, *128*, 1406–1407. [CrossRef]
53. Tsukamoto, H.; Ito, K.; Ueno, T.; Shiraishi, M.; Kondo, Y.; Doi, T. Palladium(0)-Catalyzed *Anti*-Selective Addition-Cyclizations of Alkynyl Electrophiles. *Chem. Eur. J.* **2022**, e202203068. [CrossRef]
54. Tsukamoto, H.; Nakamura, S.; Tomida, A.; Doi, T. Scalable Total Syntheses and Structure–Activity Relationships of Haouamines A, B, and Their Derivatives as Stable Formate Salts. *Chem. Eur. J.* **2020**, *26*, 12528–12532. [CrossRef]

55. Boger, D.L.; Brotherton, C.E.; Panek, J.S.; Yohannes, D. Direct Introduction of Nitriles via Use of Unstable Reissert Intermediates: Convenient Procedures for the Preparation of 2-Cyanoquinolines and 1-Cyanoisoquinolines. *J. Org. Chem.* **1984**, *49*, 4056–4058. [CrossRef]
56. Chinchilla, R.; Nájera, C. The Sonogashira Reaction: A Booming Methodology in Synthetic Organic Chemistry. *Chem. Rev.* **2007**, *107*, 874–922. [CrossRef]
57. Radhika, S.; Harry, N.A.; Neetha, M.; Anikumar, G. Recent trends and applications of the Cadiot–Chodkiewicz reaction. *Org. Biomol. Chem.* **2019**, *17*, 9081–9094. [CrossRef] [PubMed]
58. Bürgi, H.B.; Duntz, J.D.; Lehn, J.M.; Wipff, G. Stereochemistry of Reaction Paths at Carbonyl Centres. *Tetrahedron* **1974**, *30*, 1563–1572. [CrossRef]
59. Gilmore, K.; Alabugin, I.V. Cyclizations of Alkynes: Revisiting Baldwin's Rules for Ring Closure. *Chem. Rev.* **2011**, *111*, 6513–6556. [CrossRef]
60. Salimbeni, A.; Canevotti, R.; Paleari, F.; Bonaccorsi, F.; Renzetti, A.R.; Belvisi, L.; Bravi, G.; Scolastico, C. Nonpeptide Angiotensin II Receptor Antagonists. Synthesis, in Vitro Activity, and Molecular Modeling Studies of *N*-[(Heterobiaryl)methylimidazole. *J. Med. Chem.* **1994**, *37*, 3928–3938. [CrossRef]
61. Shibata, N.; Tsuchiya, T.; Hashimoto, Y.; Morita, N.; Ban, S.; Tamura, O. Thiyl radical-mediated cyclization of ω-alkynyl O-tert-butyldiphenylsilyloximes. *Org. Biomol. Chem.* **2017**, *15*, 3025–3034. [CrossRef]
62. Takahashi, K.; Honda, T. Diastereoselective Syntheses of Functionalized Five-Membered Carbocycles and Heterocycles by a SmI2-Promoted Intramolecular Coupling of Bromoalkynes and α,β-Unsaturated Esters. *Org. Lett.* **2010**, *12*, 3026–3029. [CrossRef]
63. Padín, D.; Cambeiro, F.; Fañanás-Mastral, M.; Varela, J.; Saá, A.C. [2 + 1] Cycloaddition of Catalytic Ruthenium Vinyl Carbenes: A Stereoselective Controlled Access to (Z)- and (E)-Vinyl Epoxypyrrolidines. *ACS Catal.* **2017**, *7*, 992–996. [CrossRef]
64. Ordóñez, M.; De la Cruz-Cordero, R.; Fernández-Zertuche, M.; Muñoz-Hernández, M.A.; García-Barradas, O. Diastereoselective reduction of dimethyl γ-[(N-p-toluenesulfonyl)amino]-β-ketophosphonates derived from amino acids. *Tetrahedron Asymmetry* **2004**, *15*, 3035–3043. [CrossRef]
65. Morales, S.; Guijarro, F.G.; Ruano, J.L.G.; Cid, M.B. A General Aminocatalytic Method for the Synthesis of Aldimines. *J. Am. Chem. Soc.* **2014**, *136*, 1082–1089. [CrossRef]

Disclaimer/Publisher's Note: The statements, opinions and data contained in all publications are solely those of the individual author(s) and contributor(s) and not of MDPI and/or the editor(s). MDPI and/or the editor(s) disclaim responsibility for any injury to people or property resulting from any ideas, methods, instructions or products referred to in the content.

Palladium-Catalyzed Three-Component Coupling of Benzynes, Benzylic/Allylic Bromides and 1,1-Bis[(pinacolato)boryl]methane

Zhicheng Bao [1], Chaoqiang Wu [1] and Jianbo Wang [1,2,*]

1. Beijing National Laboratory of Molecular Sciences (BNLMS), Key Laboratory of Bioorganic Chemistry and Molecular Engineering of Ministry of Education, College of Chemistry, Peking University, Beijing 100871, China
2. State Key Laboratory of Organometallic Chemistry, Chinese Academy of Sciences, Shanghai 200032, China
* Correspondence: wangjb@pku.edu.cn; Tel.: +8610-6275-7248

Abstract: We report herein a palladium-catalyzed three-component cross-coupling reaction of 2-(trimethylsilyl)phenyl trifluoromethanesulfonate, benzylic/allylic bromides and 1,1-bis[(pinacolato)boryl]methane. The reaction, which affords benzyl boronates as the products, represents the first example of using 1,1-bis[(pinacolato)boryl]methane in a cross-coupling reaction involving benzyne species.

Keywords: palladium; catalysis; benzyne; 1,1-bis[(pinacolato)boryl]methane; three-component coupling

1. Introduction

Multicomponent reactions (MCRs) have been established as an efficient strategy to rapidly build up molecular complexities [1]. MCRs have found wide applications in many areas, including organic synthesis [2–9], chemical biology [10,11], drug developments [12,13] and polymer synthesis [14–16]. Thus, it is highly desirable to develop novel MCRs for further expanding the scope of this type of reaction. In this regard, one of the major challenging issues lies in the arrangement of each component in a proper order to react one by one, especially when these components may react with each other. Arynes, as highly reactive components typically generated in situ, have been utilized as one of the reaction partners and inserted into ordinary reactions [17–19]. For example, to expand the two-component reaction of Suzuki–Miyaura coupling, Cheng and coworkers developed an π-allylpalladium-involved three-component coupling reaction using arylboronic acids as the terminating reagents (Scheme 1a) [20]. In this transformation, the π-allylpalladium species first react with highly reactive aryne intermediate. Upon carbopalladation of the aryne, the newly formed aryl palladium intermediate reacts with arylboronic acid to afford the three-component product, *o*-allylbiaryls. In another report, Cheng and co-workers developed a Ni(0)-catalyzed coupling of arynes, alkenes and boronic acids, in which a nickelacycle intermediate is formed through the reaction of Ni(0) with enone and aryne (Scheme 1b) [21]. The same group also developed a series of other three-component reactions based on palladium-catalyzed reactions involving aryne species [22–25]. Furthermore, Dong and coworkers reported a similar coupling process using a Pd(II)−Pb(II) bimetallic metal−organic framework (MOF) as an active heterogeneous catalyst [26].

On the other hand, 1,1-bis[(pinacolato)boryl]methane, as a readily available *gem*-diboronate reagent, has attracted considerable attention in recent years [27–33]. The *gem*-diboronates can be successfully employed in transition-metal-catalyzed cross-coupling reactions. In particular, Endo, Shibata and coworkers developed a palladium-catalyzed cross-coupling with 1,1-diborylalkanes with organohalides to afford alkyl boronates [34–37]. In connection to our interest in the chemistry of 1,1-diborylalkanes [38–40], we conceived to apply bis(boryl) methane as one of the substrates in the palladium-catalyzed three-component coupling reaction of arynes and halides (Scheme 1c). To the best of our knowledge, *gem*-diboronates have not been utilized as substrates in transition-metal-catalyzed reactions involving arynes. The reaction would generate substituted benzyl boronates, which are highly useful, but their preparation is not trivial [41].

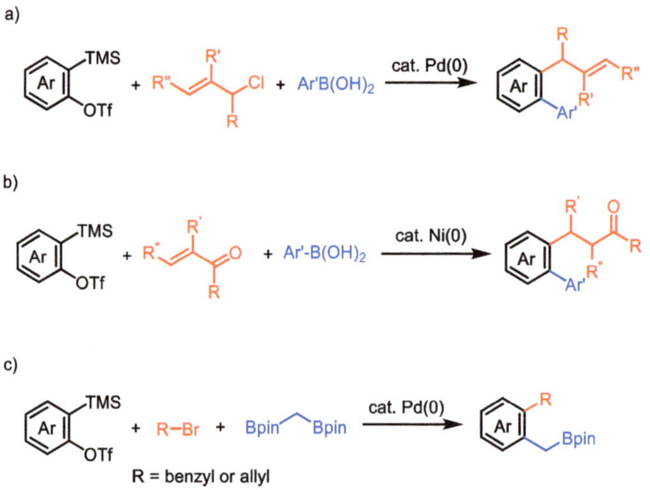

Scheme 1. Arynes as the coupling partners in transition-metal-catalyzed three-component reactions. (**a**) Pd(0)-catalyzed coupling with allyl halides and arylboronic acids [7]; (**b**) Ni(0)-catalyzed coupling with vinyl ketones and arylboronic acids [8]; (**c**) Pd(0)-catalyzed reaction with 1,1-bis[(pinacolato)boryl]methane (this work).

2. Results

The preliminary study commenced with 2-(trimethylsilyl)phenyl trifluoromethanesulfonate **1a**, (bromomethyl)benzene **2a** and 1,1-bis[(pinacolato)boryl]methane **3a** as the model substrates. Carefully screening the reaction conditions revealed that DCE was the most suitable solvent (Tables S1 and S2). However, further reaction condition optimization showed no obvious improvements, which was attributed to the low solubility of the reaction substrates in DCE. To circumvent the solubility problem, we then focused on a mixed solvent. While mixing DCE with various solvents failed to improve the reaction, a 1:1 mixture of toluene and acetonitrile afforded a better yield (Table 1, entry 1). With this mixed solvent, we then inspected the influence of catalysts and ligands (Table 1, entries 2–6). The results indicated that the combination of Pd(OAc)$_2$/PPh$_3$ gave the highest yields (Table 1, entry 6). The triarylphosphine ligand was further tuned by introducing substituents onto the para position of the aryl ring (Table 1, entries 7–9). With tris(*p*-fluorophenyl) phosphine as the ligand, the reaction was further improved. Furthermore, we found that increasing the loading of **2a** from 1 equiv to 1.4 equiv led to the optimal yield of 77% after stirring the reaction mixture for 10 h (Table 1, entry 10). Finally, it was observed that the base had a significant effect on the reaction. When the loading of CsF was reduced from 4 equiv to 3 equiv, the yield was diminished (Table 1, entry 11). No desired product could be detected when CsF was replaced by KF (Table 1, entry 12). A combination of KF and 18-crown-6 gave a trace amount of the product (Table 1, entry 13). These results suggested that the counterncations had a significant effect on the reaction.

Table 1. Reaction condition optimization [a].

Entry	[Pd]	[P]	Base	Yield (%) [b]
1	PdCl$_2$	dppe	CsF	47
2	Pd(MeCN)$_2$Cl$_2$	dppe	CsF	56
3	Pd$_2$(dba)$_3$	dppe	CsF	39
4	Pd(OAc)$_2$	dppe	CsF	61
5	Pd(OAc)$_2$	dppe	CsF	53
6	Pd(OAc)$_2$	PPh$_3$	CsF	65
7	Pd(OAc)$_2$	(p-MeOC$_6$H$_4$)$_3$P	CsF	61
8	Pd(OAc)$_2$	(p-CF$_3$C$_6$H$_4$)$_3$P	CsF	64
9	Pd(OAc)$_2$	(p-FC$_6$H$_4$)$_3$P	CsF	69
10 [c]	Pd(OAc)$_2$	(p-FC$_6$H$_4$)$_3$P	CsF	77 [d]
11 [e]	Pd(OAc)$_2$	(p-FC$_6$H$_4$)$_3$P	CsF	43
12	Pd(OAc)$_2$	(p-FC$_6$H$_4$)$_3$P	KF	0
13	Pd(OAc)$_2$	(p-FC$_6$H$_4$)$_3$P	KF/18-Crown-6	6

[a] If not otherwise noted, the reaction conditions are as follows: a solution of **3a** (0.1 mmol), **2a** (1.0 equiv), **1a** (2.0 equiv), [Pd] (5 mol), [P] (10 mol%) and base (4 equiv) in PhMe/MeCN (1:1) was stirred at 85 °C for 12 h. [b] Yields were determined by ^{1}H NMR spectroscopy analysis of the crude reaction mixture using CH$_2$Br$_2$ as the internal standard. [c] A total of 1.4 equiv benzyl bromide was used. [d] The reaction time was 10 h. [e] A total of 3 equiv CsF was used.

With optimized reaction conditions (Table 1, entry 10) in hand, we proceeded to study the substrate scope (Scheme 2). First, we investigated the substrate scope with regard to the substituents of benzyl bromides. For the model reaction with **2a**, the corresponding product **4a** could be isolated in a 75% yield. Notably, the side product due to direct coupling between **2a** and **3a** was not detected. Various substituents in the para position of benzyl bromides could be tolerant, including electron-withdrawing substituents (CN, F, CO$_2$Me) and electron-donating groups (Me, ^{t}Bu), affording the corresponding products **4a–h** in 58–75% yields. Similarly, the ortho- and meta-substituted benzyl bromides could also be utilized in the reaction, providing the corresponding products **4i–m** in moderate yields. Both 1-(bromomethyl)naphthalene and 1-(bromomethyl)naphthalene could participate in this coupling reaction. However, the yields of the products **4n** and **4o** were low to moderate, which is presumably attributed to the steric effect. 3-Bromomethylthiophene was tolerated well to afford a moderate yield (**4p**). When (1-bromoethyl)benzene was used as the substrate, none of products were produced, which might be attributed to the steric effects and the possible β-H elimination.

Subsequently, we turned our attention to allylic bromides. In the cases when the structure of π-allylpalladium was symmetrical, a single product was obtained in each case (**4q–s**). However, if an unsymmetrical π-allylpalladium was generated, a pair of isomers were obtained with essentially no selectivity (**4t** and **4t'**). Moreover, we also investigated the reaction with iodobenzene as the electrophilic reagent. The reaction worked, but only giving the product **4u** in a low yield. Other electrophiles, including ethyl bromoacetate and alkyl iodide, were found unsuitable for this coupling reaction.

For the scope of the aryne precursor, a MeO-substituted substrate was examined. The reaction gave a mixture of isomeric products **4v** and **4v'** in low yields, and the reaction was essentially nonselective. Finally, a series of substituted *gem*-diboronates were examined. However, the substituted diboronates did not participate in the coupling reaction, which was consistent with the above-mentioned observations, namely that the reaction was quite sensitive to steric hindrance.

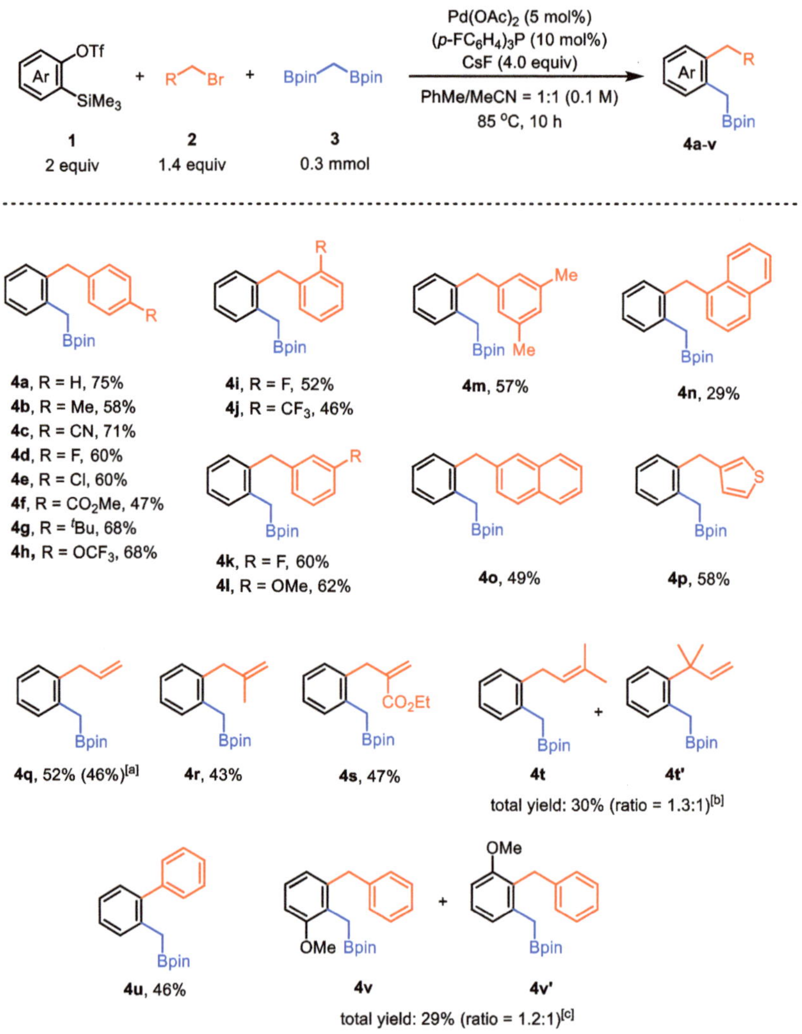

Scheme 2. Substrate scope of the three-component coupling. [a] The yield in the bracket refers to the reaction in which allyl iodide was used instead of allyl bromide. [b] 1-Bromo-3-methylbut-2-ene was used as substrate **2**. [c] 2-Methoxy-6-(trimethylsilyl)phenyl trifluoromethanesulfinate was used as substrate **1**.

The proposed mechanism is shown in Scheme 3. The reaction is initiated by the oxidative addition of benzylic bromides to Pd(0), affording benzylic-Pd(II) complex **A**. Subsequently, insertion of **A** to the in situ-generated benzyne occurs, to afford aryl-Pd(II) complex **B**. Then, transmetalation with diborylmethane generates intermediate **C**, from which reductive elimination is followed to provide the final product and regenerate the Pd(0) catalyst to restart a new catalytic cycle. In this reaction, cesium fluoride played two roles: (1) for the in situ generation of aryne; (2) for the transmetalation of diborylmethane 3. It is worth mentioning that this mechanistic proposal is tentative. Other possible pathways—for example, the formation of a benzyne-Pd(0) first and then followed by oxidation addition to generate intermediate **B**—cannot be ruled out. Further solid studies are needed to firmly establish the reaction mechanism for this three-component reaction.

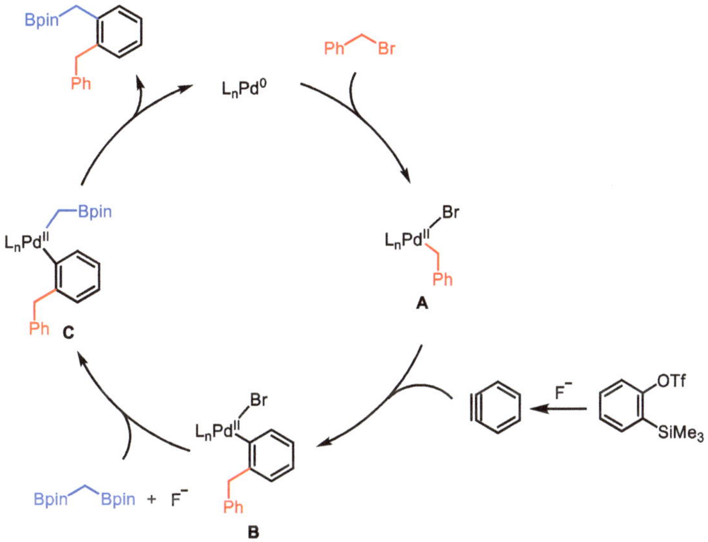

Scheme 3. Proposed mechanism of the three-component coupling reaction.

The three-component coupling reaction could provide a comparable yield of **4a** when the reaction was carried out in 4 mmol scale (Scheme 4). Given the versatility of benzyl boronates in synthetic chemistry, we further proceeded to explore some transformations with **4a**. Thus, as shown in Scheme 4, oxidation of **4a** afforded benzyl alcohol **5** [42], and fluorination gave benzyl fluoride **6** [43]. Palladium-catalyzed cross-coupling of **4a** with phenyl bromide generated 1,2-dibenzyl benzene **7** [44].

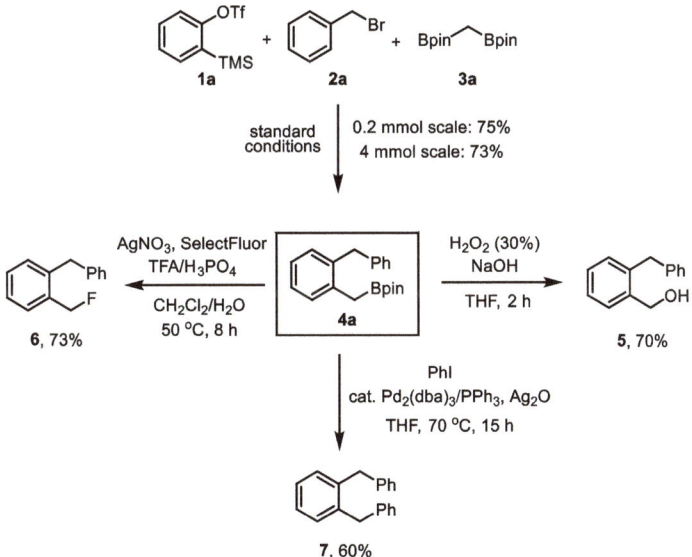

Scheme 4. Scale-up reaction and derivatization.

3. Materials and Methods

3.1. Materials

All the reactions were performed under nitrogen atmosphere in an oven-dried reaction tube. The solvents were distilled under nitrogen atmosphere prior to use. Toluene, dioxane and THF were dried over Na with benzophenone ketyl intermediate as the indicator. MeCN was dried over CaH_2. The boiling point of petroleum ether was between 60 and 70 °C. Unless otherwise noted, commercially available reagents were used as received. For chromatography, 200–300 mesh silica gel (Qingdao, China) was used. Bis(4,4,5,5-tetramethyl-1,3,2-dioxaborolan-2-yl) methane 3 was prepared according to reported procedure [45].

3.2. Methods

An oven-dried 10 mL Schlenk flask with magnetic stir bar was charged with *gem*-diboronates 3 (0.3 mmol), Pd(OAc)$_2$ (5 mol%) and tris (*p*-fluorophenyl) phosphine (10 mol%). The flask was sealed with a rubber stopper, evacuated and filled with nitrogen three times, followed by the addition of toluene (1.5 mL), MeCN (1.5 mL), 2-(trimethylsilyl)phenyl trifluoromethanesulfonate (2 equiv) and benzyl bromide (1.4 equiv). The reaction mixture was stirred at 85 °C for 10 h. Upon completion, the mixture was cooled to room temperature and filtered through a short plug of silica gel, rinsed with ethyl acetate. The filtrate was evaporated by rotary evaporation and the crude product was purified by silica gel column chromatography to afford the pure product (petroleum ether: EtOAc = 10:1).

4. Conclusions

In summary, we developed a three-component coupling of benzyne, benzylic/allylic bromide and 1,1-bis[(pinacolato)boryl]methane to afford benzyl boronates. The reaction represents the first example of using 1,1-bis[(pinacolato)boryl]methane in the palladium-catalyzed cross-coupling involving benzyne intermediate.

Supplementary Materials: The following supporting information can be downloaded at: https://www.mdpi.com/article/10.3390/catal13010126/s1, screening tables, preparation of substrates, analytical data, ^{1}H and ^{13}C NMR spectra, more detailed materials and methods. Table S1: Reaction optimization for the coupling reaction using DCE as the solvent. Table S2: Reaction optimization for the coupling reaction using PhMe/MeCN as the solvents.

Author Contributions: Conceptualization, J.W.; methodology, C.W. and J.W.; investigation, Z.B. and C.W.; writing—original draft preparation, Z.B.; writing—review and editing, Z.B. and J.W.; supervision, J.W.; funding acquisition, J.W. All authors have read and agreed to the published version of the manuscript.

Funding: We acknowledge the funding support from NSFC (Grant Nos. 21871010 and "Laboratory for Synthetic Chemistry and Chemical Biology" under the Health@InnoHK Program launched by Innovation and Technology Commission, The Government of Hong Kong Special Administrative Region of the People's Republic of China.

Data Availability Statement: All experimental data are contained in the article and Supplementary Material.

Acknowledgments: We thank Hongpei Chan for her assistance during the course of this study.

Conflicts of Interest: The authors declare no conflict of interest.

References

1. Zhu, J.; Wang, Q.; Wang, M. *Multicomponent Reactions in Organic Synthesis*; Wiley VCH: Weinheim, Germany, 2015.
2. Lesma, G.; Cecchi, R.; Crippa, S.; Giovanelli, P.; Meneghetti, F.; Musolino, M.; Sacchetti, A.; Silvani, A. Ugi 4-CR/Pictet–Spengler reaction as a short route to tryptophan-derived peptidomimetics. *Org. Biomol. Chem.* **2012**, *10*, 9004–9012. [CrossRef] [PubMed]
3. Van der Heijden, G.; Ruijter, E.; Orru, R.V.A. Efficiency, diversity, and complexity with multicomponent reactions. *Synlett* **2013**, *24*, 666–685.
4. Guo, X.; Hu, W. Novel multicomponent reactions via trapping of protic onium ylides with electrophiles. *Acc. Chem. Res.* **2013**, *46*, 2427–2440. [CrossRef] [PubMed]

5. Eberlin, L.; Tripoteau, F.; Carreaux, F.; Whiting, A.; Carboni, B. Boron-substituted 1,3-dienes and heterodienes as key elements in multicomponent processes. *Beilstein J. Org. Chem.* **2014**, *10*, 237–250. [CrossRef]
6. Koopmanschap, G.; Ruijter, E.; Orru, R.V.A. Isocyanide-based multicomponent reactions towards cyclic constrained peptidomimetics. *Beilstein J. Org. Chem.* **2014**, *10*, 544–598. [CrossRef]
7. Rotstein, B.H.; Zaretsky, S.; Rai, V.; Yudin, A.K. Small heterocycles in multicomponent reactions. *Chem. Rev.* **2014**, *114*, 8323–8359. [CrossRef]
8. Brauch, S.; van Berkel, S.S.; Westermann, B. Higher-order multicomponent reactions: Beyond four reactants. *Chem. Soc. Rev.* **2013**, *42*, 4948–4962. [CrossRef]
9. Neochoritis, C.G.; Zhao, T.; Domling, A. Tetrazoles via multicomponent reactions. *Chem. Rev.* **2019**, *119*, 1970–2042. [CrossRef]
10. Dömling, A. Recent advances in isocyanide-based multicomponent chemistry. *Curr. Opin. Chem. Biol.* **2002**, *6*, 306–313. [CrossRef]
11. Domling, A.; Wang, W.; Wang, K. Chemistry and biology of multicomponent reactions. *Chem. Rev.* **2012**, *112*, 3083–3135. [CrossRef]
12. Estevez, V.; Villacampa, M.; Menéndez, J.C. Multicomponent reactions for the synthesis of pyrroles. *Chem. Soc. Rev.* **2010**, *39*, 4402–4421. [CrossRef]
13. Estevez, V.; Villacampa, M.; Menéndez, J.C. Recent advances in the synthesis of pyrroles by multicomponent reactions. *Chem. Soc. Rev.* **2014**, *43*, 4633–4657. [CrossRef] [PubMed]
14. Blasco, E.; Sims, M.B.; Goldmann, A.S.; Sumerlin, B.S.; Barner-Kowollik, C. 50th Anniversary Perspective: Polymer Functionalization. *Macromolecules* **2017**, *50*, 5215–5252. [CrossRef]
15. Kakuchi, R. Multicomponent reactions in polymer synthesis. *Angew. Chem. Int. Ed.* **2014**, *53*, 46–48. [CrossRef] [PubMed]
16. Theato, P. *Multi-Component and Sequential Reactions in Polymer Synthesis*; Springer International Publishing: Berlin/Heidelberg, Germany, 2015; Volume 269.
17. Shi, J.; Li, L.; Li, Y. o-Silylaryl triflates: A journey of Kobayashi aryne precursors. *Chem. Rev.* **2021**, *121*, 3892–4044. [CrossRef] [PubMed]
18. Sarmah, M.; Hazarika, H.; Gogoi, P. Aryne annulations for the synthesis of carbocycles and hetero-cycles: An updated review. *Synthesis* **2022**, *54*, 4932–4962.
19. Dubrovskiy, A.V.; Markina, N.A.; Larock, R.C. Use of benzynes for the synthesis of heterocycles. *Org. Biomol. Chem.* **2013**, *11*, 191–218. [CrossRef]
20. Jayanth, T.T.; Jeganmohan, M.; Cheng, C.-H. Highly efficient route to o-allylbiaryls via palladium-catalyzed three-component coupling of benzynes, allylic halides, and aryl organometallic reagents. *Org. Lett.* **2005**, *7*, 2921–2924. [CrossRef]
21. Jayanth, T.T.; Cheng, C.-H. Nickel-catalyzed coupling of arynes, alkenes, and boronic acids: Dual role of the boronic acid. *Angew. Chem. Int. Ed.* **2007**, *46*, 5921–5924. [CrossRef]
22. Jeganmohan, M.; Cheng, C.-H. Palladium-catalyzed allylalkynylation of benzynes: A highly efficient route to substituted 1-allyl-2-alkynylbenzenes. *Org. Lett.* **2004**, *6*, 2821–2824. [CrossRef]
23. Bhuvaneswari, S.; Jeganmohan, M.; Cheng, C.-H. Carbocyclization of aromatic iodides, bicyclic alkenes, and benzynes involving a palladium-catalyzed C−H bond activation as a key step. *Org. Lett.* **2006**, *8*, 5581–5584. [CrossRef]
24. Bhuvaneswari, S.; Jeganmohan, M.; Yang, M.-C.; Cheng, C.-H. Palladium-catalyzed three-component coupling of arynes with allylic acetates or halides and terminal alkynes promoted by cuprous iodide. *Chem. Commun.* **2008**, 2158–2160. [CrossRef]
25. Jeganmohan, M.; Bhuvaneswari, S.; Cheng, C.-H. A cooperative copper- and palladium-catalyzed three-component coupling of benzynes, allylic epoxides, and terminal alkynes. *Angew. Chem. Int. Ed.* **2009**, *48*, 391–394. [CrossRef] [PubMed]
26. Dong, Y.; Li, Y.; Wei, Y.-L.; Wang, J.-C.; Ma, J.-P.; Ji, J.; Yao, B.-J.; Dong, Y.-B. A N-heterocyclic tetracarbene Pd(II) moiety containing a Pd(II)–Pb(II) bimetallic MOF for three-component cyclotrimerization via benzyne. *Chem. Commun.* **2016**, *52*, 10505–10508. [CrossRef] [PubMed]
27. Paul, S.; Das, K.K.; Aich, D.; Manna, S.; Panda, S. Recent developments in the asymmetric synthesis and functionalization of symmetrical and unsymmetrical *gem*-diborylalkanes. *Org. Chem. Front.* **2022**, *9*, 838–852. [CrossRef]
28. Lee, Y.; Han, S.; Cho, S.H. Catalytic chemo- and enantioselective transformations of *gem*-diborylalkanes and (diborylmethyl)metallic species. *Acc. Chem. Res.* **2021**, *54*, 3917–3929. [CrossRef]
29. Nallagonda, R.; Padala, K.; Masarwa, A. *gem*-Diborylalkanes: Recent advances in their preparation, transformation and application. *Org. Biomol. Chem.* **2018**, *16*, 1050–1064. [CrossRef]
30. Miralles, N.; Maza, R.J.; Fernández, E. Synthesis and reactivity of 1,1-diborylalkanes towards C–C bond formation and related mechanisms. *Adv. Synth. Catal.* **2018**, *360*, 1306–1327. [CrossRef]
31. Wu, C.; Wang, J. Geminal bis(boron) compounds: Their preparation and synthetic applications. *Tetrahedron Lett.* **2018**, *59*, 2128–2140. [CrossRef]
32. Jo, W.; Lee, J.H.; Cho, S.H. Advances in transition metal-free deborylative transformations of *gem*-diborylalkanes. *Chem. Commun.* **2021**, *57*, 4346–4353. [CrossRef]
33. Xu, L.; Zhang, S.; Li, P. Boron-selective reactions as powerful tools for modular synthesis of diverse complex molecules. *Chem. Soc. Rev.* **2015**, *44*, 8848–8858. [CrossRef] [PubMed]
34. Endo, K.; Ohkubo, T.; Hirokami, M.; Shibata, T. Chemoselective and regiospecific Suzuki coupling on a multisubstituted sp^3-carbon in 1,1-diborylalkanes at room temperature. *J. Am. Chem. Soc.* **2010**, *132*, 11033–11035. [CrossRef]

35. Endo, K.; Ohkubo, T.; Shibata, T. Chemoselective Suzuki coupling of diborylmethane for facile synthesis of benzylboronates. *Org. Lett.* **2011**, *13*, 3368–3371. [CrossRef] [PubMed]
36. Endo, K.; Ishioka, T.; Ohkubo, T.; Shibata, T. One-pot synthesis of symmetrical and unsymmetrical diarylmethanes via diborylmethane. *J. Org. Chem.* **2012**, *77*, 7223–7231. [CrossRef] [PubMed]
37. Endo, K.; Ohkubo, T.; Ishioka, T.; Shibata, T. Cross coupling between sp3-carbon and sp^3-carbon using a diborylmethane derivative at room temperature. *J. Org. Chem.* **2012**, *77*, 4826–4831. [CrossRef]
38. Li, H.; Shangguan, X.; Zhang, Z.; Huang, S.; Zhang, Y.; Wang, J. Formal carbon insertion of N-tosylhydrazone into B–B and B–Si bonds: Gem-diborylation and gem-silylborylation of sp^3 carbon. *Org. Lett.* **2014**, *16*, 448–451. [CrossRef]
39. Li, H.; Zhang, Z.; Shangguan, X.; Huang, S.; Chen, J.; Zhang, Y.; Wang, J. Palladium(0)-catalyzed cross-coupling of 1,1-diboronates with vinyl bromides and 1,1-dibromoalkenes. *Angew. Chem. Int. Ed.* **2014**, *53*, 11192–11925. [CrossRef]
40. Xu, S.; Shangguan, X.; Li, H.; Zhang, Y.; Wang, J. Pd(0)-Catalyzed cross-coupling of 1,1-diboronates with 2,2′-dibromobiphenyls: Synthesis of 9H-fluorenes. *J. Org. Chem.* **2015**, *80*, 7779–7784. [CrossRef]
41. Wu, C.; Wu, G.; Zhang, Y.; Wang, J. One-carbon homologation of arylboronic acids: A convenient approach to the synthesis of pinacol benzylboronates. *Org. Chem. Front.* **2016**, *3*, 817–822. [CrossRef]
42. Stymiest, J.; Bagutski, V.; French, R.; Aggarwal, V. Enantiodivergent conversion of chiral secondary alcohols into tertiary alcohols. *Nature* **2008**, *456*, 778–782. [CrossRef]
43. Li, Z.; Song, L.; Li, C. Silver-catalyzed radical aminofluorination of unactivated alkenes in aqueous media. *J. Am. Chem. Soc.* **2013**, *135*, 4640–4643. [CrossRef] [PubMed]
44. Crudden, C.; Ziebenhaus, C.; Rygus, J.; Ghozati, K.; Unsworth, P.; Nambo, M.; Voth, S.; Hutchinson, M.; Laberge, V.; Maekawa, Y.; et al. Iterative protecting group-free cross-coupling leading to chiral multiply arylated structures. *Nat. Commun.* **2016**, *7*, 11065–11072. [CrossRef] [PubMed]
45. Hong, K.; Liu, X.; Morken, J. Simple access to elusive α-boryl carbanions and their alkylation: An umpolung construction for organic synthesis. *J. Am. Chem. Soc.* **2014**, *136*, 10581–10584. [CrossRef] [PubMed]

Disclaimer/Publisher's Note: The statements, opinions and data contained in all publications are solely those of the individual author(s) and contributor(s) and not of MDPI and/or the editor(s). MDPI and/or the editor(s) disclaim responsibility for any injury to people or property resulting from any ideas, methods, instructions or products referred to in the content.

Article

Photocatalyzed Oxidative Decarboxylation Forming Aminovinylcysteine Containing Peptides †

Masaya Kumashiro, Kosuke Ohsawa and Takayuki Doi *

Graduate School of Pharmaceutical Sciences, Tohoku University, 6-3 Aza-Aoba, Aramaki, Aoba-ku, Sendai 980-8578, Japan
* Correspondence: doi_taka@mail.pharm.tohoku.ac.jp; Tel.: +81-22-795-6865
† This paper is dedicated to late Professor Jiro Tsuji.

Abstract: The formation of (2S,3S)-S-[(Z)-aminovinyl]-3-methyl-D-cysteine (AviMeCys) substructures was developed based on the photocatalyzed-oxidative decarboxylation of lanthionine-bearing peptides. The decarboxylative selenoetherification of the N-hydroxyphthalimide ester, generated in situ, proceeded under mild conditions at −40 °C in the presence of 1 mol% of eosin Y-Na$_2$ as a photocatalyst and the Hantzsch ester. The following β-elimination of the corresponding N,Se-acetal was operated in a one-pot operation, led to AviMeCys substructures found in natural products in moderate to good yields. The sulfide-bridged motif, and also the carbamate-type protecting groups, such as Cbz, Teoc, Boc and Fmoc groups, were tolerant under the reaction conditions.

Keywords: (2S,3S)-S-[(Z)-aminovinyl]-3-methyl-D-cysteine (AviMeCys); photocatalytic reaction; oxidative decarboxylation; ribosomally synthesized and post-translationally modified peptides (RiPPs); β-thioenamide; N-hydroxyphthalimide (NHPI) ester; Eosin Y

Citation: Kumashiro, M.; Ohsawa, K.; Doi, T. Photocatalyzed Oxidative Decarboxylation Forming Aminovinylcysteine Containing Peptides. *Catalysts* **2022**, *12*, 1615.
https://doi.org/10.3390/catal12121615

Academic Editors: Ewa Kowalska and Yuichi Kobayashi

Received: 21 November 2022
Accepted: 6 December 2022
Published: 9 December 2022

Publisher's Note: MDPI stays neutral with regard to jurisdictional claims in published maps and institutional affiliations.

Copyright: © 2022 by the authors. Licensee MDPI, Basel, Switzerland. This article is an open access article distributed under the terms and conditions of the Creative Commons Attribution (CC BY) license (https://creativecommons.org/licenses/by/4.0/).

1. Introduction

Ribosomally synthesized and post-translationally modified peptides (RiPPs) are one of the largest classes of natural products that exhibit various biological properties [1–3]. Among the diverse substructures found in RiPPs, cross-linked sulfides between two amino acid residues have been identified as an important chemical functionality for constrained conformation in the peptide backbone, providing high target specificity and biological stability [4–7]. The major components of RiPPs with thioether bonds are (2S,6R)-lanthionine (Lan) and (2S,3S,6R)-3-methyllanthionine (β-MeLan). Thioether-bridged units in Lan/β-MeLan are biosynthetically constructed through a conjugated addition of thiols of cysteine residues to dehydroalanine (Dha)/dehydrobutyrine (Dhb) [8,9]. Intriguingly, (Z)-thioenolates, which are generated by the oxidative decarboxylation of cysteine residues positioned at the C-terminal, also attack Dhb/Dha residues to produce S-[(Z)-aminovinyl]-D-cysteine (AviCys) and (2S,3S)-S-[(Z)-aminovinyl]-3-methyl-D-cysteine (AviMeCys), respectively. Owing to β-thioenamide units including sp^2 α-carbons in the peptide backbone, AviCys/AviMeCys are attractive substructures for improving the structural rigidity and drug-like properties of cyclopeptides [10–12].

Biosynthesis-inspired approaches to obtain lanthionines have been accomplished through the stereoselective conjugated addition of thiols of cysteine derivatives to Dha/Dhb derivatives [13–15], whereas the synthesis of AviCys/AviMeCys motifs in a similar manner is a difficult task due to the chemical instability of thioenolates [16]. Thus, alternative methodologies for constructing AviCys/AviMeCys have been developed to date. There have been several reports on the synthesis of AviCys substructures by condensation using primary amides and α-thioaldehyde/acetals [17,18] and the β-addition of thiyl radicals to terminal ynamides [19]. However, the methodologies for constructing AviMeCys substructures in complex natural products are limited. One of the efficient approaches is a decarboxylative olefination of carboxylic acid derivatives, as in the palladium-catalyzed reaction of allyl β-ketoesters reported by Tsuji [20]. Recently, oxidative decarboxylation/decarbonylation using lanthionine units has been reported for the AviMeCys formation. VanNieuwenhze et al. reported Z-selective AviMeCys formation via nickel(0)-promoted decarbonylation of activated thioesters in short peptide fragments during the synthesis of D-ring in mersacidin (Scheme 1a) [21]. Furthermore, they realized the direct conversion from carboxylic acids through two procedures: (i) Curtius rearrangement using diphenylphosphoryl azide (DPPA), followed by the collapse of the resulting isocyanate, and (ii) oxidative decarboxylation using lead tetraacetate, followed by the elimination of the resulting acetate (Scheme 1b) [22]. Nevertheless, there is no report on the total synthesis of any AviMeCys-containing natural peptides, due to harsh conditions and an excess amount of toxic oxidants during oxidative decarboxylation/decarbonylation steps in the late-stages of the synthesis. Considering the tolerance to functional groups in RiPPs, we focused on photocatalytic reactions mediated by visible light, which have been widely used in the modification of amino acids and peptides [23–25]. We envisioned that a radical species, generated from a N-hydroxyphthalimide (NHPI) ester of the corresponding lanthionine by oxidative decarboxylation [26], can be readily trapped in the presence of oxidation-sensitive sulfides under mild conditions. The following β-elimination with a weak base would yield the desired β-thioenamide motifs, suppressing the retro-thio-Michael reaction (Scheme 1c). Herein, we report the β-thioenamide formation through the photocatalyzed oxidative decarboxylation of lanthionine derivatives. A series of reactions were conducted in a one-pot operation under mild conditions, providing AviMeCys units with functional group compatibility.

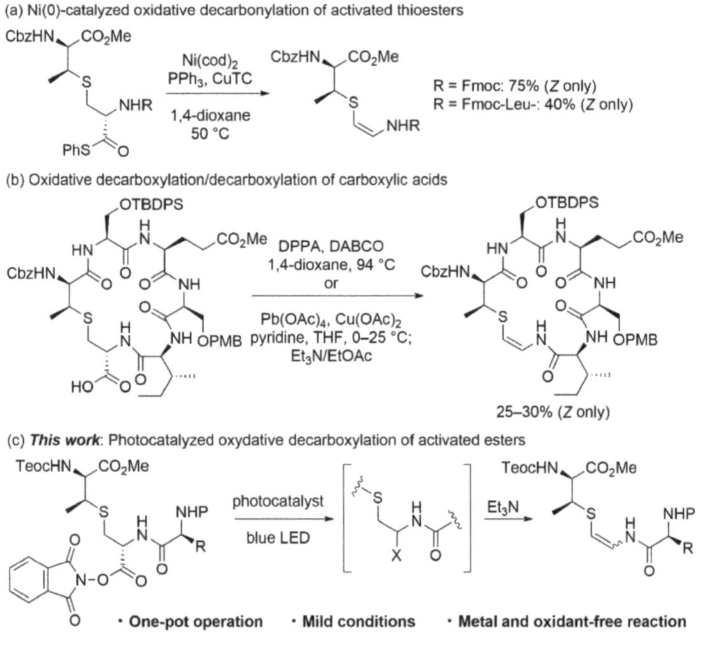

Scheme 1. AviMeCys formation via decarboxylation/decarbonylation of lanthionine derivatives.

2. Results and Discussion

Our study began with the preparation of the lanthionine-containing peptides **1**, as shown in Scheme 2. According to the procedure reported by VanNieuwenhze [27], the regioselective ring-opening of the N-{2-(trimethylsilyl)ethoxycarbonyl} (Teoc)-protected aziridine **2**, which was prepared from D-threonine, with Fmoc-Cys-OH (**3**) was performed in the presence of indium(III) chloride, providing the lanthionine derivative **4** in a 54% yield.

The subsequent protection of the carboxylic acid group with a methoxymethyl (MOM) group provided the MOM ester **5** in a 99% yield. The removal of the Fmoc group in **5** with 20% diethylamine/acetonitrile, followed by the coupling of the resulting amine with N-protected amino acids afforded the peptides **6a–f** over two steps. Finally, the MOM group in **5** was removed under acidic conditions, yielding the carboxylic acids **1a–f**.

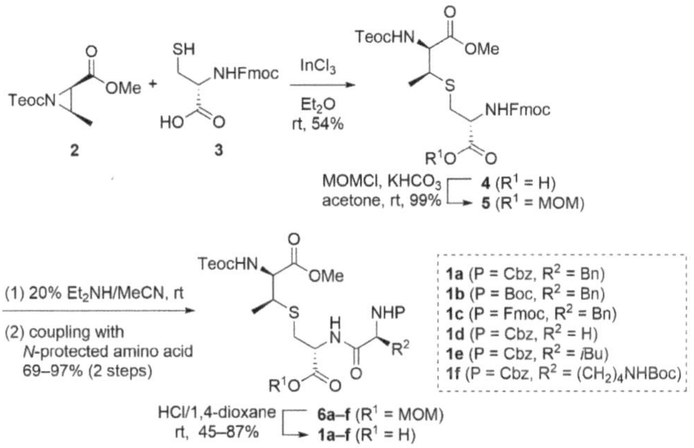

Scheme 2. Preparation of the lanthionine-bearing peptides **1**.

Next, we surveyed the formation of the AviMeCys unit using **1a** as the model lanthionine-bearing peptide as shown in Scheme 3. Given the immediate capture of the resulting radical species during the oxidative decarboxylation [28–30], diphenyl diselenide was selected as a radical trapping agent [31,32]. In addition, we envisioned that the resulting N,Se-acetal would be converted into the corresponding β-thioenamide without losing the β-methylcysteine unit due to high leaving activity of phenylselenolates [33–35]. According to the reported procedures [32,36,37], the NHPI ester **7a**, prepared from **1a** in situ, was treated with 1 mol% of [Ru(bpy)$_3$](PF$_6$)$_2$ as the photocatalyst in the presence of the Hantzsch ester and diphenyl diselenide under 40 W blue light-emitting diode (LED) irradiation. The following β-elimination of the obtained selenoether **8a** by treatment with triethylamine furnished the β-thioenamieds (Z)-**9a** and (E)-**9a** in 23% and 13% yields, respectively. The geometry of olefins in (Z)-**9a** and (E)-**9a** was determined by ^{1}H nuclear magnetic resonance (NMR) spectroscopy through the coupling constants ($^3J_{H,H}$ = 7.2 Hz for (Z)-**9a**, and 13.8 Hz for (E)-**9a**) of isolated compounds [19].

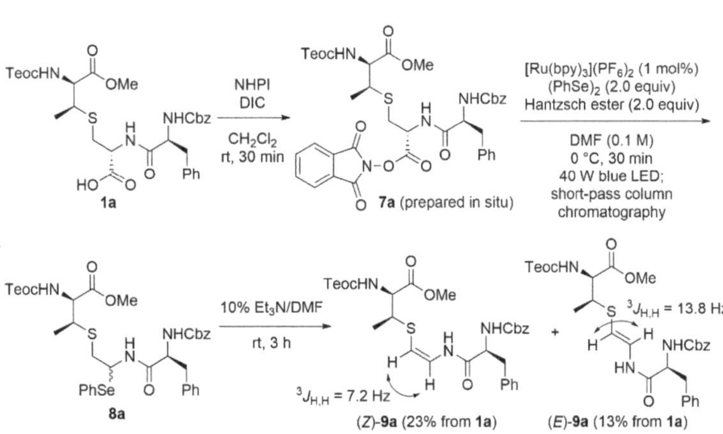

Scheme 3. Photocatalytic selenoetherification/β-elimination of the lanthionine-bearing peptide **1a**.

As a moderate yield was observed, we conducted the screening of photocatalysts, as shown in Table 1. When selenoetherification, followed by β-elimination was conducted in a one-pot operation, the combined yields of (Z)-**9a** and (E)-**9a** were slightly up to 49% (entry 1). After the optimization of metal and organic photosensitizers, eosin Y-Na$_2$ [38] promoted the transformation to increase the yields by up to 59% (entries 2–4). Intriguingly, selenoetherification of **7a** proceeded without eosin Y-Na$_2$, albeit with slightly lower yields, suggesting that the formation of the electron donor-acceptor (EDA) complex between the NHPI and Hantzsch esters should promote the reaction (entry 5) [39,40]. No product was obtained in the absence of the blue light (entry 6).

Table 1. Preliminary screening of photocatalysts for selenoetherification/β-elimination [a].

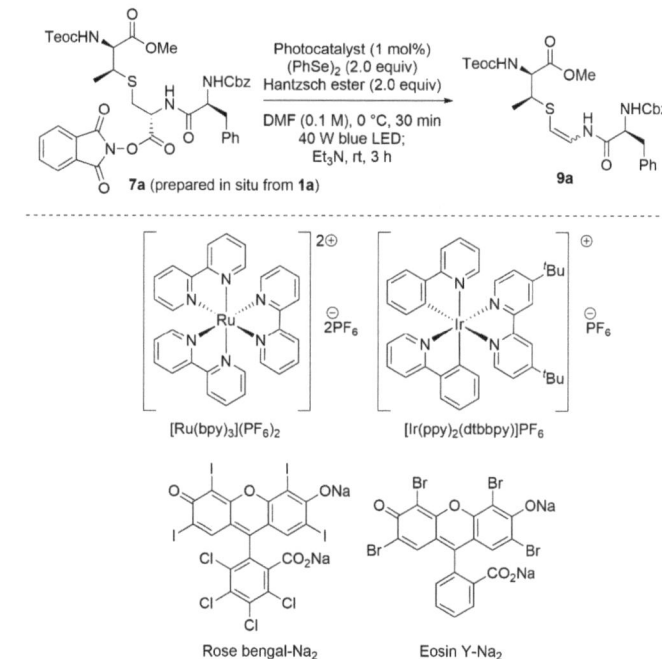

Entry	Photocatalyst	Yield [b]
1	[Ru(bpy)$_3$](PF$_6$)$_2$	49% (Z: 28%, E: 21%)
2	[Ir(dtbbpy)(ppy)$_2$]PF$_6$	48% (Z: 28%, E: 20%)
3	Rose bengal-Na$_2$	52% (Z: 30%, E: 22%)
4	Eosin Y-Na$_2$	59% (Z: 34%, E: 25%)
5	none	53% (Z: 31%, E: 22%)
6 [c]	Eosin Y-Na$_2$	-

[a] All reactions were conducted on a 0.1 mmol scale. [b] Isolated yield based on **1a**. [c] No reaction was performed in the absence of the blue light.

To further improve the yield, we optimized the reaction conditions using eosin Y-Na$_2$, and the results are summarized in Table 2. Using other solvents, such as CH$_2$Cl$_2$, MeCN, N,N-dimethylaniline (DMA) and dimethyl sulfoxide (DMSO), instead of N,N-dimethylformamide (DMF) was fruitless (entry 1 vs, entries 2–5). As reductants, 1-benzyl-1,4-dihydronicotinamide and γ-terpinene decreased the yield (entry 1 vs, entries 6 and 7). Notably, N,N-diisopropylethylamine (DIEA), widely used for photocatalytic reactions, involved the decomposition of the NHPI ester **7a** because of its strong basicity (entry 8). The yield increased to 68% when the amount of the Hantzsch ester was reduced to 1.0 equiv (entry 9). Excess amounts of the Hantzsch ester may interfere with the capture of the resulting radical species by diphenyl diselenide [41,42]. Further reduction of the Hantzsch ester decreased the yield (entries 10 and 11). With a decrease in the reaction temperature at −40 °C, the yield was up to 74% (entry 12). In contrast, selenoetherification did not complete at −78 °C (entry 13). Thus, we determined that the optimized condition is observed in entry 12. Our developed methodology was performed on a 1.0 mmol scale, giving **9a** in a moderate yield (51%, entry 14).

Table 2. Optimization of reaction conditions [a].

Entry	Reductant (Equiv)	Solvent	Temp. (°C)	Yield [b]
1	Hantzsch ester (2.0)	DMF	0	59% (Z: 34%, E: 25%)
2	Hantzsch ester (2.0)	CH_2Cl_2	0	33% (Z: 20%, E: 13%)
3	Hantzsch ester (2.0)	MeCN	0	52% (Z: 33%, E: 19%)
4	Hantzsch ester (2.0)	DMA	0	53% (Z: 30%, E: 23%)
5	Hantzsch ester (2.0)	DMSO	0	45% (Z: 26%, E: 19%)
6	1-benzyl-1,4-dihydronicotinamide (2.0)	DMF	0	38% (Z: 22%, E: 16%)
7	γ-terpinene (2.0)	DMF	0	6% (Z: 3%, E: 3%)
8	DIEA (2.0)	DMF	0	16% (Z: 9%, E: 7%)
9	Hantzsch ester (1.0)	DMF	0	68% (Z: 36%, E: 32%)
10	Hantzsch ester (0.5)	DMF	0	51% (Z: 28%, E: 23%)
11 [c]	none	DMF	0	22% (Z: 13%, E: 9%)
12	Hantzsch ester (1.0)	DMF	−40	74% (Z: 42%, E: 32%)
13	Hantzsch ester (1.0)	DMF	−78	34% (Z: 19%, E: 15%)
14 [d]	Hantzsch ester (1.0)	DMF	−40	51% (Z/E = 57:43) [e]

[a] All reactions were conducted on a 0.1 mmol scale. [b] Isolated yield based on **1a**. [c] Selenoetherification was conducted for 2 h. [d] 1.0 mmol scale. [e] The product was obtained as a Z/E mixture. The ratio was determined by ^{1}H NMR.

The substrate scope for our developed AviMeCys formation is shown in Scheme 4. Carbamate-type protecting groups, such as Cbz, Teoc, Boc and Fmoc groups, were tolerant under the reaction conditions, providing the corresponding β-thioenamides **9a–c** in moderate to good yields (37–74%). AviMeCys substructures in natural products, such as **9d** for cacaodin [43], **9e** for mersacidin [44], and **9f** for lexapeptide [45], were obtained from lanthionines **1d–1f** in 58–68% yields.

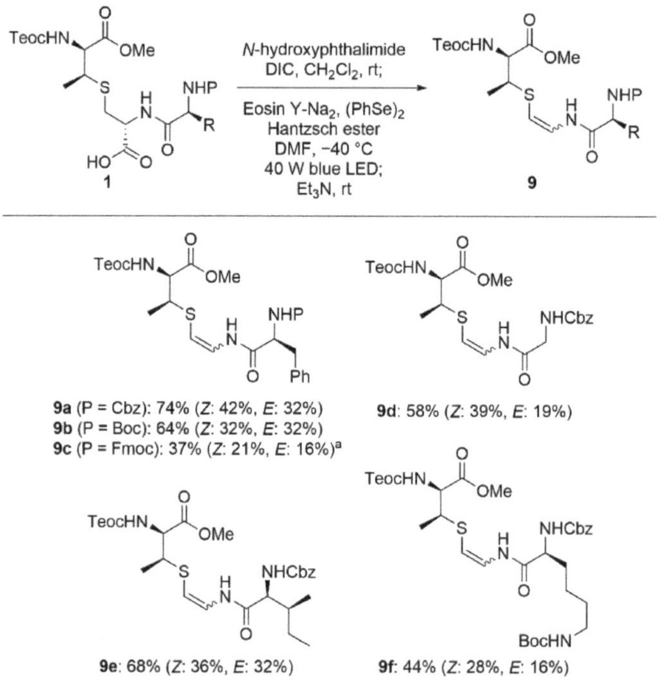

Scheme 4. Scope of lanthionine-bearing peptides in the AviMeCys formation. [a] 3.0 equiv of Et$_3$N was used.

A plausible reaction mechanism of the photocatalytic synthesis of AviMeCys is depicted in Scheme 5 according to the above results and previous reports on decarboxylative selenoetherification [31,46]. Given that the reaction proceeded without photocatalysts, we assumed the formation of an EDA complex between the NHPI and Hantzsch esters [39,40]. Photoirradiation induces intramolecular single-electron transfer, generating a phthalimide radical anion **B** with a dihydropyridine radical cation **A**. The resulting **B** undergoes decarboxylation to form a radical species **D** and a phthalimide anion **C**. The radical **D** is then captured by diphenyl diselenide to form a *N,Se*-acetal **F** with a seleno radical **E**. The β-elimination with the *N,Se*-acetal **F** in the presence of Et$_3$N affords the corresponding AviMeCys (**G**) (Scheme 5a). A (Z)-isomer will be obtained with slight priority because of the electrostatic attraction between a sulfur atom and the amide moiety [17]. The resulting **A**, **C**, and **E** are converted into phthalimide, pyridine derivative and phenylselenol, respectively, through two possible pathways. Although the radical-quenching of **A** and **E** automatically occurs (Scheme 5c), photocatalysts may mediate this step to improve the yields (Scheme 5b).

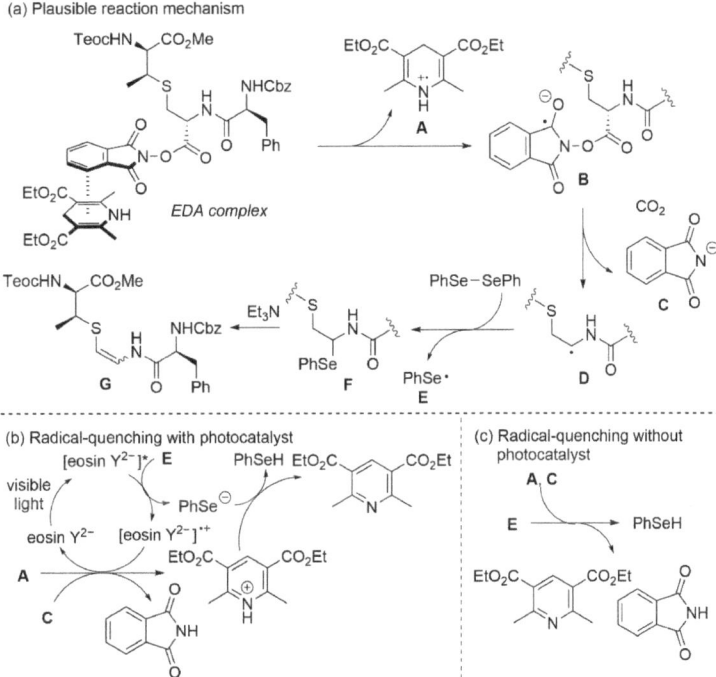

Scheme 5. Plausible reaction mechanism.

3. Materials and Methods

3.1. General Techniques

All commercially available reagents were purchased from commercial suppliers and used as received. Dry THF and CH_2Cl_2 (Kanto Chemical Co., Inc., Tokyo, Japan) were obtained by passing commercially available pre-dried, oxygen-free formulations. DMF (for peptide synthesis) was purchased from Watanabe Chemical Industries, Ltd. (Hiroshima, Japan). Photocatalyzed oxidative decarboxylation was performed with a Kessil A160WE Tuna Blue (Dicon Fiberoptic Inc., Richmond, CA, USA), as shown in Figure S1.

All reactions were monitored by TLC carried out on Merck silica gel plates (0.2 mm, 60F-254) with UV light, and visualized by *p*-anisaldehyde/H_2SO_4/EtOH solution, phosphomolybdic acid–EtOH solution or ninhydrin/AcOH/BuOH solution. Column chromatography was carried out with silica gel 60 N (Kanto Chemical Co. 100–210 μm). Preparative TLC was performed on 0.75 mm Wakogel® B-5F PLC plates (FUJIFILM Wako Pure Chemical Co., Ltd., Osaka, Japan). ^{1}H NMR spectra (400 and 600 MHz) and ^{13}C{^{1}H} NMR spectra (100 and 150 MHz) were recorded on JEOL JNM-AL400 and JEOL JNM-ECA600 spectrometers (JEOL Ltd., Tokyo, Japan) in the indicated solvent. Chemical shifts (δ) are reported in unit parts per million (ppm) relative to the signal for internal TMS (0.00 ppm for ^{1}H) for solutions in $CDCl_3$. NMR spectral data are reported as follows: $CHCl_3$ (7.26 ppm for ^{1}H) or $CDCl_3$ (77.0 ppm for ^{13}C), and DMSO (2.49 ppm for ^{1}H) or DMSO-d_6 (39.5 ppm for ^{13}C), when internal standard is not indicated. Multiplicities are reported by using standard abbreviations, and coupling constants are given in hertz.

High-resolution mass spectra (HRMS) were recorded on Thermo Scientific Exactive Plus Orbitrap Mass Spectrometer (Thermo Fisher Scientific K.K., Tokyo, Japan) for ESI or JEOL JMS-AX500 (JEOL Ltd., Tokyo, Japan) for FAB. IR spectra were recorded on a JASCO FTIR-4100 spectrophotometer (JASCO Co., Tokyo, Japan). Only the strongest and/or structurally important absorption are reported as the IR data afforded in wavenumbers (cm^{-1}). Optical rotations were measured on a JASCO P-1010 polarimeter (JASCO Co.,

Tokyo, Japan). Melting points were measured with Round Science Inc. RFS-10 (J-SCIENCE LAB Co., Ltd., Kyoto, Japan), and are not corrected.

3.2. Synthesis of the Lanthionine 5

3.2.1. 2-Methyl 1-(2-(Trimethylsilyl)ethyl) (2R,3R)-3-methylaziridine-1,2-dicarboxylate (2)

To a solution of D-threonine (5.00 g, 42.0 mmol, 1.0 equiv) in MeOH (150 mL) was added $SOCl_2$ (15.3 mL, 210 mmol, 5.0 equiv) dropwise at 0 °C, and the mixture was stirred at the same temperature for 30 min. After being stirred at reflux in an oil bath for 12 h, the reaction mixture was cooled to room temperature, and concentrated in vacuo. The resulting crude methyl ester was used for the next reaction without further purification.

To a solution of the crude amine in dry CH_2Cl_2 (150 mL) were added Et_3N (14.6 mL, 105 mmol, 2.5 equiv) and TrtCl (11.7 g, 42.0 mmol, 1.0 equiv) at 0 °C under an argon atmosphere. After being stirred at room temperature for 43 h, the reaction mixture was washed with 10% aqueous citric acid, saturated aqueous $NaHCO_3$ and brine, dried over $MgSO_4$, and filtered. The filtrate was concentrated in vacuo, and the resulting crude N-Trt amine was used for the next reaction without further purification.

To a solution of the crude alcohol in dry THF (120 mL) were added Et_3N (14.6 mL, 105 mmol, 2.5 equiv) and MsCl (3.6 mL, 46.2 mmol, 1.1 equiv) at 0 °C under an argon atmosphere, and the mixture was stirred at the same temperature for 30 min. After being stirred at reflux in an oil bath for 72 h, the reaction mixture was concentrated in vacuo to remove THF. The resulting residue was diluted with EtOAc, and the organic layer was washed with 10% aqueous citric acid and saturated aqueous $NaHCO_3$, dried over $MgSO_4$, and filtered. The filtrate was concentrated in vacuo, and the resulting crude aziridine was used for next reaction without further purification.

To a solution of the crude N-Trt aziridine in dry CH_2Cl_2 (150 mL) were added dry MeOH (2.6 mL, 63.0 mmol, 1.5 equiv) and TFA (6.5 mL, 84.0 mmol, 2.0 equiv) at 0 °C under an argon atmosphere. After being stirred at the same temperature for 1 h, the reaction mixture was basified by Et_3N (20.5 mL, 147 mmol, 3.5 equiv). TeocOSu (10.9 g, 42.0 mmol, 1.0 equiv) was then added to the above mixture at 0 °C. After being stirred at room temperature for 19 h, the reaction mixture was washed with 10% aqueous citric acid and saturated aqueous $NaHCO_3$, dried over $MgSO_4$, and filtered. The filtrate was concentrated in vacuo, and the resulting residue was suspended in CH_2Cl_2/MeOH. The suspension was filtered through a pad of Celite®, and the filtrate was concentrated in vacuo. The resulting residue was purified by column chromatography on silica gel (eluted with hexane/EtOAc = 4:1) to afford the N-Teoc aziridine **2** (5.98 g, 23.0 mmol, 55% in 4 steps) as a colorless oil. $[\alpha]^{22}_D$ +64 (c 1.0, $CHCl_3$); ^{1}H NMR (400 MHz, $CDCl_3$) δ 4.18–4.22 (m, 2H), 3.79 (s, 3H), 3.15 (d, 1H, J = 6.8 Hz), 2.77–2.82 (m, 1H), 1.35 (d, 3H, J = 6.4 Hz), 1.00–1.04 (m, 2H), 0.00 (s, 9H); ^{13}C{^{1}H} NMR (100 MHz, $CDCl_3$) δ 167.7, 161.8, 65.3, 52.2, 39.7, 38.7, 17.4, 12.9, −1.5; IR (neat) 2955, 1756, 1729, 1442, 1425, 1285, 1252, 1201, 1181, 1081, 1038, 860, 838 cm^{-1}; HRMS [ESI] m/z calcd for $C_{11}H_{21}NO_4SiNa$ [M+Na]$^+$ 282.1132, found 282.1131.

3.2.2. Fmoc-Cys-OH (3)

To a solution of L-cystine (5.00 g, 20.8 mmol, 1.0 equiv) in 1,4-dioxane (90 mL) were added a solution of Na_2CO_3 (6.62 g, 62.4 mmol, 3.0 equiv) in water (60 mL) and a solution of FmocOSu (14.0 g, 41.6 mmol, 2.0 equiv) in 1,4-dioxane (90 mL) at 0 °C. After being stirred at the room temperature for 15 h, the reaction mixture was concentrated in vacuo to remove 1,4-dioxane. The aqueous layer was acidified with 6 M aqueous HCl until pH1, and extracted three times with CH_2Cl_2. The combined organic layers were dried over Na_2SO_4, and filtered. The filtrate was concentrated in vacuo, and the resulting residue was suspended in Et_2O. The white precipitate was filtered, and dried under vacuum to afford the N-Fmoc amine (12.7 g, 18.5 mmol, 89%) as a white solid. The spectral data of synthetic compound were in good agreement with those of reported [47]. Mp 153–154 °C [lit. 149–151 °C]; $[\alpha]^{20}_D$ −89 (c 1.2, MeOH) [lit. $[\alpha]^{24}_D$ −87.1 (c 1.0, MeOH)]; ^{1}H NMR (400 MHz, DMSO-d_6, rotamer mixture) δ 13.0 (s, 1H), 7.87 (d, 2H, J = 7.5 Hz), 7.78 (d, 1H,

J = 7.5 Hz), 7.69 (d, 2H, J = 7.5 Hz), 7.37–7.41 (m, 2H), 7.30 (t, 2H, J = 7.5 Hz), 4.26–4.31 (m, 3H), 4.21 (dd, 1H, J = 12.6, 5.6 Hz), 3.16 (dd, 1H, J = 13.5, 3.9 Hz), 2.94 (dd, 1H, J = 13.5, 10.3 Hz); ^{13}C{^{1}H} NMR (100 MHz, DMSO-d_6, rotamer mixture) δ 172.2, 156.0, 143.8, 143.7, 140.7, 127.6, 127.1, 125.3, 125.2, 120.1, 65.8, 53.0, 46.6, 39.1; IR (neat) 1696, 1515, 1448, 1331, 1228, 1049, 758, 739 cm^{-1}; HRMS [FAB] m/z calcd for $C_{36}H_{33}N_2O_8S_2$ [M+H]$^+$ 685.1673, found 685.1690.

To a solution of the disulfide (14.0 g, 20.4 mmol, 1.0 equiv) in dry THF (70 mL) were added 1 M aqueous HCl (70 mL) and activated zinc dust (4.00 g, 61.2 mmol, 3.0 equiv) at 0 °C. After being stirred at the room temperature for 30 min, the reaction mixture was filtered through of a pad of Celite®. The filtrate was concentrated in vacuo, and the resulting residue was diluted with 1 M aqueous HCl. The aqueous layer was extracted with CH_2Cl_2. The organic layer was dried over Na_2SO_4, and filtered. The filtrate was concentrated in vacuo, and the resulting residue was suspended in CH_2Cl_2/hexane. The precipitate was filtered and dried under vacuum to afford the thiol 3 (10.7 g, 31.3 mmol, 77%) as a white solid. The spectral data of synthetic compound were in good agreement with those of reported [48]. Mp 119–123 °C; $[\alpha]^{20}_D$ -5.7 (c 0.93, MeOH); ^{1}H NMR (400 MHz, DMSO-d_6) δ 12.9 (s, 1H), 7.88 (d, 2H, J = 7.5 Hz), 7.69–7.73 (m, 3H), 7.41 (t, 2H, J = 7.5 Hz), 7.32 (t, 2H, J = 7.5 Hz), 4.29–4.31 (m, 2H), 4.23 (t, 1H, J = 7.0 Hz), 4.14 (dt, 1H, J = 8.3, 4.3 Hz), 2.89–2.92 (m, 1H), 2.71–2.78 (m, 1H); ^{13}C{^{1}H} NMR (100 MHz, DMSO-d_6) δ 171.8, 156.0, 143.8, 140.7, 127.6, 127.1, 125.2, 120.1, 65.7, 56.5, 46.6, 25.4; IR (neat) 3314, 1694, 1536, 1476, 1447, 1418, 1230, 1103, 1047, 756, 736, 620 cm^{-1}; HRMS [FAB] m/z calcd for $C_{18}H_{18}NO_4S$ [M+H]$^+$ 344.0951, found 344.0942.

3.2.3. The Lanthionine 4

To a solution of the aziridine 2 (4.62 g, 17.8 mmol, 2.0 equiv) in dry Et_2O (90 mL) were added Fmoc-Cys-OH (3) (3.06 g, 8.91 mmol, 1.0 equiv) and $InCl_3$ (788 mg, 3.56 mmol, 0.4 equiv) at room temperature under an argon atmosphere. After being stirred at the same temperature for 18 h, the reaction mixture was quenched with water. The organic layer was separated, and aqueous layer was extracted three times with EtOAc. The combined organic layers were dried over Na_2SO_4, and filtered. The filtrate was concentrated in vacuo, and the resulting residue was purified by column chromatography on silica gel (eluted with CH_2Cl_2/MeCN = 1:1) to afford the lanthionine 4 (2.89 g, 4.79 mmol, 54%) as a white amorphous solid. $[\alpha]^{24}_D$ $+2.3$ (c 1.0, CHCl$_3$); ^{1}H NMR (400 MHz, DMSO-d_6) δ 7.89 (d, 2H, J = 7.5 Hz), 7.71–7.73 (m, 3H), 7.41 (t, 2H, J = 7.5 Hz), 7.31–7.33 (m, 3H), 4.21–4.34 (m, 4H), 4.09–4.15 (m, 1H), 4.04–4.06 (m, 2H), 3.65 (s, 3H), 3.21–3.24 (m, 1H), 2.95 (dd, 1H, J = 13.6, 4.7 Hz), 2.74 (dd, 1H, J = 13.5, 9.4 Hz), 1.21 (d, 3H, J = 7.0 Hz), 0.91–0.93 (m, 2H), 0.00 (s, 9H); ^{13}C{^{1}H} NMR (100 MHz, DMSO-d_6) δ 172.0, 170.8, 156.2, 155.9, 143.7, 140.7, 127.6, 127.0, 125.2, 120.0, 65.7, 62.2, 58.4, 54.0, 51.9, 46.6, 41.6, 32.1, 18.8, 17.3, -1.6; IR (neat) 3327, 3019, 2953, 1720, 1513, 1478, 1449, 1338, 1249, 1213, 1080, 1049, 859, 837, 758, 740 cm^{-1}; HRMS [ESI] m/z calcd for $C_{29}H_{38}N_2O_8NaSSi$ [M+Na]$^+$ 625.2010, found 625.2007.

3.2.4. The MOM Ester 5

To a solution of the carboxylic acid 4 (2.89 g, 4.79 mmol, 1.0 equiv) in dry acetone (90 mL) were added $KHCO_3$ (1.20 g, 12.0 mmol, 2.5 equiv) and MOMCl (437 µL, 5.75 mmol, 1.2 equiv) at room temperature under an argon atmosphere. After being stirred at the same temperature for 15 h, the reaction mixture was concentrated in vacuo to remove acetone. The resulting residue was diluted with EtOAc. The organic layer was washed with saturated aqueous $NaHCO_3$, dried over Na_2SO_4, and filtered. The filtrate was concentrated in vacuo, and the resulting residue was purified by column chromatography on silica gel (eluted with hexane/EtOAc = 1:1) to afford the MOM ester 5 (3.06 g, 4.73 mmol, 99%) as a white amorphous solid. $[\alpha]^{23}_D$ -16 (c 1.3, CHCl$_3$); ^{1}H NMR (400 MHz, CDCl$_3$) δ 7.74 (d, 2H, J = 7.5 Hz), 7.57–7.59 (m, 2H), 7.38 (t, 2H, J = 7.5 Hz), 7.29 (d, 2H, J = 7.5 Hz), 5.66 (d, 1H, J = 6.8 Hz), 5.41 (d, 1H, J = 8.5 Hz), 5.31 (d, 1H, J = 5.6 Hz), 5.27 (d, 1H, J = 5.6 Hz), 4.58–4.61 (m, 1H), 4.51 (d, 1H, J = 7.5 Hz), 4.36–4.44 (m, 2H), 4.22 (t, 1H, J = 6.9 Hz), 4.11–4.16 (m,

2H), 3.72 (s, 3H), 3.44–3.47 (m, 4H), 3.03 (dd, 1H, J = 13.4, 3.7 Hz), 2.91 (dd, 1H, J = 13.4, 5.4 Hz), 1.31 (d, 3H, J = 7.0 Hz), 0.95–0.97 (m, 2H), −0.01 (s, 9H); $^{13}C\{^{1}H\}$ NMR (100 MHz, CDCl$_3$) δ 171.1, 170.0, 156.7, 155.7, 143.7, 141.3, 127.7, 127.0, 125.0, 120.0, 91.8, 67.2, 63.7, 58.1, 58.0, 53.8, 52.5, 47.1, 43.7, 33.5, 19.4, 17.6, −1.6; IR (neat) 3335, 2953, 1722, 1511, 1450, 1339, 1249, 1210, 1157, 1081, 1047, 994, 928, 859, 837, 759, 741 cm^{-1}; HRMS[ESI] m/z calcd for C$_{31}$H$_{42}$N$_2$O$_9$NaSSi [M+Na]$^+$ 669.2272, found 669.2280.

3.3. Synthesis of the Tripeptide 1 by Solution-Phase Peptide Synthesis

3.3.1. The Tripeptide 1a

To a solution of the N-Fmoc-amine **5** (3.06 g, 4.73 mmol, 1.0 equiv) in dry MeCN (40 mL) was added Et$_2$NH (10 mL) at room temperature under an argon atmosphere. After being stirred at the same temperature for 40 min, the reaction mixture was concentrated in vacuo. The resulting residue was azeotroped three times with MeCN to remove Et$_2$NH, and the resulting crude amine was used for the next reaction without further purification.

To a solution of the crude amine in dry CH$_2$Cl$_2$ (50 mL) were added DIEA (1.7 mL, 9.46 mmol, 2.0 equiv), Cbz-Phe-OH (1.70 g, 5.68 mmol, 1.2 equiv), HOBt (773 mg, 5.68 mmol, 1.2 equiv) and EDCI·HCl (1.09 g, 5.68 mmol, 1.2 equiv) at 0 °C under an argon atmosphere. After being stirred at room temperature for 13 h, the reaction mixture was diluted with CH$_2$Cl$_2$. The organic layer was washed with 10% aqueous citric acid and saturated aqueous NaHCO$_3$, dried over Na$_2$SO$_4$, and filtered. The filtrate was concentrated in vacuo, and the resulting residue was purified by column chromatography on silica gel (eluted with hexane/EtOAc = 1:1) to afford the tripeptide **6a** (2.88 g, 4.08 mmol, 86% in 2 steps) as a white amorphous solid. [α]$^{23}_D$ −25 (c 0.90, CHCl$_3$); ^{1}H NMR (400 MHz, CDCl$_3$) δ 7.20–7.27 (m, 10H), 6.73 (d, 1H, J = 6.5 Hz), 5.45 (d, 1H, J = 9.2 Hz), 5.38 (d, 1H, J = 5.8 Hz), 5.29 (d, 1H, J = 5.8 Hz), 5.26 (d, 1H, J = 5.8 Hz), 5.08 (s, 2H), 4.72–4.74 (m, 1H), 4.48–4.50 (m, 2H), 4.15–4.17 (m, 2H), 3.75 (s, 3H), 3.48 (s, 3H), 3.37–3.41 (m, 1H), 3.00–3.18 (m, 3H), 2.84 (dd, 1H, J =13.9, 5.9 Hz), 1.27 (d, 3H, J = 7.0 Hz), 0.96–1.01 (m, 2H), 0.02 (s, 9H); $^{13}C\{^{1}H\}$ NMR (100 MHz, CDCl$_3$) δ 171.1, 171.0, 169.5, 156.6, 155.9, 136.2, 136.1, 129.3, 128.6, 128.4, 128.1, 127.9, 127.0, 91.7, 67.0, 63.7, 58.1, 57.9, 56.0, 52.5, 52.2, 43.3, 38.1, 32.8, 19.2, 17.6, −1.6; IR (neat) 3314, 3030, 2953, 1721, 1519, 1454, 1338, 1249, 1215, 1155, 1083, 1048, 931, 860, 837, 750, 699 cm^{-1}; HRMS [ESI] m/z calcd for C$_{33}$H$_{47}$N$_3$O$_{10}$NaSSi [M+Na]$^+$ 728.2644, found 728.2650.

To a solution of the MOM ester **6a** (2.88 g, 4.08 mmol, 1.0 equiv) in 1,4-dioxane (30 mL) was added 4 M HCl/1,4-dioxane (10 mL) at 0 °C under argon atmosphere. After being stirred at room temperature for 1 h, the reaction mixture was concentrated in vacuo. The resulting residue was purified by column chromatography on silica gel (eluted with CH$_2$Cl$_2$/MeOH = 50:1) to afford the carboxylic acid **1a** (2.35 g, 3.55 mmol, 87%) as a white amorphous solid. [α]$^{23}_D$ −13 (c 1.1, CHCl$_3$); ^{1}H NMR (400 MHz, DMSO-d_6) δ 12.9 (brs, 1H), 8.38 (d, 1H, J = 7.7 Hz), 7.46 (d, 1H, J = 8.7 Hz), 7.19–7.31 (m, 11H), 4.93 (s, 2H), 4,25–4.43 (m, 3H), 4.03–4.05 (m, 2H), 3.64 (s, 3H), 3.25–3.31 (m, 1H), 2.96–3.03 (m, 2H), 2.70–2.80 (m, 2H), 1.21 (d, 3H, J = 6.8 Hz), 0.90–0.92 (m, 2H), 0.00 (s, 9H); $^{13}C\{^{1}H\}$ NMR (100 MHz, DMSO-d_6) δ 171.7, 171.6, 170.8, 156.2, 155.7, 138.0, 136.9, 129.2, 128.2, 128.0, 127.6, 127.3, 126.2, 65.1, 62.2, 58.5, 55.9, 52.2, 51.9, 41.8, 37.4, 32.1, 18.7, 17.3, −1.5; IR (neat) 3315, 3064,3030, 2953, 1721, 1518, 1454, 1439, 1340, 1287, 1250, 1215, 1180, 1081, 1050, 860, 837, 753, 698 cm^{-1}; HRMS [ESI] m/z calcd for C$_{31}$H$_{43}$N$_3$O$_9$NaSSi [M+Na]$^+$ 684.2381, found 684.2391.

3.3.2. The Tripeptide 1b

Compound **6b** was prepared from the N-Fmoc-amine **5** (248 mg, 383 µmol) according to the procedure above described for **6a**, and obtained in 69% yield (178 mg, 265 µmol) as a white amorphous solid after purification by column chromatography on silica gel (eluted with hexane/EtOAc = 1:1). [α]$^{22}_D$ −17 (c 1.3, CHCl$_3$); ^{1}H NMR (400 MHz, CDCl$_3$) δ 7.18–7.28 (m, 5H), 6.75 (d, 1H, J = 7.2 Hz), 5.40 (d, 1H, J = 8.5 Hz), 5.26 (d, 1H, J = 5.8 Hz), 5.22 (d, 1H, J = 5.8 Hz), 5.01–5.03 (m, 1H), 4.69–4.71 (m, 1H), 4.39–4.46 (m, 2H), 4.13–4.15 (m, 2H), 3.74 (s, 3H), 3.44 (s, 3H), 3.35–3.37 (m, 1H), 3.11 (dd, 1H, J = 14.2, 5.8 Hz), 2.99–3.02 (m, 2H), 2.82 (dd, 1H, J = 13.8, 6.3 Hz), 1.36 (s, 9H), 1.26 (d, 3H, J = 7.2 Hz), 0.95–0.97 (m,

2H), 0.00 (s, 9H); ^{13}C{^{1}H} NMR (100 MHz, CDCl$_3$) δ 171.4, 171.1, 169.6, 156.7, 155.3, 136.5, 129.3, 128.7, 127.0, 91.7, 80.3, 63.7, 58.1, 58.0, 55.8, 52.5, 52.3, 43.5, 38.1, 33.1, 28.2, 19.3, 17.7, −1.5; IR (neat) 3317, 2953, 1719, 1510, 1454, 1366, 1339, 1249, 1210, 1168, 1086, 1048, 933, 860, 837, 776, 699 cm$^{−1}$; HRMS[ESI] m/z calcd for C$_{30}$H$_{49}$N$_3$O$_{10}$NaSSi [M+Na]$^+$ 650.2800, found 650.2819.

To a solution of the MOM ester **6b** (158 mg, 235 µmol, 1.0 equiv) in 1,4-dioxane (4.20 mL) was added 4 M HCl/1,4-dioxane (0.6 mL) at 0 °C under argon atmosphere. After being stirred at room temperature for 2.5 h, the reaction mixture was concentrated in vacuo. The resulting residue was purified by column chromatography on silica gel (eluted with CH$_2$Cl$_2$/MeOH = 10:1) to afford the carboxylic acid **1b** (109 mg, 173 µmol, 74%) as a white amorphous solid. [α]^{21}$_D$ −12 (c 1.0, CHCl$_3$); ^{1}H NMR (400 MHz, DMSO-d_6) δ 8.21 (d, 1H, J = 7.2 Hz), 7.20–7.36 (m, 5H), 7.14–7.20 (m, 1H), 6.84 (d, 1H, J = 8.5 Hz), 4.36–4.45 (m, 1H), 4.16–4.35 (m, 2H), 3.99–4.09 (m, 2H), 3.64 (s, 3H), 3.19–3.28 (m, 1H), 2.90–3.06 (m, 2H), 2.66–2.83 (m, 2H), 1.27 (s, 9H), 1.21 (d, 3H, J = 6.5 Hz), 0.89–0.94 (m, 2H), 0.00 (s, 9H); ^{13}C{^{1}H} NMR (100 MHz, DMSO-d_6) δ 171.7, 170.8, 156.2, 155.1, 138.0, 129.2, 127.9, 126.1, 78.0, 66.2, 58.4, 55.6, 52.0, 51.9, 41.8, 37.4, 32.2, 28.1, 18.7, 17.3, −1.5; IR (neat) 3320, 2954, 1721, 1512, 1453, 1367, 1339, 1249, 1170, 1080, 1049, 859, 837, 699 cm$^{−1}$; HRMS[ESI] m/z calcd for C$_{28}$H$_{45}$N$_3$O$_9$NaSSi [M+Na]$^+$ 650.2538, found 650.2546.

3.3.3. The Tripeptide **1c**

Compound **6c** was prepared from the *N*-Fmoc-amine **5** (367 mg, 568 µmol) according to the procedure above described for **6a**, and obtained in 86% yield (388 mg, 488 µmol) as a white amorphous solid after purification by column chromatography on silica gel (eluted with hexane/EtOAc = 1:1). [α]^{20}$_D$ −28 (c 1.0, CHCl$_3$); ^{1}H NMR (400 MHz, CDCl$_3$) δ 7.73 (d, 2H, J = 7.5 Hz), 7.50 (t, 2H, J = 7.5 Hz), 7.37 (t, 2H, J = 7.5 Hz), 7.21–7.29 (m, 7H), 6.70–6.82 (m, 1H), 5.44 (d, 2H, J = 8.5 Hz), 5.23–5.26 (m, 2H), 4.71–4.72 (m, 1H), 4.46–4.48 (m, 2H), 4.39–4.42 Hz (m, 1H), 4.27–4.30 (m, 1H), 4.13–4.17 (m, 3H), 3.71 (s, 3H), 3.43 (s, 3H), 3.34–3.40 (m, 1H), 2.98–3.10 Hz (m, 3H), 2.81–2.85 (m, 1H), 1.24 (d, 3H, J = 7.1 Hz), 0.95–0.97 (m, 2H), −0.01 (s, 9H); ^{13}C{^{1}H} NMR (100 MHz, CDCl$_3$) δ 171.1, 171.0, 169.5, 156.6, 155.9, 143.71, 143.66, 141.2, 136.2, 129.3, 128.7, 127.7, 127.0, 125.0, 119.9, 91.7, 77.2, 67.1, 63.7, 58.0, 52.5, 52.2, 47.0, 43.3, 38.2, 32.8, 29.2, 19.2, 17.6, −1.6; IR (neat) 3310, 3064, 3025, 2953, 1719, 1670, 1517, 1450, 1412, 1381, 1338, 1287, 1249, 1217, 1154, 1084, 1047, 932, 860, 837, 757, 742, 700 cm$^{−1}$; HRMS[ESI] m/z calcd for C$_{40}$H$_{51}$N$_3$O$_{10}$NaSSi [M+Na]$^+$ 816.2957, found 816.2960.

Compound **1c** was prepared from the MOM ester **6c** (340 mg, 428 µmol) according to the procedure above described for **1a**, and obtained in 85% yield (272 mg, 362 µmol) as a white amorphous solid after purification by column chromatography on silica gel (eluted with CH$_2$Cl$_2$/MeOH = 50:1). [α]^{24}$_D$ −14 (c 0.96, CHCl$_3$); ^{1}H NMR (400 MHz, DMSO-d_6) δ 8.37 (d, 1H, J = 7.5 Hz), 7.87 (d, 2H, J = 7.5 Hz), 7.60–7.64 (m, 3H), 7.16–7.42 (m, 10H), 4.25–4.40 (m, 3H), 4.09–4.16 (m, 3H), 4.03–4.05 (m, 2H), 3.64 (s, 3H), 3.25–3.26 (m, 1H), 2.95–3.03 (m, 2H), 2.78–2.80 (m, 2H), 1.19 (d, 3H, J = 7.0 Hz), 0.90–0.92 (m, 2H), 0.00 (s, 9H); ^{13}C{^{1}H} NMR (100 MHz, DMSO-d_6) δ 171.8, 171.5, 170.8, 156.2, 155.6, 143.7, 143.6, 140.6, 138.0, 129.2, 127.9, 127.5, 127.0, 126.1, 125.23, 125.17, 120.0, 65.6, 62.2, 58.4, 56.0, 52.3, 51.8, 46.5, 41.7, 37.4, 32.2, 18.7, 17.3, −1.6; IR (neat) 3313, 3064, 3028, 2953, 1721, 1516, 1450, 1338, 1249, 1216, 1180, 1081, 1048, 859, 837, 757, 742, 699 cm$^{−1}$; HRMS [ESI] m/z calcd for C$_{38}$H$_{47}$N$_3$O$_9$NaSSi [M+Na]$^+$ 772.2694, found 772.2715.

3.3.4. The Tripeptide **1d**

Compound **6d** was prepared from the *N*-Fmoc-amine **5** (480 mg, 742 µmol) according to the procedure above described for **6a**, and obtained in 75% yield (344 mg, 558 µmol) as a white amorphous solid after purification by column chromatography on silica gel (eluted with hexane/EtOAc = 1:1 to hexane/acetone = 3:1). [α]^{23}$_D$ −16 (c 1.1, CHCl$_3$); ^{1}H NMR (400 MHz, CDCl$_3$) δ 7.25–7.32 (m, 5H), 7.01 (s, 1H), 5.67 (s, 1H), 5.48 (d, 1H, J = 9.2 Hz), 5.28 (d, 1H, J = 5.8 Hz), 5.23 (d, 1H, J = 5.8 Hz), 5.10 (s, 2H), 4.77 (dt, 1H, J = 6.3, 5.6 Hz), 4.48 (dd, 1H, J = 8.8, 3.0 Hz), 4.12–4.15 (m, 2H), 3.92 (d, 2H, J = 5.6 Hz), 3.72 (s, 3H), 3.36–3.44 (m,

4H), 3.03 (dd, 1H, J = 13.6, 4.0 Hz), 2.86 (dd, 1H, J = 13.9, 6.2 Hz), 1.27 (d, 3H, J = 7.2 Hz), 0.95–0.97 (m, 2H), −0.01 (s, 9H); ^{13}C{^{1}H} NMR (100 MHz, CDCl$_3$) δ 171.1, 169.8, 169.2, 156.6, 156.5, 136.1, 128.5, 128.1, 128.0, 91.7, 77.2, 67.1, 63.7, 57.9, 52.5, 52.1, 44.4, 43.3, 32.7, 19.1, 17.6, −1.6; IR (neat) 3325, 2953, 1723, 1515, 1453, 1381, 1339, 1249, 1156, 1088, 1048, 927, 860, 837, 754, 698 cm^{-1}; HRMS[ESI] m/z calcd for C$_{26}$H$_{41}$N$_3$O$_{10}$NaSSi [M+Na]$^+$ 638.2174, found 638.2180.

Compound **1d** was prepared from the MOM ester **6d** (215 mg, 349 μmol) according to the procedure above described for **1a**, and obtained in 45% yield (89.6 mg, 157 μmol) as a white amorphous solid after purification by column chromatography on silica gel (eluted with CH$_2$Cl$_2$/MeOH = 50:1 to 10:1). [α]^{24}$_D$ +2.7 (c 0.95, CHCl$_3$); ^{1}H NMR (400 MHz, DMSO-d_6) δ 12.9 (brs, 1H), 8.18 (d, 1H, J = 8.0 H), 7.30–7.44 (m, 7H), 5.02 (s, 2H), 4.37–4.40 (m, 1H), 4.23 (dd, 1H, J = 8.0, 6.0 Hz), 4.03–4.05 (m, 2H), 3.66 (d, 2H, J = 6.3 Hz), 3.63 (s, 3H), 3.18–3.20 (m, 1H), 2.91 (dd, 1H, J = 13.5, 5.1 Hz), 2.72 (dd, 1H, J = 13.5, 8.1 Hz), 1.18 (d, 3H, J = 6.8 Hz), 0.90–0.93 (m, 2H), 0.00 (s, 9H); ^{13}C{^{1}H} NMR (100 MHz, DMSO-d_6) δ 171.7, 170.7, 169.0, 156.4, 156.2, 137.0, 128.2, 127.7, 127.6, 65.4, 62.2, 58.4, 51.9, 51.8, 43.2, 41.8, 32.3, 18.6, 17.3, −1.6; IR (neat) 3327, 2953, 1722, 1523, 1453, 1438, 1340, 1249, 1081, 1050, 860, 837, 697 cm^{-1}; HRMS [ESI] m/z calcd for C$_{24}$H$_{37}$N$_3$O$_9$NaSSi [M+Na]$^+$ 594.1912, found 594.1924.

3.3.5. The Tripeptide 1e

To a solution of *N*-Fmoc-amine **5** (359 mg, 554 μmol, 1.0 equiv) in dry MeCN (4.4 mL) was added Et$_2$NH (1.1 mL) at room temperature under an argon atmosphere. After being stirred at the same temperature for 1.5 h, the reaction mixture was concentrated in vacuo. The resulting residue was azeotroped three times with MeCN, and the resulting crude amine was used for the next reaction without further purification.

To a solution of the crude amine in dry CH$_2$Cl$_2$ (5.5 mL) were added DIEA (193 μL, 1.11 mmol, 2.0 equiv), Cbz-Ile-OH (177 mg, 665 μmol, 1.2 equiv) and HATU (253 mg, 665 μmol, 1.2 equiv) at 0 °C under an argon atmosphere. After being stirred at room temperature for 7 h, the reaction mixture was diluted with CH$_2$Cl$_2$. The organic layer was washed with 10% aqueous citric acid and saturated aqueous NaHCO$_3$, dried over Na$_2$SO$_4$, and filtered. The filtrate was concentrated in vacuo, and the resulting residue was purified by column chromatography on silica gel (eluted with hexane/EtOAc = 1:1) to afford the tripeptide **6e** (373 mg, 539 mmol, 97% in 2 steps) as a yellowish amorphous solid. [α]^{23}$_D$ −19 (c 0.98, CHCl$_3$); ^{1}H NMR (400 MHz, CDCl$_3$) δ 7.29–7.34 (m, 5H), 6.82 (d, 1H, J = 6.5 Hz), 5.51 (d, 1H, J = 9.2 Hz), 5.43 (d, 1H, J = 8.7 Hz), 5.32 (d, 1H, J = 5.8 Hz), 5.27 (d, 1H, J = 5.8 Hz), 5.11 (s, 2H), 4.78–4.80 (m, 1H), 4.52 (dd, 1H, J = 9.3, 3.0 Hz), 4.14–4.17 (m, 3H), 3.75 (s, 3H), 3.48 (s, 3H), 3.37–3.48 (m, 1H), 3.04 (dd, 1H, J = 13.8, 4.3 Hz), 2.89 (dd, 1H, J = 13.8, 6.0 Hz), 1.89–1.91 (m, 1H), 1.60–1.70 (m, 1H) 1.30 (d, 3H, J = 7.0 Hz), 1.07–1.26 (m, 1H), 0.90–1.01 (m, 8H), 0.03 (s, 9H); ^{13}C{^{1}H} NMR (100 MHz, CDCl$_3$) δ 171.4, 171.2, 169.7, 156.7, 156.2, 136.2, 128.5, 128.1, 128.0, 91.7, 67.1, 63.7, 59.6, 58.1, 58.0, 52.5, 52.0, 43.5, 37.4, 32.9, 24.6, 19.3, 17.6, 15.4, 11.3, −1.6; IR (neat) 3311, 2959, 1723, 1666, 1524, 1454, 1382, 1339, 1284, 1248, 1156, 1087, 1045, 932, 860, 837, 697 cm^{-1}; HRMS[ESI] m/z calcd for C$_{30}$H$_{49}$N$_3$O$_{10}$NaSSi [M+Na]$^+$ 694.2800, found 694.2814.

Compound **1e** was prepared from the MOM ester **6e** (373 mg, 539 μmol) according to the procedure above described for **1a**, and obtained in 67% yield (228 mg, 362 μmol) as a white amorphous solid after purification by column chromatography on silica gel (eluted with CH$_2$Cl$_2$/MeCN = 3:1 to CH$_2$Cl$_2$/MeOH = 20:1). [α]^{21}$_D$ −6.5 (c 0.99, CHCl$_3$); ^{1}H NMR (400 MHz, DMSO-d_6) δ 12.8 (brs, 1H), 8.21 (d, 1H, J = 7.7 Hz), 7.24–7.33 (m, 7H), 5.02 (s, 2H), 4.35–4.38 (m, 1H), 4.25 (dd, 1H, J = 7.4, 5.7 Hz), 4.03–4.05 (m, 2H), 3.91–3.99 (m, 1H), 3.62–3.64 (m, 3H), 3.21 (s, 1H), 2.92 (dd, 1H, J = 13.0, 5.1 Hz), 2.72 (dd, 1H, J = 13.0, 8.2 Hz), 1.63–1.82 (m, 1H), 1.31–1.49 (m, 1H), 1.03–1.23 (m, 4H), 0.72–0.96 (m, 8H), 0.00 (s, 9H); ^{13}C{^{1}H} NMR (100 MHz, DMSO-d_6) δ 171.7, 171.2, 170.8, 156.2, 155. 9, 137.0, 128.2, 127.6, 127.5, 65.3, 62.2, 59.0, 58.4, 51.90, 51.85, 41.6, 36.6, 32.1, 24.2, 18.6, 17.3, 15.2, 10.8, −1.5; IR (neat) 3316, 2959, 1721, 1666, 1517, 1454, 1340, 1286, 1249, 1216, 1179, 1082, 1045, 859, 837 cm^{-1}; HRMS [ESI] m/z calcd for C$_{28}$H$_{45}$N$_3$O$_9$NaSSi [M+Na]$^+$ 650.2538, found 650.2551.

3.3.6. The Tripeptide **1f**

Compound **6f** was prepared from the *N*-Fmoc-amine **5** (268 mg, 414 μmol) according to the procedure above described for **6a**, and obtained in 74% yield (242 mg, 307 μmol) as a white amorphous solid after purification by column chromatography on silica gel (eluted with hexane/EtOAc = 1:1). $[\alpha]^{22}_D$ −21 (*c* 1.2, CHCl$_3$); ^{1}H NMR (400 MHz, CDCl$_3$) δ 7.25–7.32 (m, 5H), 6.92–6.94 (m, 1H), 5.47–5.49 (m, 2H), 5.30 (d, 1H, *J* = 5.8 Hz), 5.23 (d, 1H, *J* = 5.8 Hz), 5.08 (s, 2H), 4.74 (dt, 1H, *J* = 6.8, 5.4 Hz), 4.65 (brs, 1H), 4.48 (dd, 1H, *J* = 8.7, 2.4 Hz), 4.21–4.22 (m, 1H), 4.13–4.15 (m, 2H), 3.73 (s, 3H), 3.37–3.43 (m, 4H), 3.02–3.04 (m, 3H), 2.86 (dd, 1H, *J* = 13.2, 5.9 Hz), 1.25–1.85 (m, 18H), 0.95–0.97 (m, 2H), −0.01 (s, 9H); ^{13}C{^{1}H} NMR (100 MHz, CDCl$_3$) δ171.8, 171.2, 169.7, 156.7, 156.2, 156.1, 136.2, 128.4, 128.1, 128.0, 91.7, 79.0, 67.0, 63.7, 58.1, 57.9, 54.7, 52.5, 52.1, 43.3, 39.8, 32.7, 31.9, 29.6, 28.4, 22.3, 19.2, 17.6, −1.6; IR (neat) 3319. 2952, 1714, 1511, 1454, 1365, 1339, 1249, 1169, 1086, 1046, 860, 837 cm^{-1}; HRMS[ESI] *m*/*z* calcd for C$_{35}$H$_{58}$N$_4$O$_{12}$NaSSi [M+Na]$^+$ 809.3433, found 809.3434.

Compound **1f** was prepared from the MOM ester **6f** (215 mg, 279 μmol) according to the procedure above described for **1b**, and obtained in 87% yield (180 mg, 242 μmol) as a white amorphous solid after purification by column chromatography on silica gel (eluted with CH$_2$Cl$_2$/MeCN = 1:1 to CH$_2$Cl$_2$/MeOH = 10:1). $[\alpha]^{24}_D$ −13 (*c* 1.1, CHCl$_3$); ^{1}H NMR (400 MHz, DMSO-*d*$_6$) δ 7.91 (d, 1H, *J* = 7.2 Hz), 7.28–7.32 (m, 5H), 7.11–7.12 (m, 1H), 6.90–6.91 (m, 1H), 6.47–6.48 (m, 1H), 5.02 (s, 2H), 4.32–4.37 (m, 1H), 4.23 (dd, 1H, *J* = 8.1, 5.9 Hz), 3.99–4.06 (m, 3H), 3.63 (s, 3H), 3.19–3.25 (m, 1H), 2.93–3.02 (m, 3H), 2,74 (dd, 1H, *J* = 13.4, 7.6 Hz), 1.61–1.64 (m, 1H), 1.51–1.53 (m, 1H), 1.13–1.37 (m, 16H), 0.90–0.93 (m, 2H), 0.00 (s, 9H); ^{13}C{^{1}H} NMR (100 MHz, DMSO-*d*$_6$) δ 171.8, 171.7, 170.7, 156.1, 155.8, 155.4, 136.9, 128.2, 127.6, 127.5, 66.2, 65.3, 62.1, 58.4, 54.5, 52.0, 51.7, 41.6, 32.2, 31.6, 29.1, 28.1, 27.8, 22.6, 18.6, 17.2, −1.6; IR (neat) 3325, 2953, 1713, 1515, 1453, 1411, 1391, 1366, 1340, 1249, 1214, 1172, 1081, 1047, 860, 837, 754, 697 cm^{-1}; HRMS [ESI] *m*/*z* calcd for C$_{33}$H$_{54}$N$_4$O$_{11}$NaSSi [M+Na]$^+$ 765.3171, found 765.3179.

3.4. The Photocatalytic AviMeCys Formation Using 1

3.4.1. The β-Thioenamides (*Z*)-**9a** and (*E*)-**9a**

To a solution of the carboxylic acid **6a** (66.6 mg, 100 μmol, 1.0 equiv) and *N*-hydroxyphthalimide (18.1 mg, 110 μmol, 1.1 equiv) in dry CH$_2$Cl$_2$ (1.0 mL) was added DIC (17.3 μL, 110 μmol, 1.1 equiv) at room temperature under an argon atmosphere, and the mixture was stirred at the same temperature for 30 min. After complete consumption of **6a** (monitored by TLC analysis), the reaction mixture was cooled to −40 °C. A solution of eosin Y-Na$_2$ (0.7 mg, 1.00 μmol, 1 mol%), Hantzsch ester (25.5 mg, 100 μmol, 1.0 equiv) and diphenyl diselenide (62.8 mg, 200 μmol, 2.0 equiv) in dry DMF (1.5 mL, used immediately after freeze-pump-thaw cycling) was then added to the above mixture at −40 °C. After being stirred at the same temperature for 30 min under irradiated Blue LEDs, Et$_3$N (250 μL, 1.79 mmol, 18 equiv) was added to the solution at −40 °C. After being stirred at room temperature for 3 h, the reaction mixture was quenched with saturated aqueous NaHCO$_3$, and stirred for 12 h. The aqueous layer was extracted three times with Et$_2$O. The combined organic layers were dried over Na$_2$SO$_4$, and filtered. The filtrate was concentrated in vacuo, and the resulting residue was purified by preparative TLC (eluted with hexane/EtOAc = 7:2) to afford the β-thioenamide (*Z*)-**9a** (26.2 mg, 42.5 μmol, 42%) as a white amorphous solid and the β-thioenamide (*E*)-**9a** (19.6 mg, 31.8 μmol, 32%) as a white amorphous solid. (*Z*)-**9a**: $[\alpha]^{24}_D$ −41 (*c* 1.1, CHCl$_3$); ^{1}H NMR (600 MHz, CDCl$_3$) δ 8.42 (d, 1H, *J* = 10.6 Hz), 7.19–7.34 (m, 10H), 7.15 (dd, 1H, *J* = 10.6, 7.2 Hz), 5.49–5.50 (m, 1H), 5.40–5.41 (m, 1H), 5.32 (d, 1H, *J* = 7.2 Hz), 5.07–5.09 (m, 2H), 4.56–4.58 (m, 1H), 4.51 (dd, 1H, *J* = 9.2, 3.9 Hz), 4.14–4.15 (m, 2H), 3.65 (s, 3H), 3.31–3.32 (m, 1H), 3.19 (dd, 1H, *J* = 13.9, 6.2 Hz), 3.09–3.11 (m, 1H), 1.32 (d, 3H, *J* = 7.2 Hz), 0.95–0.97 (m, 2H), 0.00 (s, 9H); ^{13}C{^{1}H} NMR (150 MHz, CDCl$_3$) δ 171.3, 168.9, 156.8, 156.2, 136.2, 136.1, 129.7, 129.3, 128.9, 128.6, 128.3, 128.2, 127.2, 99.7, 67.3, 63.9, 58.9, 56.4, 52.6, 45.8, 38.1, 18.3, 17.8, −1.4; IR (neat) 3310, 3064, 3031, 2952, 1697, 1628, 1498, 1455, 1380, 1337, 1248, 1178, 1080, 1046, 860, 837, 742, 698 cm^{-1}; HRMS[ESI] *m*/*z* calcd for C$_{30}$H$_{41}$N$_3$O$_7$NaSSi [M+Na]$^+$ 638.2327, found 638.2330. (*E*)-**9a**: $[\alpha]^{23}_D$ −3.5

(*c* 0.89, CHCl$_3$); ^{1}H NMR (600 MHz, CDCl$_3$) δ 8.01 (m, 1H), 7.14–7.32 (m, 10H), 6.99 (dd, 1H, *J* = 13.7, 6.6 Hz), 5.59 (d, 1H, *J* = 13.7 Hz), 5.42–5.44 (m, 2H), 5.05 (s, 2H), 4.48 (dd, 1H, *J* = 5.5, 2.7 Hz), 4.40–4.42 (m, 1H), 4.13–4.17 (m, 2H), 3.64 (s, 3H), 3.34–3.38 (m, 1H), 3.06 (d, 2H, *J* = 6.8 Hz), 1.29 (d, 3H, *J* = 7.5 Hz), 1.00–1.01 (m, 2H), 0.03 (s, 9H); ^{13}C{^{1}H} NMR (150 MHz, CDCl$_3$) δ 171.2, 168.3, 156.8, 156.3, 136.0, 135.9, 129.4, 129.3, 129.0, 128.7, 128.5, 128.2, 127.4, 104.1, 67.5, 63.9, 57.9, 56.4, 52.5, 45.3, 38.2, 18.6, 17.8, −1.4; IR (neat) 3303, 2953, 1725, 1688, 1669, 1629, 1505, 1454, 1341, 1288, 1261, 1246, 1209, 1176, 1079, 1046, 938, 862, 835, 750, 697 cm^{-1}; HRMS[ESI] *m/z* calcd for C$_{30}$H$_{41}$N$_3$O$_7$NaSSi [M+Na]$^+$ 638.2327, found 638.2334.

3.4.2. The β-Thioenamides (*Z*)-9b and (*E*)-9b

Compounds (*Z*)-9b and (*E*)-9b were prepared from the carboxylic acid 6b (63.7 mg, 101 µmol) according to the procedure above described for 9a, and purified by preparative TLC (eluted with hexane/EtOAc = 7:2, hexane/IPA = 20:1) to be obtained in 32% yield (18.6 mg, 32.0 µmol) as a white amorphous solid and 32% yield (18.9 mg, 32.5 µmol) as a white amorphous solid, respectively. (*Z*)-9b: [α]$^{24}_D$ −49 (*c* 1.2, CHCl$_3$); ^{1}H NMR (600 MHz, CDCl$_3$) δ 8.42 (d, 1H, *J* = 10.9 Hz), 7.28–7.30 (m, 2H), 7.22–7.24 (m, 3H), 7.12–7.15 (m, 1H), 5.45–5.47 (m, 1H), 5.28–5.29 (m, 1H), 5.14–5.16 (m, 1H), 4.52 (d, 2H, *J* = 10.3 Hz), 4.14–4.17 (m, 2H), 3.70 (s, 3H), 3.34–3.36 (m, 1H), 3.18 (dd, 1H, *J* = 14.0, 5.8 Hz), 3.04–3.06 (m, 1H), 1.39 (s, 9H), 1.32 (d, 3H, *J* = 6.8 Hz), 0.98–0.99 (m, 2H), 0.01 (s, 9H); ^{13}C{^{1}H} NMR (150 MHz, CDCl$_3$) δ 171.3, 169.3, 156.8, 155.6, 136.5, 129.5, 129.3, 128.8, 127.1, 99.6, 80.6, 63.9, 58.7, 55.8, 52.6, 45.8, 37.9, 28.3, 18.4, 17.8, −1.4; IR (neat) 3324, 2953, 1690, 1627, 1499, 1366, 1338, 1285, 1249, 1171, 1080, 1047, 860, 837, 754, 699 cm^{-1}; HRMS [ESI] *m/z* calcd for C$_{27}$H$_{43}$N$_3$NaO$_7$SSi [M+Na]$^+$ 604.2483, found 604.2490. (*E*)-9b: [α]$^{24}_D$ −6.8 (*c* 1.1, CHCl$_3$); ^{1}H NMR (600 MHz, CDCl$_3$) δ 8.04–8.05 (m, 1H), 7.28 (t, 2H, *J* = 7.5 Hz), 7.22 (t, 1H, *J* = 7.5 Hz), 7.15 (d, 2H, *J* = 7.5 Hz), 7.00 (dd, 1H, *J* = 13.7, 10.9 Hz), 5.59 (d, 1H, *J* = 13.7 Hz), 5.42 (d, 1H, *J* = 6.0 Hz), 5.06 (d, 1H, *J* = 8.2 Hz), 4.47 (dd, 1H, *J* = 9.2, 3.1 Hz), 4.33 (s, 1H), 4.15–4.16 (m, 2H), 3.65 (s, 3H), 3.37–3.38 (m, 1H), 3.05–3.06 (m, 1H), 2.99–3.00 (m, 1H), 1.37 (s, 9H), 1.28 (d, 3H, *J* = 6.8 Hz), 0.98–1.00 (m, 2H), 0.04 (s, 9H); ^{13}C{^{1}H} NMR (150 MHz, CDCl$_3$) δ 171.2, 168.7, 156.8, 155.8, 136.3, 129.6, 129.3, 128.8, 127.2, 103.6, 80.8, 63.5, 57.9, 56.0, 52.5, 45.3, 38.0, 28.3, 18.5, 17.8, −1.4; IR (neat) 3310, 2954, 1723, 1681, 1628, 1513, 1454, 1391, 1367, 1339, 1289, 1248, 1211, 1169, 1079, 1047, 941, 860, 837, 754, 698 cm^{-1}; HRMS [ESI] *m/z* calcd for C$_{27}$H$_{43}$N$_3$NaO$_7$SSi [M+Na]$^+$ 604.2483, found 604.2487.

3.4.3. The β-Thioenamides (*Z*)-9c and (*E*)-9c

To a solution of the carboxylic acid 6c (75.2 mg, 100 µmol, 1.0 equiv) and *N*-hydroxyphthalimide (18.0 mg, 110 µmol, 1.1 equiv) in dry CH$_2$Cl$_2$ (1.0 mL) was added DIC (17.3 µL, 110 µmol, 1.1 equiv) at room temperature under an argon atmosphere, and the mixture was stirred at the same temperature for 5 h. After complete consumption of 5c (monitored by TLC analysis), the reaction mixture was cooled to −40 °C. A solution of Eosin Y (696 µg, 1.00 µmol, 1 mol%), Hantzsch ester (25.4 mg, 100 µmol, 1.0 equiv) and diphenyl diselenide (62.6 mg, 200 µmol, 2.0 equiv) in dry DMF (1.5 mL, used immediately after Freeze-Pump-Thaw cycling) was then added to the above mixture at −40 °C. After being stirred at the same temperature for 30 min under irradiated Blue LEDs, Et$_3$N (41.9 µL, 300 µmol, 3.0 equiv) was added to the solution at −40 °C. After being stirred at room temperature for 3 h, the reaction mixture was quenched with saturated aqueous NaHCO$_3$, diluted with Et$_2$O, and stirred for 12 h. The organic layer was separated, and the aqueous layer was extracted three times with Et$_2$O. The combined organic layers were dried over Na$_2$SO$_4$, and filtered. The filtrate was concentrated in vacuo, and the resulting residue was purified by preparative TLC (eluted with hexane/EtOAc = 7:2, hexane/IPA = 10:1) to afford the β-thioenamide (*Z*)-9c (14.6 mg, 20.7 µmol, 21%) as a white amorphous solid and the β-thioenamide (*E*)-9c (11.1 mg, 15.8 µmol, 16%) as a white amorphous solid. (*Z*)-9c: [α]$^{24}_D$ −25 (*c* 0.83, CHCl$_3$); ^{1}H NMR (600 MHz, CDCl$_3$) δ 8.41–8.42 (m, 1H), 7.75 (d, 2H, *J* = 7.5 Hz), 7.49–7.52 (m, 2H), 7.39 (t, 2H, *J* = 7.5 Hz), 7.25–7.28 (m, 7H), 7.17 (dd, 1H, *J* = 10.8, 7.6 Hz) 5.53–5.55 (m, 1H), 5.38–5.40 (m, 1H), 5.33 (d, 1H, *J* = 4.6 Hz), 4.59–4.61 (m, 1H), 4.50–4.52

(m, 1H), 4.42 (dd, 1H, J = 10.8, 7.5 Hz), 4.34–4.35 (m, 1H), 4.12–4.18 (m, 3H), 3.68 (s, 3H), 3.31–3.33 (m, 1H), 3.12–3.19 (m, 2H), 1.32 (d, 3H, J = 6.8 Hz), 0.93–0.95 (m, 2H), 0.01 (s, 9H); ^{13}C{^{1}H} NMR (150 MHz, CDCl$_3$) δ 171.4, 168.9, 156.8, 156.2, 143.83, 143.80, 141.4, 136.2, 129.8, 129.4, 129.0, 127.9, 127.4, 127.2, 125.2, 120.1, 99.8, 67.4, 64.0, 59.0, 56.4, 52.7, 47.2, 45.9, 38.3, 18.3, 17.8, −1.4; IR (neat) 3308, 3064, 3028, 2952, 1696, 1628, 1499, 1451, 1336, 1248, 1178, 1080, 1045, 859, 837, 757, 740, 699 cm^{-1}; HRMS [ESI] m/z calcd for C$_{37}$H$_{45}$N$_3$O$_7$NaSSi [M+Na]$^+$ 726.2640, found 726.2649. (E)-**9c**: [α]^{24}$_D$ −16 (c 0.59, CHCl$_3$); ^{1}H NMR (600 MHz, CDCl$_3$) δ 7.76 (d, 2H, J = 7.5 Hz), 7.50 (t, 2H, J = 7.5 Hz), 7.40 (t, 2H, J = 7.5 Hz), 7.25–7.29 (m, 8H), 6.95–6.99 (m, 1H), 5.58 (d, 1H, J = 13.7 Hz), 5.39 (d, 1H, J = 9.6 Hz), 5.26–5.28 (m, 1H), 4.45–4.47 (m, 2H), 4.34–4.36 (m, 2H), 4.15–4.17 (m, 3H), 3.63 (s, 3H), 3.36–3.37 (m, 1H), 3.06–3.07 (m, 2H), 1.28 (d, 3H, J = 6.8 Hz), 0.99–1.00 (m, 2H), 0.04 (s, 9H); ^{13}C{^{1}H} NMR (150 MHz, CDCl$_3$) δ 171.1, 168.0, 156.7, 156.3, 143.6, 141.4, 136.0, 129.3, 129.0, 127.9, 127.4, 127.2, 125.0, 120.1, 104.2, 67.3, 63.8, 57.9, 56.3, 52.5, 47.1, 45.2, 38.0, 18.5, 17.8, −1.4; IR (neat) 3316, 3064, 3027, 2950, 1691, 1626, 1509, 1450, 1338, 1289, 1247, 1216, 1079, 1045, 937, 859, 837, 757, 739, 700 cm^{-1}; HRMS [ESI] m/z calcd for C$_{37}$H$_{45}$N$_3$O$_7$NaSSi [M+Na]$^+$ 726.2640, found 726.2643.

3.4.4. The β-Thioenamides (Z)-**9d** and (E)-**9d**

Compounds (Z)-**9d** and (E)-**9d** were prepared from the carboxylic acid **6d** (57.1 mg, 99.9 µmol) according to the procedure above described for **9a**, and purified by preparative TLC (eluted with hexane/EtOAc = 7:2, toluene/acetone = 8:1, hexane/IPA = 10:1) to be obtained in 39% yield (20.2 mg, 38.4 µmol) as a white amorphous solid and 19% yield (9.9 mg, 19 µmol) as a white amorphous solid, respectively. (Z)-**9d**: [α]^{23}$_D$ +3.0 (c 1.0, CHCl$_3$); ^{1}H NMR (600 MHz, CDCl$_3$) δ 8.57–8.59 (m, 1H), 7.25–7.34 (m, 5H), 7.14 (dd, 1H, J = 10.9, 7.5 Hz), 5.75–5.77 (m, 1H), 5.59–5.61 (m, 1H), 5.32 (d, 1H, J = 7.5 Hz), 5.18 (d, 1H, J = 12.3 Hz), 5.14 (d, 1H, J = 12.3 Hz), 4.54 (dd, 1H, J = 8.9, 2.7 Hz), 4.12–4.14 (m, 2H), 3.98 (d, 2H, J = 5.5 Hz), 3.64 (s, 3H), 3.35–3.37 (m, 1H), 1.34 (d, 3H, J = 6.8 Hz), 0.93–0.95 (m, 2H), −0.01 (s, 9H); ^{13}C{^{1}H} NMR (150 MHz, CDCl$_3$) δ 171.3, 167.1, 156.9, 156.8, 136.1, 129.3, 128.6, 128.4, 128.2, 100.1, 67.5, 63.9, 58.7, 52.5, 46.4, 44.9, 18.5, 17.8, −1.4; IR (neat) 3320, 2952, 1700, 1629, 1517, 1338, 1248, 1176, 1081, 1046, 860, 837, 738, 697 cm^{-1}; HRMS [ESI] m/z calcd for C$_{23}$H$_{35}$N$_3$O$_7$NaSSi [M+Na]$^+$ 548.1857, found 548.1864. (E)-**9d**: [α]^{23}$_D$ +5.0 (c 0.50, CHCl$_3$); ^{1}H NMR (600 MHz, CDCl$_3$) δ 8.31–8.32 (m, 1H), 7.31–7.36 (m, 5H), 7.05 (dd, 1H, J = 13.7, 10.9 Hz), 5.72 (d, 1H, J = 13.7 Hz), 5.55–5.56 (m, 1H), 5.45 (d, 1H, J = 8.9 Hz), 5.12 (s, 2H), 4.48 (dd, 1H, J = 8.9, 3.4 Hz), 4.15–4.16 (m, 2H), 3.87 (d, 2H, J = 5.5 Hz), 3.70 (s, 3H), 3.37–3.39 (m, 1H), 1.29 (d, 3H, J = 6.8 Hz), 0.98–1.00 (m, 2H), 0.03, (s, 9H); ^{13}C{^{1}H} NMR (150 MHz, CDCl$_3$) δ 171.3, 166.4, 156.9, 156.7, 135.9, 129.5, 128.7, 128.5, 128.2, 103.7, 67.6, 63.8, 57.9, 52.6, 45.3, 44.8, 18.4, 17.8, −1.4; IR (neat) 3311, 2953, 1696, 1627, 1512, 1454, 1338, 1248, 1174, 1080, 1047, 860, 837, 697 cm^{-1}; HRMS [ESI] m/z calcd for C$_{23}$H$_{35}$N$_3$O$_7$NaSSi [M+Na]$^+$ 548.1857, found 548.1866.

3.4.5. The β-Thioenamides (Z)-**9e** and (E)-**9e**

Compounds (Z)-**9e** and (E)-**9e** were prepared from the carboxylic acid **6e** (62.2 mg, 99.1 µmol) according to the procedure above described for **9a**, and purified by preparative TLC (eluted with hexane/EtOAc = 7:2) to be obtained in 36% yield (20.5 mg, 35.2 µmol) as a white amorphous solid and 32% yield (18.3 mg, 31.5 µmol) as a white amorphous solid, respectively. (Z)-**9e**: [α]^{24}$_D$ −45 (c 1.1, CHCl$_3$); ^{1}H NMR (600 MHz, CDCl$_3$) δ 8.52 (d, 1H, J = 10.6 Hz), 7.30–7.33 (m, 5H), 7.19 (dd, 1H, J = 10.6, 7.5 Hz), 5.66 (d, 1H, J = 8.2 Hz), 5.41 (d, 1H, J = 10.0 Hz), 5.34 (d, 1H, J = 7.5 Hz), 5.15 (d, 1H, J = 12.3 Hz), 5.10 (d, 1H, J = 12.3 Hz), 4.55 (dd, 1H, J = 10.0, 5.0 Hz), 4.26–4.27 (m, 1H), 4.11–4.15 (m, 2H), 3.61 (s, 3H), 3.33–3.34 (m, 1H), 2.03–2.04 (m, 1H), 1.48–1.49 (m, 1H), 1.37 (d, 3H, J = 6.8 Hz), 1.18–1.24 (m, 1H), 0.97 (d, 3H, J = 7.6 Hz), 0.94–0.97 (m, 2H), 0.91 (t, 3H, J = 7.6 Hz), 0.01 (s, 9H); ^{13}C{^{1}H} NMR (150 MHz, CDCl$_3$) δ 171.5, 169.3, 156.8, 156.5, 136.2, 130.0, 128.6, 128.34, 128.26, 99.1, 67.4, 63.9, 60.1, 59.2, 52.6, 45.6, 37.3, 24.6, 18.2, 17.8, 15.7, 11.6, −1.4; IR (neat) 3314, 3065, 3033, 2959, 1720, 1695, 1628, 1512, 1381, 1337, 1283, 1248, 1178, 1127, 1080, 1043, 938, 860, 837,

773, 738, 696 cm^{-1}; HRMS [ESI] m/z calcd for C$_{27}$H$_{43}$N$_3$NaO$_7$SSi [M+Na]$^+$ 604.2483, found 604.2487. (E)-**9e**: [α]^{24}$_D$ +8.8 (c 0.78, CHCl$_3$); ^{1}H NMR (600 MHz, CDCl$_3$) δ 8.35–8.37 (m, 1H), 7.30–7.35 (m, 5H), 7.04 (dd, 1H, J = 13.7, 10.3 Hz), 5.71 (d, 1H, J = 13.7 Hz), 5.46–5.47 (m, 2H), 5.10 (d, 1H, J = 12.0 Hz), 5.05 (d, 1H, J = 12.0 Hz), 4.48 (dd, 1H, J = 8.9, 3.4 Hz), 4.14–4.16 (m, 2H), 4.01 (t, 1H, J = 7.9 Hz), 3.65 (s, 3H), 3.40–3.41 (m, 1H), 1.85–1.93 (m, 1H), 1.48–1.50 (m, 1H), 1.29 (d, 3H, J = 6.8 Hz), 1.08–1.11 (m, 1H), 0.98–1.00 (m, 2H), 0.91 (d, 3H, J = 6.8 Hz), 0.87 (t, 3H, J = 7.5 Hz), 0.02 (s, 9H); ^{13}C{^{1}H} NMR (150 MHz, CDCl$_3$) δ 171.3, 168.8, 156.8, 156.7, 136.0, 129.5, 128.7, 128.4, 128.1, 103.9, 67.4, 63.8, 59.9, 57.9, 52.5, 45.4, 37.1, 24.8, 18.6, 17.8, 15.6, 11.3, −1.4; IR (neat) 3297, 2960, 2929, 1747, 1714, 1692, 1664, 1630, 1515, 1455, 1380, 1335, 1283, 1242, 1211, 1176, 1081, 1041, 862, 836, 697 cm^{-1}; HRMS [ESI] m/z calcd for C$_{27}$H$_{43}$N$_3$NaO$_7$SSi [M+Na]$^+$ 604.2483, found 604.2485.

3.4.6. The β-Thioenamides (Z)-**9f** and (E)-**9f**

Compounds (Z)-**9f** and (E)-**9f** were prepared from the carboxylic acid **6f** (74.9 mg, 101 μmol) according to the procedure above described for **9a**, and purified by preparative TLC (eluted with hexane/EtOAc = 3.5:1 to 2:1, hexane/IPA = 20:1, toluene/acetone = 20:1, hexane/tBuOH = 9:1) to be obtained in 28% yield (19.9 mg, 28.6 μmol) as a white amorphous solid and 16% yield (11.5 mg, 16.5 μmol) as a white amorphous solid, respectively. (Z)-**9f**: [α]^{23}$_D$ −7.6 (c 1.0, CHCl$_3$); ^{1}H NMR (600 MHz, CDCl$_3$) δ 8.58 (d, 1H, J = 10.3 Hz), 7.30–7.34 (m, 5H), 7.14 (dd, 1H, J = 10.3, 6.8 Hz), 5.63–5.65 (m, 2H), 5.31 (d, 1H, J = 6.8 Hz), 5.15 (d, 1H, J = 12.3 Hz), 5.10 (d, 1H, J = 12.3 Hz), 4.62–4.61 (m, 1H), 4.55 (dd, 1H, J = 9.6, 3.4 Hz), 4.24–4.25 (m, 1H), 4.12–4.14 (m, 2H), 3.63 (s, 3H), 3.38–3.40 (m, 1H), 3.16–3.18 (m, 1H), 3.07–3.08 (m, 1H), 1.96–1.97 (m, 1H), 1.82–1.83 (m, 1H), 1.69–1.70 (m, 1H), 1.41–1.48 (m, 12H), 1.35 (d, 3H, J = 7.5 Hz), 0.95–0.96 (m, 2H), 0.03 (s, 9H); ^{13}C{^{1}H} NMR (150 MHz, CDCl$_3$) δ 171.4, 169.7, 156.9, 156.6, 156.4, 136.2, 129.6, 128.6, 128.3, 99.9, 79.3, 67.4, 63.8, 58.8, 55.3, 52.5, 46.1, 39.7, 31.4, 29.7, 28.5, 22.5, 18.5, 17.8, −1.4; IR (neat) 3325, 2952, 1698, 1628, 1522, 1365, 1337, 1249, 1172, 1080, 1045, 860, 837, 754, 697 cm^{-1}; HRMS [ESI] m/z calcd for C$_{32}$H$_{52}$N$_4$O$_9$NaSSi [M+Na]$^+$ 719.3116, found 719.3137. (E)-**9f**: [α]^{23}$_D$ +0.1 (c 0.58, CHCl$_3$); ^{1}H NMR (600 MHz, CDCl$_3$) δ 8.57–8.58 (m, 1H), 7.30–7.35 (m, 5H), 7.04 (dd, 1H, J = 13.3, 10.6 Hz), 5.71 (d, 1H, J = 13.3 Hz), 5.60–5.62 (m, 1H), 5.44 (d, 1H, J = 10.6 Hz), 5.08 (s, 2H), 4.68–4.70 (m, 1H), 4.47 (dd, 1H, J = 8.9, 3.4 Hz), 4.14–4.16 (m, 3H), 3.66 (s, 3H), 3.38–3.40 (m, 1H), 3.06–3.08 (m, 2H), 1.96–1.98 (m, 1H), 1.83–1.85 (m, 1H), 1.63–1.65 (m, 1H), 1.28–1.48 (m, 3H), 1.40 (s, 9H), 1.29 (d, 3H, J = 6.8 Hz), 0.98–1.00 (m, 2H), 0.02 (s, 9H); ^{13}C{^{1}H} NMR (150 MHz, CDCl$_3$) δ 171.3, 169.3, 156.8, 156.7, 156.5, 136.0, 129.9, 128.6, 128.4, 128.2, 103.5, 79.5, 67.4, 63.8, 57.9, 54.8, 52.5, 45.4, 39.4, 31.3, 29.5, 28.5, 22.3, 18.6, 17.8, −1.4; IR (neat) 3313, 2952, 1694, 1628, 1513, 1454, 1365, 1337, 1248, 1172, 1081, 1045, 860, 837, 754 cm^{-1}; HRMS [ESI] m/z calcd for C$_{32}$H$_{52}$N$_4$O$_9$NaSSi [M+Na]$^+$ 719.3116, found 719.3125.

4. Conclusions

In summary, we have demonstrated the formation of AviMeCys using lanthionine-bearing peptides **1**. The decarboxylative selenoetherification of the NHPI esters **7** smoothly proceeded at a low temperature (−40 °C) in the presence of 1 mol% of eosin Y-Na$_2$ as a photocatalyst, and subsequent β-elimination in a one-pot operation resulted in the corresponding β-thioenamides **9** in moderate to good yields without losing the sulfide-bridged motif. The carbamate-type protecting groups, such as Cbz, Teoc, Boc and Fmoc groups, were tolerant under the reaction conditions, suggesting that the reaction should be useful in synthesizing structurally complicated RiPPs. The synthesis of neothioviridamide [49,50] and its related natural products [51–54] is underway.

Supplementary Materials: The following supporting information can be downloaded at: https://www.mdpi.com/article/10.3390/catal12121615/s1, Figure S1: Experimental setup for photocatalyzed oxidative decarboxylation; Figure S2. Emission spectrum for Kessil A160WE Tuna Blue; ^{1}H and ^{13}C{^{1}H} NMR spectra of synthetic compounds.

Author Contributions: Conceptualization, M.K., K.O. and T.D.; methodology, M.K., K.O. and T.D.; formal analysis, M.K. and K.O.; investigation, M.K. and K.O.; data curation, M.K. and K.O.; writing—original draft preparation, M.K., K.O. and T.D.; writing—review and editing, M.K., K.O. and T.D.; supervision, T.D.; project administration, T.D.; funding acquisition, M.K., K.O. and T.D. All authors have read and agreed to the published version of the manuscript.

Funding: This research was funded by JSPS KAKENHI Grant Number JP21K15216 and 22H02740, and JST SPRING Grant Number JPMJSP2114 (Tohoku University Advanced Graduate School Pioneering Research Support Project for PhD Students). This research was partially supported by Research Support Project for Life Science and Drug Discovery (Basis for Supporting Innovative Drug Discovery and Life Science Research (BINDS)) from AMED under Grant Number JP22ama121038.

Data Availability Statement: Not applicable.

Conflicts of Interest: The authors declare no conflict of interest.

References

1. Montalbán-López, M.; Scott, T.A.; Ramesh, S.; Rahman, I.R.; Van Heel, A.J.; Viel, J.H.; Bandarian, V.; Dittmann, E.; Genilloud, O.; Goto, Y.; et al. New developments in RiPP discovery, enzymology and engineering. *Nat. Prod. Rep.* **2021**, *38*, 130–239. [CrossRef] [PubMed]
2. Cao, L.; Do, T.; Link, A.J. Mechanisms of action of ribosomally synthesized and posttranslationally modified peptides (RiPPs). *J. Ind. Microbiol. Biotechnol.* **2021**, *1*, 1524–1540. [CrossRef] [PubMed]
3. Hetrick, K.J.; van der Donk, W.A. Ribosomally synthesized and post-translationally modified peptide natural product discovery in the genomic era. *Curr. Opin. Chem. Biol.* **2017**, *38*, 36–44. [CrossRef]
4. Katz, B.A.; Johnson, C.; Cass, R.T. Structure-based design of high affinity streptavidin binding cyclic peptide ligands containing thioether crosslinks. *J. Am. Chem. Soc.* **1995**, *117*, 8541–8547. [CrossRef]
5. Aimetti, A.A.; Shoemaker, R.K.; Lin, C.-C.; Anseth, K.S. On-resin peptide macrocyclization using thiol–ene click chemistry. *Chem. Commun.* **2010**, *46*, 4061–4063. [CrossRef]
6. Galande, A.K.; Bramlett, K.S.; Burris, T.P.; Wittliff, J.L.; Spatola, A.F. Thioether side chain cyclization for helical peptide formation: Inhibitors of estrogen receptor-coactivator interactions. *J. Pept. Res.* **2004**, *63*, 297–302. [CrossRef]
7. Rathman, B.M.; Del Valle, J.R. Late-Stage Sidechain-to-Backbone Macrocyclization of N-Amino Peptides. *Org. Lett.* **2022**, *24*, 1536–1540. [CrossRef]
8. Chatterjee, C.; Paul, M.; Xie, L.; van der Donk, W.A. Biosynthesis and Mode of Action of Lantibiotics. *Chem. Rev.* **2005**, *105*, 633–684. [CrossRef]
9. Smith, L.; Hillman, J.D. Therapeutic potential of type A (I) lantibiotics, a group of cationic peptide antibiotics. *Curr. Opin. Macrobiol.* **2008**, *11*, 401–408. [CrossRef]
10. Sit, C.S.; Yoganathan, S.; Vederas, J.C. Biosynthesis of aminovinyl-cysteine-containing peptides and its application in the production of potential drug candidates. *Acc. Chem. Res.* **2011**, *44*, 261–268. [CrossRef]
11. Grant-Mackie, E.S.; Williams, E.T.; Harris, P.W.R.; Brimble, M.A. Aminovinyl Cysteine Containing Peptides: A Unique Motif That Imparts Key Biological Activity. *JACS Au* **2021**, *1*, 1527–1540. [CrossRef]
12. De Leon Rodriguez, L.M.; Williams, E.T.; Brimble, M.A. Chemical Synthesis of Bioactive Naturally Derived Cyclic Peptides Containing Ene-Like Rigidifying Motifs. *Chem. Eur. J.* **2018**, *24*, 17869–17880. [CrossRef] [PubMed]
13. Aydillo, C.; Avenoza, A.; Busto, J.H.; Jiménez-Osés, G.; Pergrina, J.M.; Zurbano, M.M. A biomimetic approach to lanthionines. *Org. Lett.* **2012**, *14*, 334–337. [CrossRef] [PubMed]
14. Gutiérrez-Jiménez, M.I.; Aydillo, C.; Navo, C.D.; Avenoza, A.; Corzana, F.; Jiménez-Osés, G.; Zurbano, M.M.; Busto, J.H.; Peregrina, J.M. Bifunctional Chiral Dehydroalanines for Peptide Coupling and Stereoselective S-Michael Addition. *Org. Lett.* **2016**, *18*, 2796–2799. [CrossRef] [PubMed]
15. Zhou, H.; van der Donk, W.A. Biomimetic Stereoselective Formation of Methyllanthionine. *Org. Lett.* **2002**, *4*, 1335–1338. [CrossRef] [PubMed]
16. Sikandar, A.; Lopatniuk, M.; Luzhetskyy, A.; Müller, R.; Koehnke, J. Total In Vitro Biosynthesis of the Thioamitide Thioholgamide and Investigation of the Pathway. *J. Am. Chem. Soc.* **2022**, *144*, 5136–5144. [CrossRef]
17. Lutz, J.; Don, V.S.; Kumar, R.; Taylor, C.M. Influence of sulfur on acid-mediated enamide formation. *Org. Lett.* **2017**, *19*, 5146–5149. [CrossRef]
18. Lutz, J.; Taylor, C.M. Synthesis of the Aminovinylcysteine-Containing C-Terminal Macrocycle of the Linaridins. *Org. Lett.* **2020**, *22*, 1874–1877. [CrossRef]
19. Banerjee, B.; Litvinov, D.N.; Kang, J.; Bettale, J.D.; Castle, S.L. Stereoselective additions of thiyl radicals to terminal ynamides. *Org. Lett.* **2010**, *12*, 2650–2652. [CrossRef]
20. Shimizu, I.; Tsuji, J. Palladium-Catalyzed Decarboxylation-Dehydrogenation of Allyl β-Keto Carboxylates and Allyl Enol Carbonates as a Novel Synthetic Method for α-Substituted α,β-Unsaturated Ketones. *J. Am. Chem. Soc.* **1982**, *104*, 5844–5846. [CrossRef]

21. García-Reynaga, P.; Carrillo, K.A.; VanNieuwenhze, S.M. Decarbonylative Approach to the Synthesis of Enamides from Amino Acids: Stereoselective Synthesis of the (Z)-Aminovinyl-D-Cysteine Unit of Mersacidin. *Org. Lett.* **2012**, *14*, 1030–1033. [CrossRef] [PubMed]
22. Carrillo, K.A.; Van Nieuwenhze, S.M. Synthesis of the AviMeCys-Containing D-Ring of Mersacidin. *Org. Lett.* **2012**, *14*, 1034–1037. [CrossRef] [PubMed]
23. King, T.A.; Mandrup Kandemir, J.; Walsh, S.J.; Spring, D.R. Photocatalytic methods for amino acid modification. *Chem. Soc. Rev.* **2021**, *50*, 39–57. [CrossRef] [PubMed]
24. Bottecchia, C.; Noël, T. Photocatalytic Modification of Amino Acids, Peptides, and Proteins. *Chem. Eur. J.* **2019**, *25*, 26–42. [CrossRef] [PubMed]
25. Malins, L.R. Decarboxylative couplings as versatile tools for late-stage peptide modifications. *Pep. Sci.* **2018**, *110*, e24049. [CrossRef]
26. Murarka, S. N-(Acyloxy)phthalimides as Redox-Active Esters in Cross-Coupling Reactions. *Adv. Synth. Catal.* **2018**, *360*, 1735–1753. [CrossRef]
27. Li, Z.; Gentry, Z.; Murphy, B.; Vannieuwenhze, M.S. Scalable synthesis of orthogonally protected β-methyllanthionines by indium(III)-mediated ring opening of aziridines. *Org. Lett.* **2019**, *21*, 2200–2203. [CrossRef]
28. Russell, G.A.; Tashtoush, H. Free-radical chain-substitution reactions of alkylmercury halides. *J. Am. Chem. Soc.* **1983**, *105*, 1398–1399. [CrossRef]
29. Perkins, M.J.; Turner, E.S. S_H2 reactions of diphenyl diselenide; preparation and reactions of bridgehead selenides. *J. Chem. Soc. Chem. Commun.* **1981**, *3*, 139–140. [CrossRef]
30. Russell, G.A.; Ngoviwatchai, P.; Tashtoush, H.I.; Pla-Dalmau, A.; Khanna, R.K. Reactions of alkylmercurials with heteroatom-centered acceptor radicals. *J. Am. Chem. Soc.* **1988**, *110*, 3530–3538. [CrossRef]
31. Jiang, M.; Yang, H.; Fu, H. Visible-Light Photoredox Synthesis of Chiral α-Selenoamino Acids. *Org. Lett.* **2016**, *18*, 1968–1971. [CrossRef] [PubMed]
32. Huang, Z.; Lumb, J.P. Mimicking oxidative radical cyclizations of lignan biosynthesis using redox-neutral photocatalysis. *Nat. Chem.* **2021**, *13*, 24–32. [CrossRef] [PubMed]
33. Mautner, H.G.; Chu, S.-H.; Gunther, W.H.H. The Aminolysis of Thioacyl and Selenoacyl Analogs. *J. Am. Chem. Soc.* **1963**, *85*, 3458–3462. [CrossRef]
34. Durek, T.; Alewood, P.F. Preformed selenoesters enable rapid native chemical ligation at intractable sites. *Angew. Chem. Int. Ed.* **2011**, *50*, 12042–12045. [CrossRef]
35. Raj, M.; Wu, H.; Blosser, S.L.; Vittoria, M.A.; Arora, P.S. Aldehyde capture ligation for synthesis of native peptide bonds. *J. Am. Chem. Soc.* **2015**, *137*, 6932–6940. [CrossRef]
36. Okada, K.; Okamoto, K.; Morita, N.; Okubo, K.; Oda, M. Photosensitized Decarboxylative Michael Addition through N-(Acyloxy)phthalimides via an Electron-Transfer Mechanism. *J. Am. Chem. Soc.* **1991**, *113*, 9401–9402. [CrossRef]
37. Pratsch, G.; Lackner, G.L.; Overman, L.E. Constructing Quaternary Carbons from N-(Acyloxy)phthalimide Precursors of Tertiary Radicals Using Visible-Light Photocatalysis. *J. Org. Chem.* **2015**, *80*, 6025–6036. [CrossRef]
38. Srivastava, V.; Singh, P.P. Eosin Y Catalysed Photoredox Synthesis: A Review. *RSC Adv.* **2017**, *7*, 31377–31392. [CrossRef]
39. Zhang, J.; Li, Y.; Xu, R.; Chen, Y. Donor–Acceptor Complex Enables Alkoxyl Radical Generation for Metal-Free $C(sp^3)$–$C(sp^3)$ Cleavage and Allylation/Alkenylation. *Angew. Chem. Int. Ed.* **2017**, *56*, 12619–12623. [CrossRef]
40. Crisenza, G.E.M.; Mazzarella, D.; Melchiorre, P. Synthetic Methods Driven by the Photoactivity of Electron Donor-Acceptor Complexes. *J. Am. Chem. Soc.* **2020**, *142*, 5461–5476. [CrossRef]
41. Van Bergen, T.J.; Hedstrand, D.M.; Kruizinga, W.H.; Kellogg, R.M. Hydride Transfer from 1,4-Dihydropyridine to sp^3-Hybridized Carbon in Sulfonium Salts and Activated Halides. Studies with NAD(P)H Models. *J. Org. Chem.* **1979**, *44*, 4953–4962. [CrossRef]
42. Nakamura, K.; Fujii, M.; Mekata, H.; Oka, S.; Ohno, A. Desulfonylation of β-keto sulfones with the hantzsch ester, an nad(p)h model. *Chem. Lett.* **1986**, *15*, 87–88. [CrossRef]
43. Ortiz-López, F.J.; Carretero-Molina, D.; Sánchez-Hidalgo, M.; Martín, J.; González, I.; Román-Hurtado, F.; de la Cruz, M.; García-Fernández, S.; Reyes, F.; Deisinger, J.P.; et al. Cacaoidin, First Member of the New Lanthidin RiPP Family. *Angew. Chem. Int. Ed.* **2020**, *59*, 12654–12658. [CrossRef] [PubMed]
44. Chatterjee, S.; Chatterjee, D.K.; Jani, R.H.; Blumbach, J.; Ganguli, B.N.; Klesel, N.; Limbert, M.; Seibert, G. Mersacidin, a New Antibiotic from *Bacillus*. *J. Antibiot.* **1992**, *45*, 839–845. [CrossRef]
45. Xu, M.; Zhang, F.; Cheng, Z.; Bashiri, G.; Wang, J.; Hong, J.; Wang, Y.; Xu, L.; Chen, X.; Huang, S.-X.; et al. Functional Genome Mining Reveals a Class V Lanthipeptide Containing a D-Amino Acid Introduced by an $F_{420}H_2$-Dependent Reductase. *Angew. Chem. Int. Ed.* **2020**, *59*, 18029–18035. [CrossRef]
46. Elkhalifa, M.; Elbaum, M.B.; Chenoweth, D.M.; Molander, G.A. Solid-Phase Photochemical Decarboxylative Hydroalkylation of Peptides. *Org. Lett.* **2021**, *23*, 8219–8223. [CrossRef]
47. Milewska, K.D.; Malins, L.R. Synthesis of Amino Acid α-Thioethers and Late-Stage Incorporation into Peptides. *Org. Lett.* **2022**, *24*, 3680–3685. [CrossRef]
48. Schwarz, M.K.; Tumelty, D.; Gallop, M.A. Solid-Phase Synthesis of 3,5-Disubstituted 2,3-Dihydro-1,5-benzothiazepin-4(5H)-ones. *J. Org. Chem.* **1999**, *64*, 2219–2231. [CrossRef]

49. Kawahara, T.; Izumikawa, M.; Kozone, I.; Hashimoto, J.; Kagaya, N.; Koiwai, H.; Komatsu, M.; Fujie, M.; Sato, N.; Ikeda, H.; et al. Neothioviridamide, a Polythioamide Compound Produced by Heterologous Expression of a *Streptomyces* sp. Cryptic RiPP Biosynthetic Gene Cluster. *J. Nat. Prod.* **2018**, *81*, 264–269. [CrossRef]
50. Kjaerulff, L.; Sikandar, A.; Zaburannyi, N.; Adam, S.; Herrmann, J.; Koehnke, J.; Müller, R. Thioholgamides: Thioamide-Containing Cytotoxic RiPP Natural Products. *ACS Chem. Biol.* **2017**, *12*, 2837–2841. [CrossRef]
51. Hayakawa, Y.; Sasaki, K.; Adachi, H.; Furihata, K.; Nagai, K.; Shin-ya, K. Thioviridamide, a Novel Apoptosis Inducer in Transformed Cells from *Streptomyces olivoviridis*. *J. Antibiot.* **2006**, *59*, 1–5. [CrossRef] [PubMed]
52. Izumikawa, M.; Kozone, I.; Hashimoto, J.; Kagaya, N.; Takagi, M.; Koiwai, H.; Komatsu, M.; Fujie, M.; Satoh, N.; Ikeda, H.; et al. Novel thioviridamide derivative—JBIR-140: Heterologous expression of the gene cluster for thioviridamide biosynthesis. *J. Antibiot.* **2015**, *68*, 533–536. [CrossRef] [PubMed]
53. Frattaruolo, L.; Lacret, R.; Cappello, A.R.; Truman, A.W. A Genomics-Based Approach Identifies a Thioviridamide-Like Compound with Selective Anticancer Activity. *ACS Chem. Biol.* **2017**, *12*, 2815–2822. [CrossRef] [PubMed]
54. Frattaruolo, L.; Fiorillo, M.; Brindisi, M.; Curcio, R.; Dolce, V.; Lacret, R.; Truman, A.W.; Sotgia, F.; Lisanti, M.P.; Cappello, A.R. Thioalbamide, A Thioamidated Peptide from *Amycolatopsis alba*, Affects Tumor Growth and Stemness by Inducing Metabolic Dysfunction and Oxidative Stress. *Cells* **2019**, *8*, 1408. [CrossRef]

Communication

Rhodium-Catalyzed Dynamic Kinetic Resolution of Racemic Internal Allenes towards Chiral Allylated Triazoles and Tetrazoles

Simon V. Sieger, Ilja Lubins and Bernhard Breit *

Institut für Organische Chemie, Albert-Ludwigs-Universität Freiburg, Albertstraße 21, 79104 Freiburg im Breisgau, Germany
* Correspondence: bernhard.breit@chemie.uni-freiburg.de; Tel.: +49-761-203-6051

Abstract: A general Rh-catalyzed addition reaction of nitrogen containing heterocycles to internal allenes is reported. Starting from racemic internal allenes a dynamic kinetic resolution (DKR) provides N-allylated triazoles and tetrazoles. Simultaneous control of N^1/N^x-position selectivity, enantioselectivity and olefin geometry gives access to important building blocks of target-oriented synthesis. The synthetic utility is demonstrated by a gram-scale reaction and a broad substrate scope tolerating multiple functional groups. Deuterium labeling experiments and experiments with enantioenriched allenes as starting material support a plausible reaction mechanism.

Keywords: asymmetric catalysis; dynamic kinetic resolution; internal allenes; rhodium; hydroamination

Citation: Sieger, S.V.; Lubins, I.; Breit, B. Rhodium-Catalyzed Dynamic Kinetic Resolution of Racemic Internal Allenes towards Chiral Allylated Triazoles and Tetrazoles. *Catalysts* **2022**, *12*, 1209. https://doi.org/10.3390/catal12101209

Academic Editors: Ewa Kowalska and Yuichi Kobayashi

Received: 13 September 2022
Accepted: 7 October 2022
Published: 11 October 2022

Publisher's Note: MDPI stays neutral with regard to jurisdictional claims in published maps and institutional affiliations.

Copyright: © 2022 by the authors. Licensee MDPI, Basel, Switzerland. This article is an open access article distributed under the terms and conditions of the Creative Commons Attribution (CC BY) license (https://creativecommons.org/licenses/by/4.0/).

1. Introduction

5-membered nitrogen containing heterocycles are found in a variety of bioactive compounds. Molecules containing triazoles exhibit antifungal [1], anxiolytic [2], antibacterial [3], and anticancerogenic [4] activity. Tetrazoles display antifungal [5], immunosuppressive [6], and antiviral [7] properties. In addition, the isosterism of tetrazoles to carboxylic acids makes them a suitable motif in drug design [8]. In particular, the subclass of α-chiral N-alkylated triazoles and tetrazoles features a broad range of biological activity (Figure 1) [5,9–12].

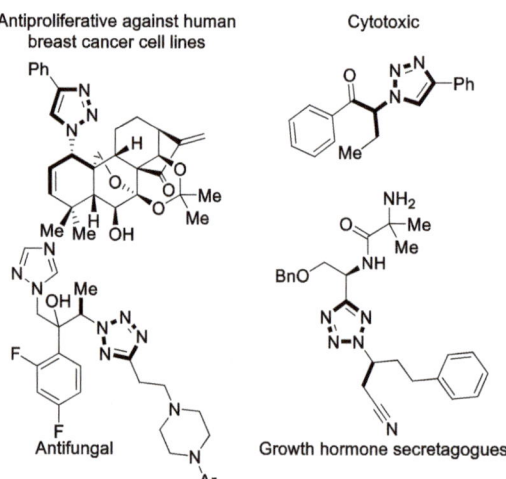

Figure 1. Bioactive compounds possessing an α-chiral triazole or tetrazole scaffold [5,9–11].

Due to the importance of these moieties, different strategies for obtaining them have been developed in the past (Scheme 1). Zhao and coworkers published the transition metal catalyzed allylic substitution with sodium benzotriazolide [13,14]. Despite the excellent enantioselectivities in some cases, the method lacks regioselectivity and requires deprotonation of the triazole in a previous step. Khan recently reported the allylic substitution of cyclic carbonates with tetrazoles [15]. However, allylic substitutions in general are not in accordance with the principle of atom economy by releasing stoichiometric amounts of waste [16]. A series of organocatalyzed aza-Michael additions overcomes this shortcoming and in some cases convinces with high selectivities, but the scope of these transformations is limited [17–22]. Furthermore, a three-component reaction of tetrazoles, aldehydes, and acid anhydrides should be mentioned, which uses a Dynamic Kinetic Resolution to synthesize α-chiral tetrazoles as hemiaminals [23,24].

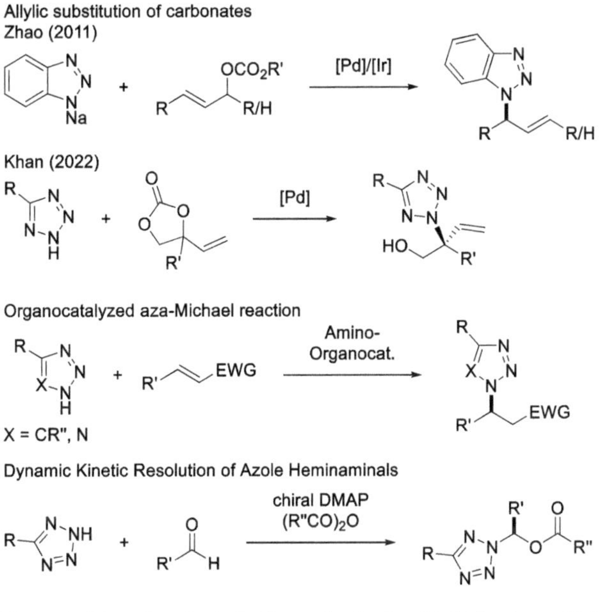

Scheme 1. Selected methods for the synthesis of α-chiral triazoles and tetrazoles [13–24].

In the last decade, our group developed the transition metal-catalyzed addition of a variety of pronucleophiles to allenes and alkynes [25,26]. C- [27–29], N- [30–32], O- [33–35], and S-pronucleophiles [36,37] can thus be allylated enantioselectively, which creates an atom economic analogon to the Tsuji-Trost allylation [38–45] and allylic oxidations [46–48]. Furthermore, the synthetic utility of these reactions has already been demonstrated in several total syntheses [49–51]. We recently reported on the rhodium-catalyzed addition of pyrazoles to internal allenes [52]. Due to the importance of triazoles and tetrazoles, we also wanted to extend the Rh-catalysis for these heterocycles. Hence an optimized catalytic protocol for the synthesis of such frameworks is presented using a DKR, a powerful and important method in asymmetric catalysis and synthesis [53–57].

2. Results and Discussion

Coupling benzotriazole (**2**) with our screening allene, **1**, resulted in 96% yield and perfect enantio-, regio-, and E/Z-selectivity by using [Rh(COD)Cl]$_2$, (R,R)-DIOP, and 50 mol% of PPTS at 60 °C. Phenyltetrazole (**4**) was coupled at 40 °C and with 20 mol% PPTS with ideal regio- and Z/E-selectivity and good enantiomeric excess (Scheme 2) (For optimization and screening tables, see the Supporting Information).

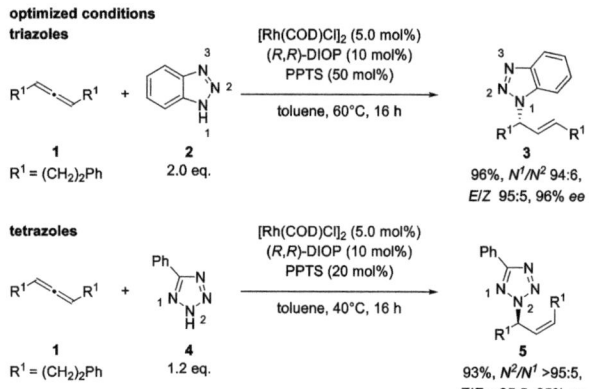

Scheme 2. Optimized reaction conditions of the Rh-catalyzed hydroamination of racemic 1,3-disubstituted allenes. Scale: 0.25 mmol; 1.25 mL of toluene (0.2 M). Yields reported for isolated products. Enantioselectivities were determined by HPLC analysis using a chiral stationary phase. COD = cyclooctadien, PPTS = pyridinium p-toluenesulfonate. Gram-scale catalysis (4.03 mmol allene): benzotriazole: 96% yield, N^1/N^2 96:4, E/Z > 95:5, 95% ee; phenyltetrazole: 84% yield, Z/E > 95:5, 85% ee. For determination of absolute configuration, see the Supporting Information.

After identifying optimized reaction conditions, the broad applicability is demonstrated by several scopes. Initially we subjected a variety of triazoles to the screening allene **1** (Figure 2).

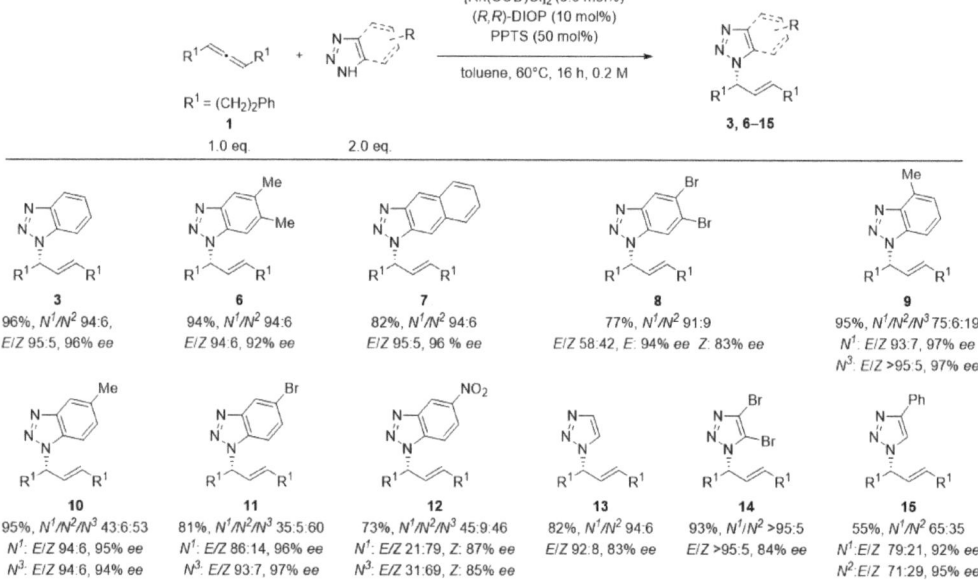

Figure 2. Scope of the addition of Triazoles to rac-1,7-diphenyl-hepta-3,4-diene (**1**). Reactions were performed at 0.25 mmol scale. Cumulative yield for all isolated isomers. E/Z-selectivities determined by ^{1}H-NMR. Enantioselectivities were determined by HPLC analysis using a chiral stationary phase. Unless specified, ee-values refer to main-stereoisomer of the main-regioisomer. Shown regioisomer is designated as an N^1-product.

The benzotriazol-derived substrates (**6–12**) showed up to 96% yield and enantioselectivities up to 97% *ee*. While N^1/N^2-selectivity was consistently greater than 91:9 for symmetrically substituted triazoles (**3, 6–8, 13, 14**), asymmetric substrates added the challenge of N^1/N^3-regioselectivity. 5-substituted benzotriazoles (**10–12**) provided the corresponding products in an almost statistical N^1/N^3-distribution. Placing the methyl-group in 4- position (**9**) and thus closer to the reactive site increased N^1/N^3-selectivity to 4:1. Notably, there was a decrease of *E/Z*-selectivity for heterosubstituted benzotriazoles (**8, 11, 12**), whereas a nearly perfect *E*-selectivity was present in all other cases. The nitro-substituted benzotriazole (**12**) even led to an inversed olefine geometry, with the *Z*-product being preferred.

We subsequently tested several allenes for compatibility in the reaction with benzotriazole **2** (Figure 3). Dialkyl substituted allenes (**16–18**) yielded perfect regioselectivity and enantiomeric excess of up to 98% *ee*. Even a sterically more demanding (**19**) and a macrocyclic allene (**20**) were tolerated as well. The method has its limitations for sterically very demanding and aryl-substituted allenes (**21, 22**), as well as for a trisubstituted allene (**23**). For these substrates, yield and selectivity decreased. Next, we subjected several tetrazoles to the corresponding reaction conditions (Figure 4).

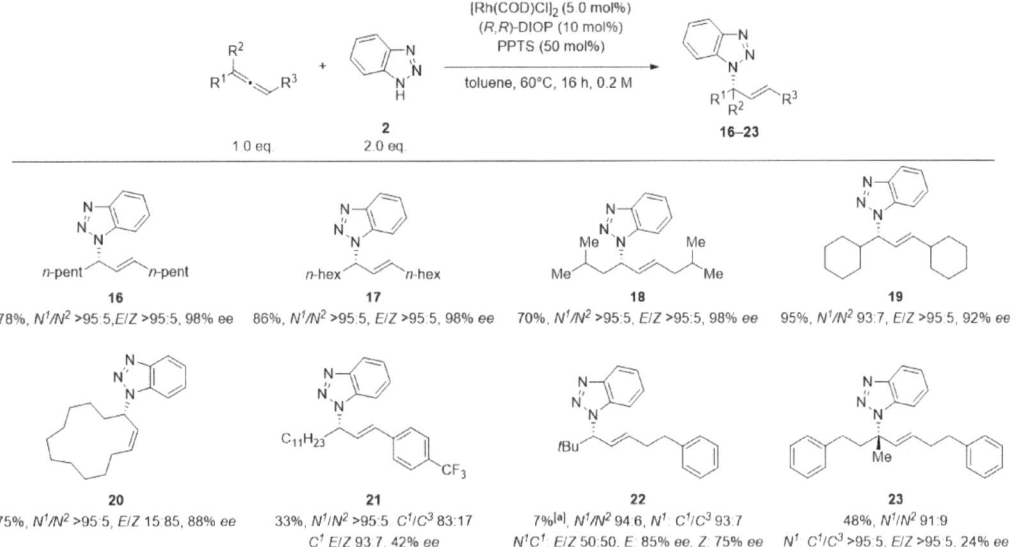

Figure 3. Scope of the addition of Benzotriazole (**2**) to different racemic allenes. Reactions were performed at 0.25 mmol scale. Cumulative yield for all isolated isomers. *E/Z*-selectivities determined by ^{1}H-NMR. Enantioselectivities were determined by HPLC analysis using a chiral stationary phase. Unless specified, *ee*-values refer to main-stereoisomer of the main-regioisomer. [a] Reaction was heated for 72 h.

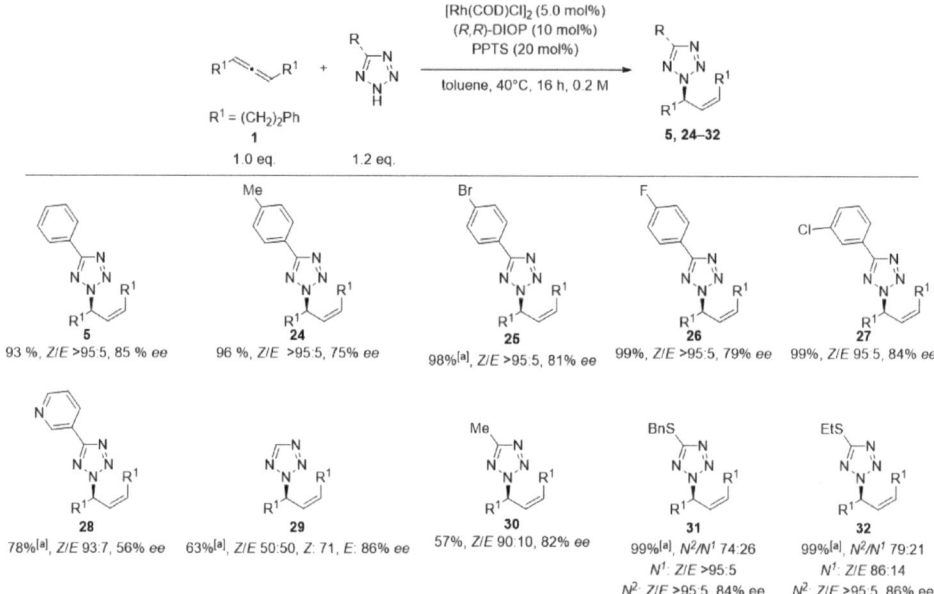

Figure 4. Scope of the addition of Tetrazoles to rac-1,7-diphenyl-hepta-3,4-diene. (**1**) Reactions were performed at 0.25 mmol scale. Cumulative yield for all isolated isomers. Z/E-selectivities were determined by ¹H-NMR. Enantioselectivities were determined by HPLC analysis using a chiral stationary phase. Unless specified, *ee*-values refer to main-stereoisomer of the main-regioisomer. Unless specified full N^2-selectivity was obtained. [a] Reaction was performed at 60 °C.

Halogen-substituted phenyltetrazoles were coupled in nearly quantitative yields (**25–27**). Tetrazoles containing thioethers (**31**, **32**) were the only ones to yield products with reduced N^2/N^1-regioselectivity (about 3:1); however, maintaining quantitative yield and consistently high Z/E- and enantioselectivity. Eventually, we subjected several allenes to the reaction with 5-phenyl-tetrazole (**4**) (Figure 5).

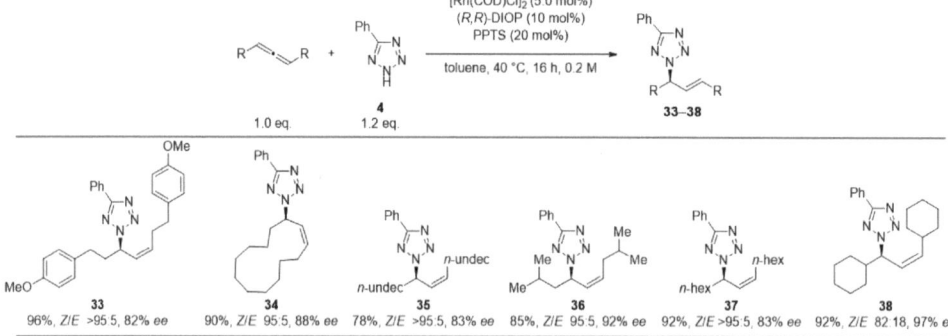

Figure 5. Scope of the addition of 5-phenyl-2H-tetrazole to different racemic allenes. Reactions were performed at 0.25 mmol scale. Cumulative yield for all isolated isomers. E/Z-selectivities determined by ¹H-NMR. Enantioselectivities were determined by HPLC analysis using a chiral stationary phase. Unless specified, *ee*-values refer to main-stereoisomer of the main-regioisomer. Unless specified full N^2-selectivity was obtained.

Except for one sterically more challenging substrate (**38**), all products were obtained in perfect Z-selectivity. Only the di-cyclohexyl substituted substrate (**38**) showed shares of the E-product, indicating the limitation of the catalyst controlling the olefin geometry for sterically more demanding allenes. Some mechanistic control experiments were carried out (Scheme 3).

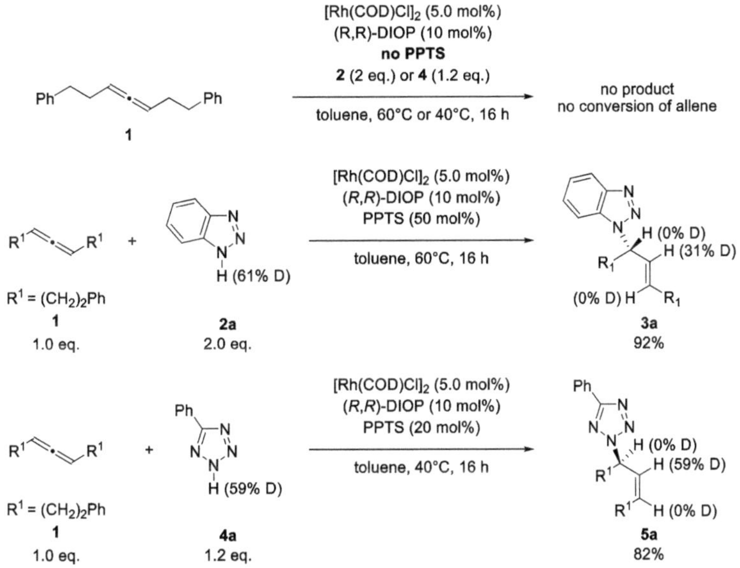

Scheme 3. Control experiments: Without PPTS, only the allene could be reisolated. After reactions with deuterated pronucleophiles, the deuterium was observed exclusively at the former central atom of the allene.

If the catalysis was performed without PPTS, no reaction occurred, revealing the important role of the additive for the reaction. Furthermore, we prepared deuterated nucleophiles, which we subjected to the respective catalysis conditions. The deuterium atoms were found exclusively at the former central atom of the allene. This is in contrast to earlier results with terminal allenes, where deuterium was found at several positions [58–60]. Using enantiomerically enriched allenes in the catalysis, both enantiomers of allene yielded the (S)-product for the triazole and the (R)-product for the tetrazole regardless of the configuration of the allene used (Table 1).

When the achiral dppb ligand was used instead of the chiral DIOP ligand in the reactions with enantiomerically enriched allene, the racemic products were obtained, indicating a racemization step in the catalytic cycle. These findings lead us to propose the following reaction mechanism (Scheme 4).

A Rh^I-DIOP complex is first formed from [Rh(COD)Cl]$_2$ and (R,R)-DIOP (inner box of Scheme 4). With the help of PPTS, the pronucleophile is then added by oxidative addition to form the Rh-hydride species **A**. Now, *syn*-hydrometallation from the sterically less demanding side occurs at the respective allene enantiomers. This occurs in such a way that the hydrogen atom results at the former central atom of the allene. Initially, two diastereomeric Z-configured σ-complexes **B** and *dia*-**B** are formed, which are in equilibrium with their corresponding *syn*, *anti*-configured π-complexes **C** and *dia*-**C**. Finally, a σ-π-σ-isomerization and a bond rotation lead to the identical pseudo-*meso*-π-allyl complex **D** for both diastereomers. Dynamic kinetic resolution is ensured via this *syn*, *syn*-configured π-complex, since the initial hydrometallation products **B** and *dia*-**B** can be converted into each other via **D**. If a triazole is the pronucleophile in the catalysis, reductive elimination occurs of pseudo-meso-π-allyl complex **D**, with the chiral ligand ensuring enantioselectivity. In contrast, when a tetrazole is used as a pronucleophile, reductive elimination preferentially occurs from *syn-anti*-π-complex **C**. This could explain why the (*S*)-*E* product is formed for the triazole and the (*R*)-*Z* product for the tetrazole.

Table 1. Control experiments with enantioenriched allene.

#	Allene	Ligand	Product
		[Rh(COD)Cl]$_2$ (5.0 mol%) ligand (10 mol%) PPTS (50 mol%) **2** (2 eq.) toluene, 60 °C, 16 h	
1	(*S*)-**1** 99% *ee*	(*R,R*)-DIOP	(*S*)-**3**, 91%, N^1/N^2 95:5, *E/Z* 95:5, 94% *ee*
2	(*R*)-**1** 90% *ee*	(*R,R*)-DIOP	(*S*)-**3**, 89%, N^1/N^2 94:6, *E/Z* 95:5, 96% *ee*
3	(*S*)-**1** 99% *ee*	dppb	rac-**3**, 87%, N^1/N^2 88:12, *E/Z* 95:5
		[Rh(COD)Cl]$_2$ (5.0 mol%) ligand (10 mol%) PPTS (20 mol%) **4** (1.2 eq.) toluene, 40 °C, 16 h	
#	Allene	Ligand	Product
4	(*S*)-**1** 99% *ee*	(*R,R*)-DIOP	(*R*)-**5**, 95%, N^2/N^1 >95:5, *Z/E* >95:5, 84% *ee*
5	(*R*)-**1** 90% *ee*	(*R,R*)-DIOP	(*R*)-**5**, 93%, N^2/N^1 >95:5, *Z/E* >95:5, 81% *ee*
6	(*S*)-**1** 99% *ee*	dppb	rac-**5**, 95%, N^2/N^1 >95:5, *Z/E* >95:5

Control experiments with enantiomerically enriched allenes led to the products with the same absolute configuration for (*R,R*)-DIOP or to the racemic products for the achiral dppb ligand.

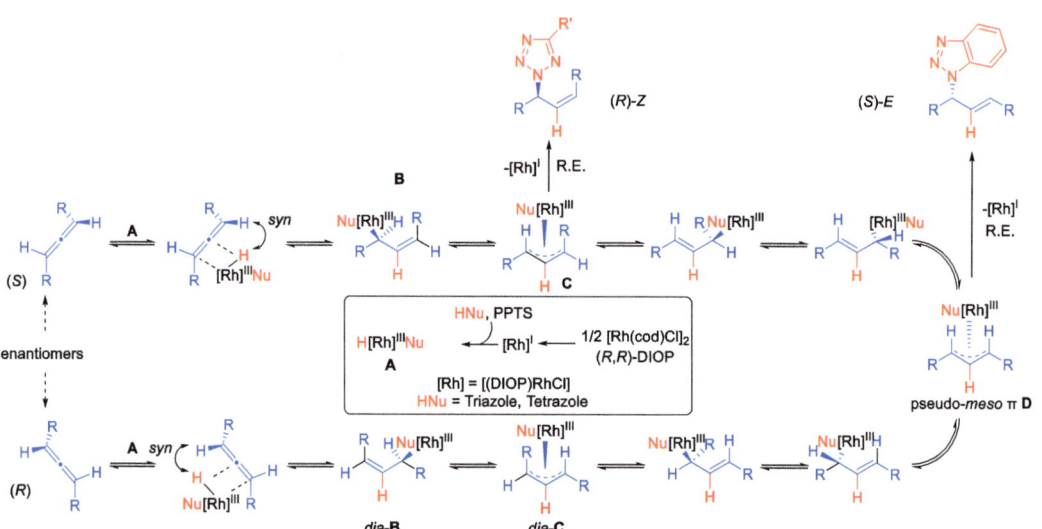

Scheme 4. Proposed reaction mechanism of Rh-catalyzed hydroamination of internal allenes with triazoles and tetrazoles. R.E. = Reductive elimination.

3. Materials and Methods

3.1. Materials

Toluene was freshly distilled over Sodium/Benzophenone and degassed with argon prior to use. Solvents employed for work-up and column chromatography were purchased in technical grade quality and distilled by rotary evaporator before use. [Rh(COD)Cl]$_2$ was purchased from Sigma-Aldrich (St. Louis, MO, USA). (*R,R*)-DIOP was prepared according to the procedure shown in the Supplementary Materials.

3.2. Methods

A 20 mL screw-cap Schlenk tube was dried under a vacuum, backfilled with argon (Argon 5.0 Sauerstoffwerke Friedrichshafen, Friedrichshafen, Germany), and cooled to room temperature using a standard Schlenk line apparatus. The tube was filled with [Rh(COD)Cl]$_2$ (6.2 mg, 0.013 mmol, 5.0 mol%), (R,R)-DIOP (12.5 mg, 0.025 mmol, 10.0 mol%), PPTS (31.5 mg, 0.125 mmol, 50.0 mol% or 12.6 mg, 0.05 mmol, 20.0 mol%), and triazole (0.500 mmol, 2.0 equiv.) or tetrazole (0.30 mmol, 1.2 equiv.). The tube was put on a vacuum and backfilled with argon again. Freshly distilled toluene (1.25 mL) and allene (0.25 mmol, 1.0 equiv.) were added by syringe under a flow of argon, and then the tube was sealed by a screw cap. The mixture was stirred at 60 °C (for triazoles) or 40 °C (for tetrazoles) for 16 h. The tube was cooled to room temperature, the solvent was removed under reduced pressure and the residue was purified by flash column chromatography using AcOEt and hexanes as eluent on silica gel. In the case of the synthesis of racemic samples dppb was used as the catalyst.

4. Conclusions

To conclude, we developed a highly selective Rh-catalysis to add triazoles and tetrazoles to internal allenes. In only one step, allylated triazoles and tetrazoles were constructed in an enantio-, stereo-, and regioselective fashion. Mechanistic studies with deuterated pronucleophiles and enantioenriched substrates led us to propose a DKR mechanism. Further investigations in terms of compatible pronucleophiles that can be subjected to this catalytic system is the goal of future research in our laboratories.

Supplementary Materials: The following supporting information can be downloaded at: https://www.mdpi.com/article/10.3390/catal12101209/s1, screening tables, preparation of ligand and substrates, analytical data, ^{1}H and ^{13}C NMR spectra, determination of absolute configuration, HPLC chromatograms of products and enantioenriched allenes, more detailed materials and methods [61–66].

Author Contributions: Conceptualization, B.B.; methodology, S.V.S. and B.B.; investigation, S.V.S. and I.L.; writing—original draft preparation, S.V.S.; writing—review and editing, S.V.S., I.L. and B.B.; supervision, B.B.; funding acquisition, B.B. All authors have read and agreed to the published version of the manuscript.

Funding: This research was funded by the Fonds der Chemischen Industrie. S.V.S. is grateful for a Ph.D. fellowship from the Fonds der Chemischen Industrie.

Data Availability Statement: All experimental data is contained in the article and supplementary material.

Acknowledgments: Jan Klauser, José Candia, and Julia Schmidt (Albert-Ludwigs Universität Freiburg) are acknowledged for their motivated technical assistance. Sincere thanks are given to Manfred Keller for NMR experiments, analysis of isomers and Joshua Emmerich for HPLC separations (Analytical Department of Albert-Ludwigs Universität Freiburg).

Conflicts of Interest: The authors declare no conflict of interest.

References

1. Rezaei, Z.; Khabnadideh, S.; Pakshir, K.; Hossaini, Z.; Amiri, F.; Assadpour, E. Design, synthesis, and antifungal activity of triazole and benzotriazole derivatives. *Eur. J. Med. Chem.* **2009**, *44*, 3064–3067. [CrossRef] [PubMed]
2. Paluchowska, M.H.; Bugno, R.; Charakchieva-Minol, S.; Bojarski, A.J.; Tatarczyńska, E.; Chojnacka-Wójcik, E. Conformational restriction in novel NAN-190 and MP3022 analogs and their 5-HT(1A) receptor activity. *Arch. Pharm.* **2006**, *339*, 498–506. [CrossRef] [PubMed]
3. Letavic, M.A.; Bronk, B.S.; Bertsche, C.D.; Casavant, J.M.; Cheng, H.; Daniel, K.L.; George, D.M.; Hayashi, S.F.; Kamicker, B.J.; Kolosko, N.L.; et al. Synthesis and Activity of a Novel Class of Tribasic Macrocyclic Antibiotics: The Triamilides. *Bioorg. Med. Chem. Lett.* **2002**, *12*, 2771–2774. [CrossRef]
4. Sambasiva Rao, P.; Kurumurthy, C.; Veeraswamy, B.; Santhosh Kumar, G.; Poornachandra, Y.; Ganesh Kumar, C.; Vasamsetti, S.B.; Kotamraju, S.; Narsaiah, B. Synthesis of novel 1,2,3-triazole substituted-N-alkyl/aryl nitrone derivatives, their anti-inflammatory and anticancer activity. *Eur. J. Med. Chem.* **2014**, *80*, 184–191. [CrossRef] [PubMed]
5. Upadhyaya, R.S.; Jain, S.; Sinha, N.; Kishore, N.; Chandra, R.; Arora, S.K. Synthesis of novel substituted tetrazoles having antifungal activity. *Eur. J. Med. Chem.* **2004**, *39*, 579–592. [CrossRef] [PubMed]
6. Wagner, R.; Mollison, K.W.; Liu, L.; Henry, C.L.; Rosenberg, T.A.; Bamaung, N.; Tu, N.; Wiedeman, P.E.; Or, Y.; Luly, J.R.; et al. Rapamycin analogs with reduced systemic exposure. *Bioorg. Med. Chem. Lett.* **2005**, *15*, 5340–5343. [CrossRef] [PubMed]
7. Johansson, A.; Poliakov, A.; Åkerblom, E.; Wiklund, K.; Lindeberg, G.; Winiwarter, S.; Danielson, U.; Samuelsson, B.; Hallberg, A. Acyl sulfonamides as potent protease inhibitors of the hepatitis C virus full-Length NS3 (Protease-Helicase/NTPase): A comparative study of different C-terminals. *Bioorg. Med. Chem.* **2003**, *11*, 2551–2568. [CrossRef]
8. Herr, R. 5-Substituted-1H-tetrazoles as carboxylic acid isosteres: Medicinal chemistry and synthetic methods. *Bioorg. Med. Chem.* **2002**, *10*, 3379–3393. [CrossRef]
9. Vantikommu, J.; Palle, S.; Reddy, P.S.; Ramanatham, V.; Khagga, M.; Pallapothula, V.R. Synthesis and cytotoxicity evaluation of novel 1,4-disubstituted 1,2,3-triazoles via CuI catalysed 1,3-dipolar cycloaddition. *Eur. J. Med. Chem.* **2010**, *45*, 5044–5050. [CrossRef]
10. Ding, C.; Zhang, Y.; Chen, H.; Wild, C.; Wang, T.; White, M.A.; Shen, Q.; Zhou, J. Overcoming synthetic challenges of oridonin A-ring structural diversification: Regio- and stereoselective installation of azides and 1,2,3-triazoles at the C-1, C-2, or C-3 position. *Org. Lett.* **2013**, *15*, 3718–3721. [CrossRef] [PubMed]
11. Hernández, A.S.; Swartz, S.G.; Slusarchyk, D.; Yan, M.; Seethala, R.K.; Sleph, P.; Grover, G.; Dickinson, K.; Giupponi, L.; Harper, T.W.; et al. Optimization of 1H-tetrazole-1-alkanenitriles as potent orally bioavailable growth hormone secretagogues. *Bioorg. Med. Chem. Lett.* **2008**, *18*, 2067–2072. [CrossRef]
12. Zhang, Y.-C.; Jiang, F.; Shi, F. Organocatalytic Asymmetric Synthesis of Indole-Based Chiral Heterocycles: Strategies, Reactions, and Outreach. *Acc. Chem. Res.* **2020**, *53*, 425–446. [CrossRef]
13. Liu, W.; Zhang, D.; Zheng, S.; Yue, Y.; Liu, D.; Zhao, X. Enantioselective Palladium-Catalyzed Allylic Substitution of Sodium Benzotriazolide. *Eur. J. Org. Chem.* **2011**, *2011*, 6288–6293. [CrossRef]
14. Zhang, M.; Guo, X.-W.; Zheng, S.-C.; Zhao, X.-M. Enantioselective iridium-catalyzed allylation of sodium benzotriazolide: An efficient way to chiral allylbenzotriazoles. *Tetrahedron Lett.* **2012**, *53*, 6995–6998. [CrossRef]
15. Khan, S.; Wang, Y.; Zhang, M.-N.; Perveen, S.; Zhang, J.; Khan, A. Regio- and enantioselective formation of tetrazole-bearing quaternary stereocenters via palladium-catalyzed allylic amination. *Org. Chem. Front.* **2022**, *9*, 456–461. [CrossRef]
16. Trost, B.M. The atom economy—A search for synthetic efficiency. *Science* **1991**, *254*, 1471–1477. [CrossRef]
17. Dinér, P.; Nielsen, M.; Marigo, M.; Jørgensen, K.A. Enantioselective organocatalytic conjugate addition of N heterocycles to alpha, beta-unsaturated aldehydes. *Angew. Chem. Int. Ed.* **2007**, *46*, 1983–1987. [CrossRef]
18. Gandelman, M.; Jacobsen, E.N. Highly enantioselective catalytic conjugate addition of N-heterocycles to alpha, beta-unsaturated ketones and imides. *Angew. Chem. Int. Ed.* **2005**, *44*, 2393–2397. [CrossRef]
19. Luo, G.; Zhang, S.; Duan, W.; Wang, W. Enantioselective Conjugate Addition of N-Heterocycles to α,β-Unsaturated Ketones Catalyzed by Chiral Primary Amines. *Synthesis* **2009**, *2009*, 1564–1572. [CrossRef]
20. Uria, U.; Vicario, J.L.; Badía, D.; Carrillo, L. Organocatalytic enantioselective aza-Michael reaction of nitrogen heterocycles and alpha, beta-unsaturated aldehydes. *Chem. Commun.* **2007**, *24*, 2509–2511. [CrossRef]
21. Wang, J.; Zu, L.; Li, H.; Xie, H.; Wang, W. Cinchona Alkaloid Based Thiourea Promoted Enantioselective Conjugate Addition of N-Heterocycles to Enones. *Synthesis* **2007**, *2007*, 2576–2580. [CrossRef]
22. Wang, J.; Li, H.; Zu, L.; Wang, W. Enantioselective organocatalytic Michael addition reactions between N-heterocycles and nitroolefins. *Org. Lett.* **2006**, *8*, 1391–1394. [CrossRef]
23. Piotrowski, D.W.; Kamlet, A.S.; Dechert-Schmitt, A.-M.R.; Yan, J.; Brandt, T.A.; Xiao, J.; Wei, L.; Barrila, M.T. Regio- and Enantioselective Synthesis of Azole Hemiaminal Esters by Lewis Base Catalyzed Dynamic Kinetic Resolution. *J. Am. Chem. Soc.* **2016**, *138*, 4818–4823. [CrossRef]
24. Kinens, A.; Sejejs, M.; Kamlet, A.S.; Piotrowski, D.W.; Vedejs, E.; Suna, E. Development of a Chiral DMAP Catalyst for the Dynamic Kinetic Resolution of Azole Hemiaminals. *J. Org. Chem.* **2017**, *82*, 869–886. [CrossRef]
25. Koschker, P.; Breit, B. Branching Out: Rhodium-Catalyzed Allylation with Alkynes and Allenes. *Acc. Chem. Res.* **2016**, *49*, 1524–1536. [CrossRef]

26. Haydl, A.M.; Breit, B.; Liang, T.; Krische, M.J. Alkynes as Electrophilic or Nucleophilic Allylmetal Precursors in Transition-Metal Catalysis. *Angew. Chem. Int. Ed.* **2017**, *56*, 11312–11325. [CrossRef]
27. Berthold, D.; Klett, J.; Breit, B. Rhodium-catalyzed asymmetric intramolecular hydroarylation of allenes: Access to functionalized benzocycles. *Chem. Sci.* **2019**, *10*, 10048–10052. [CrossRef]
28. Grugel, C.P.; Breit, B. Rhodium-Catalyzed Enantioselective Cyclization of 3-Allenyl-indoles: Access to Functionalized Tetrahydrocarbazoles. *Org. Lett.* **2019**, *21*, 5798–5802. [CrossRef]
29. Hilpert, L.J.; Breit, B. Rhodium-Catalyzed Parallel Kinetic Resolution of Racemic Internal Allenes Towards Enantiopure Allylic 1,3-Diketones. *Angew. Chem. Int. Ed.* **2019**, *58*, 9939–9943. [CrossRef]
30. Schmidt, J.P.; Breit, B. Transition metal catalyzed stereodivergent synthesis of syn- and anti-δ-vinyl-lactams: Formal total synthesis of (-)-cermizine C and (-)-senepodine G. *Chem. Sci.* **2019**, *10*, 3074–3079. [CrossRef]
31. Berthold, D.; Geissler, A.G.A.; Giofré, S.; Breit, B. Rhodium-Catalyzed Asymmetric Intramolecular Hydroamination of Allenes. *Angew. Chem.* **2019**, *131*, 10099–10102. [CrossRef]
32. Berthold, D.; Breit, B. Asymmetric Total Syntheses of (-)-Angustureine and (-)-Cuspareine via Rhodium-Catalyzed Hydroamination. *Org. Lett.* **2020**, *22*, 565–568. [CrossRef] [PubMed]
33. Koschker, P.; Lumbroso, A.; Breit, B. Enantioselective synthesis of branched allylic esters via rhodium-catalyzed coupling of allenes with carboxylic acids. *J. Am. Chem. Soc.* **2011**, *133*, 20746–20749. [CrossRef] [PubMed]
34. Liu, Z.; Breit, B. Rhodium-Catalyzed Enantioselective Intermolecular Hydroalkoxylation of Allenes and Alkynes with Alcohols: Synthesis of Branched Allylic Ethers. *Angew. Chem. Int. Ed.* **2016**, *55*, 8440–8443. [CrossRef] [PubMed]
35. Schmidt, J.P.; Breit, B. Rhodium-Catalyzed Cyclization of Terminal and Internal Allenols: An Atom Economic and Highly Stereoselective Access Towards Tetrahydropyrans. *Angew. Chem. Int. Ed.* **2020**, *59*, 23485–23490. [CrossRef] [PubMed]
36. Pritzius, A.B.; Breit, B. Asymmetric rhodium-catalyzed addition of thiols to allenes: Synthesis of branched allylic thioethers and sulfones. *Angew. Chem.* **2015**, *127*, 3164–3168. [CrossRef]
37. Khakyzadeh, V.; Wang, Y.-H.; Breit, B. Rhodium-catalyzed addition of sulfonyl hydrazides to allenes: Regioselective synthesis of branched allylic sulfones. *Chem. Commun.* **2017**, *53*, 4966–4968. [CrossRef]
38. Evans, P.A.; Leahy, D.K. Regio- and enantiospecific rhodium-catalyzed allylic etherification reactions using copper(I) alkoxides: Influence of the copper halide salt on selectivity. *J. Am. Chem. Soc.* **2002**, *124*, 7882–7883. [CrossRef]
39. Hayashi, T.; Okada, A.; Suzuka, T.; Kawatsura, M. High enantioselectivity in rhodium-catalyzed allylic alkylation of 1-substituted 2-propenyl acetates. *Org. Lett.* **2003**, *5*, 1713–1715. [CrossRef]
40. Lu, Z.; Ma, S. Metal-catalyzed enantioselective allylation in asymmetric synthesis. *Angew. Chem.* **2008**, *120*, 264–303. [CrossRef]
41. Madrahimov, S.T.; Li, Q.; Sharma, A.; Hartwig, J.F. Origins of Regioselectivity in Iridium Catalyzed Allylic Substitution. *J. Am. Chem. Soc.* **2015**, *137*, 14968–14981. [CrossRef] [PubMed]
42. Petrone, D.A.; Isomura, M.; Franzoni, I.; Rössler, S.L.; Carreira, E.M. Allenylic Carbonates in Enantioselective Iridium-Catalyzed Alkylations. *J. Am. Chem. Soc.* **2018**, *140*, 4697–4704. [CrossRef] [PubMed]
43. Qu, J.; Helmchen, G. Applications of Iridium-Catalyzed Asymmetric Allylic Substitution Reactions in Target-Oriented Synthesis. *Acc. Chem. Res.* **2017**, *50*, 2539–2555. [CrossRef] [PubMed]
44. Trost, B.M.; Crawley, M.L. Asymmetric transition-metal-catalyzed allylic alkylations: Applications in total synthesis. *Chem. Rev.* **2003**, *103*, 2921–2944. [CrossRef]
45. Trost, B.M.; van Vranken, D.L. Asymmetric Transition Metal-Catalyzed Allylic Alkylations. *Chem. Rev.* **1996**, *96*, 395–422. [CrossRef]
46. Liu, G.; Stahl, S.S. Two-faced reactivity of alkenes: Cis- versus trans-aminopalladation in aerobic Pd-catalyzed intramolecular aza-Wacker reactions. *J. Am. Chem. Soc.* **2007**, *129*, 6328–6335. [CrossRef]
47. Ma, R.; Young, J.; Promontorio, R.; Dannheim, F.M.; Pattillo, C.C.; White, M.C. Synthesis of anti-1,3 Amino Alcohol Motifs via Pd(II)/SOX Catalysis with the Capacity for Stereodivergence. *J. Am. Chem. Soc.* **2019**, *141*, 9468–9473. [CrossRef]
48. Liu, G.; Wu, Y. Palladium-catalyzed allylic C-H bond functionalization of olefins. *Top. Curr. Chem.* **2010**, *292*, 195–209. [CrossRef]
49. Steib, P.; Breit, B. Concise Total Synthesis of (-)-Vermiculine through a Rhodium-Catalyzed C2-Symmetric Dimerization Strategy. *Chem. Eur. J.* **2019**, *25*, 3532–3535. [CrossRef]
50. Schotes, C.; Ostrovskyi, D.; Senger, J.; Schmidtkunz, K.; Jung, M.; Breit, B. Total synthesis of (18S)- and (18R)-homolargazole by rhodium-catalyzed hydrocarboxylation. *Chem. Eur. J.* **2014**, *20*, 2164–2168. [CrossRef]
51. Brosowsky, J.; Lutterbeck, M.; Liebich, A.; Keller, M.; Herp, D.; Vogelmann, A.; Jung, M.; Breit, B. Syntheses of Thailandepsin B Pseudo-Natural Products: Access to New Highly Potent HDAC Inhibitors via Late-Stage Modification. *Chem. Eur. J.* **2020**, *26*, 16241–16245. [CrossRef] [PubMed]
52. Hilpert, L.J.; Sieger, S.V.; Haydl, A.M.; Breit, B. Palladium- and Rhodium-Catalyzed Dynamic Kinetic Resolution of Racemic Internal Allenes Towards Chiral Pyrazoles. *Angew. Chem.* **2019**, *131*, 3416–3419. [CrossRef]
53. Hang, Q.-Q.; Wu, S.-F.; Yang, S.; Wang, X.; Zhong, Z.; Zhang, Y.-C.; Shi, F. Design and catalytic atroposelective synthesis of axially chiral isochromenone-indoles. *Sci. China Chem.* **2022**, *65*, 1929–1937. [CrossRef]
54. Wu, P.; Yu, L.; Gao, C.-H.; Cheng, Q.; Deng, S.; Jiao, Y.; Tan, W.; Shi, F. Design and synthesis of axially chiral aryl-pyrroloindoles via the strategy of organocatalytic asymmetric (2 + 3) cyclization. *Fundam. Res.* **2022**; *in press.* [CrossRef]
55. Deska, J.; Del Pozo Ochoa, C.; Bäckvall, J.-E. Chemoenzymatic dynamic kinetic resolution of axially chiral allenes. *Chem. Eur. J.* **2010**, *16*, 4447–4451. [CrossRef]

56. Huerta, F.F.; Minidis, A.B.E.; Bäckvall, J.-E. Racemisation in asymmetric synthesis. Dynamic kinetic resolution and related processes in enzyme and metal catalysis. *Chem. Soc. Rev.* **2001**, *30*, 321–331. [CrossRef]
57. Ward, R.S. Dynamic kinetic resolution. *Tetrahedron Asymm.* **1995**, *6*, 1475–1490. [CrossRef]
58. Cooke, M.L.; Xu, K.; Breit, B. Enantioselective rhodium-catalyzed synthesis of branched allylic amines by intermolecular hydroamination of terminal allenes. *Angew. Chem. Int. Ed.* **2012**, *51*, 10876–10879. [CrossRef]
59. Xu, K.; Thieme, N.; Breit, B. Atom-economic, regiodivergent, and stereoselective coupling of imidazole derivatives with terminal allenes. *Angew. Chem. Int. Ed.* **2014**, *53*, 2162–2165. [CrossRef]
60. Haydl, A.M.; Xu, K.; Breit, B. Regio- and enantioselective synthesis of N-substituted pyrazoles by rhodium-catalyzed asymmetric addition to allenes. *Angew. Chem. Int. Ed.* **2015**, *54*, 7149–7153. [CrossRef]
61. Fürstner, A.; Wuchrer, M. Concise approach to the "higher sugar" core of the nucleoside antibiotic hikizimycin. *Chem. Eur. J.* **2005**, *12*, 76–89. [CrossRef]
62. Kielbasinski, P.; Albrycht, M.; Mikolajczyk, M.; Wieczorek, M.W.; Majzner, W.R.; Filipczak, A.; Ciolkiewicz, P. Synthesis of chiral hydroxythiolanes as potential catalysts for asymmetric organozinc additions to carbonyl compounds. *Heteroatom Chem.* **2005**, *16*, 93–103. [CrossRef]
63. Hobbs, C.F.; Knowles, W.S. Asymmetric hydroformylation of vinyl acetate with DIOP-type ligands. *J. Org. Chem.* **1981**, *46*, 4422–4427. [CrossRef]
64. Johansson, M.J.; Gorin, D.J.; Staben, S.T.; Toste, F.D. Gold(I)-catalyzed stereoselective olefin cyclopropanation. *J. Am. Chem. Soc.* **2005**, *127*, 18002–18003. [CrossRef] [PubMed]
65. Berthold, D.; Breit, B. Chemo-, Regio-, and Enantioselective Rhodium-Catalyzed Allylation of Triazoles with Internal Alkynes and Terminal Allenes. *Org. Lett.* **2018**, *20*, 598–601. [CrossRef] [PubMed]
66. Xu, K.; Raimondi, W.; Bury, T.; Breit, B. Enantioselective formation of tertiary and quaternary allylic C-N bonds via allylation of tetrazoles. *Chem. Commun.* **2015**, *51*, 10861–10863. [CrossRef]

Review

Transition-Metal-Catalyzed C–C Bond Macrocyclization via Intramolecular C–H Bond Activation

Xiao Wang [1], Ming-Zhu Lu [2,3,*] and Teck-Peng Loh [1,2,3,*]

1. Institute of Advanced Synthesis, School of Chemistry and Molecular Engineering, Nanjing Tech University, Nanjing 211816, China
2. College of Advanced Interdisciplinary Science and Technology, Henan University of Technology, Zhengzhou 450001, China
3. School of Chemistry, Chemical Engineering and Biotechnology, Nanyang Technological University, Singapore 637371, Singapore
* Correspondence: lmzhu1988@163.com (M.-Z.L.); teckpeng@ntu.edu.sg (T.-P.L.)

Abstract: Macrocycles are commonly synthesized via late-stage macrolactamization and macrolactonization. Strategies involving C–C bond macrocyclization have been reported, and examples include the transition-metal-catalyzed ring-closing metathesis and coupling reactions. In this minireview, we summarize the recent progress in the direct synthesis of polyketide and polypeptide macrocycles using a transition-metal-catalyzed C–H bond activation strategy. In the first part, rhodium-catalyzed alkene–alkene ring-closing coupling for polyketide synthesis is described. The second part summarizes the synthesis of polypeptide macrocycles. The activation of indolyl and aryl C(sp^2)–H bonds followed by coupling with various coupling partners such as aryl halides, arylates, and alkynyl bromide is then documented. Moreover, transition-metal-catalyzed C–C bond macrocyclization reactions via alkyl C(sp^3)–H bond activation are also included. We hope that this mini-review will inspire more researchers to explore new and broadly applicable strategies for C–C bond macrocyclization via intramolecular C–H activation.

Keywords: transition-metal catalysis; C–H activation; macrocyclization; polyketide macrocycles; polypeptide macrocycles

Citation: Wang, X.; Lu, M.-Z.; Loh, T.-P. Transition-Metal-Catalyzed C–C Bond Macrocyclization via Intramolecular C–H Bond Activation. *Catalysts* **2023**, *13*, 438. https://doi.org/10.3390/catal13020438

Academic Editors: Ewa Kowalska and Yuichi Kobayashi

Received: 17 January 2023
Revised: 14 February 2023
Accepted: 15 February 2023
Published: 17 February 2023

Copyright: © 2023 by the authors. Licensee MDPI, Basel, Switzerland. This article is an open access article distributed under the terms and conditions of the Creative Commons Attribution (CC BY) license (https://creativecommons.org/licenses/by/4.0/).

1. Introduction

Macrocycles have been defined as cyclic molecules that contain 12 or more covalent connected atoms, which are featured widely in numerous biologically active natural products and pharmaceutically relevant molecules (Figure 1) [1,2]. Among them, polypeptide and polyketide macrocycles are the most common. For the synthesis of this important class of compounds, one of the crucial steps is the late-stage macrocyclization reaction [3–10]. As these precursors leading to the molecules may have different conformational structures, the key step involving intramolecular the macrocyclization reaction can been challenging. The common strategies reported so far include macrolactonization [11–13], macrolactamization [14–16], and macroaldolization [17–19] (Figure 2). Over the decades, transition-metal-catalyzed intramolecular coupling and ring-closing metathesis (RCM) have also been shown to be effective strategies for constructing large rings [20–23]. However, all of the reported methods have some disadvantages, such as the need to use expensive catalysts, difficulty in accessing the substrates, and difficulty in introducing alkene functionality with high selectivity. Furthermore, many of these methods are not atom-economical and sometimes produce substantial amounts of hazardous wastes that are difficult to treat.

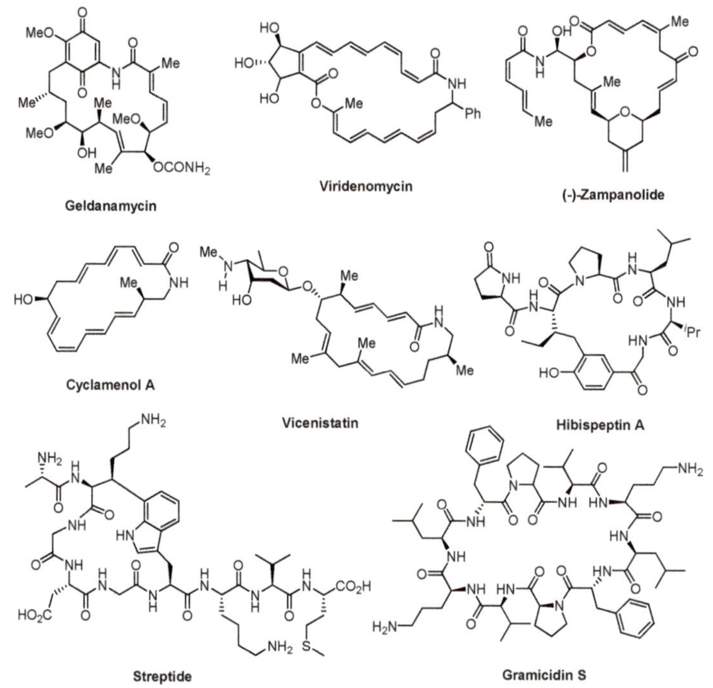

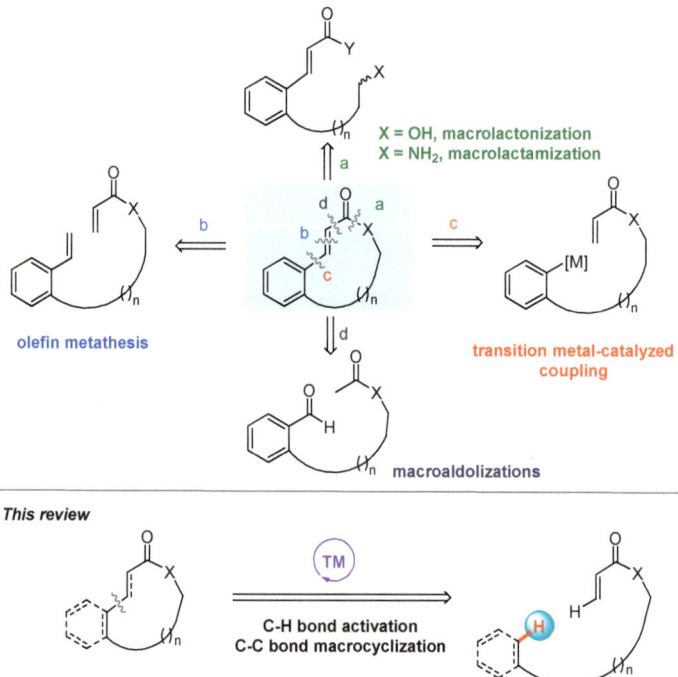

Figure 1. Representative examples of bioactive macrocyclic natural products.

Figure 2. General strategies for the direct synthesis of macrocycles.

In recent years, transition-metal-catalyzed C–H activation reactions have attracted increasing attention [24–31]. In contrast to the classical cross-coupling reactions, such as Suzuki coupling, Negishi coupling, and Stille coupling, the C–H bond functionalization coupling reactions do not need to use organometallic or organohalide reagents. Due to these remarkable advantages, the C–H bond functionalization reactions have been utilized widely in organic synthesis, and a large number of excellent reviews on this topic have also been reported [32–43]. While the intermolecular versions of metal-catalyzed, group-controlled $C(sp^2)$–H for aryl groups and vinyl groups and $C(sp^3)$–H for alkyl groups have been well studied, the application of the intramolecular version of macrocyclization to construct various macrocycles has remained relatively unexplored [44–47]. The lengthy synthesis of the precursors for the risky macrocyclization step may be one of the main reasons. In recent years, we have witnessed more activities in the polypeptide area as compared to polyketide synthesis. In this mini-review, we will mainly focus on the direct synthesis of polyketide and polypeptide macrocycles involving carbon–carbon-based cyclization reactions via intramolecular C–H bond macrocyclization reactions (Figure 2) [48].

2. Polyketide Macrocycles

Macrolides containing a diene moiety are found in many bioactive natural products such as vicenistatin [49–51], geldanamycin [52,53], and cyclamenol A [54]. Therefore, the development of robust methods to construct structurally diverse macrocycles via C–C bond macrocyclization containing a fixed configuration of the diene moiety can be challenging. Extending the reactions they developed for the intermolecular version [55], Loh's group in 2018 first reported an efficient method to construct macrolactams containing diene functionality with high Z and E selectivity via an alkenyl sp^2 C–H bond olefination using a Cp*Rh(III) catalyst (Figure 3) [56]. This strategy provides easy access to a wide variety of macrocodes with different ring sizes in an atom-economical manner. Mechanistically, this ring-closing alkene–alkene coupling reaction was proposed to proceed via a Z-olefinic vinylrhodium(III) intermediate of the acrylamide derivative (Figure 4). Initially, the anion exchange of the [RhCp*Cl$_2$]$_2$ catalyst with the NaBARF additive and Cu(OAc)$_2$•H$_2$O produced the highly reactive species **A**, which then selectively activated the Z-olefinic $C(sp^2)$–H bond of acrylamides to generate the vinylrhodium(III) intermediate **B**. Subsequent coordination and further migratory insertion occurred smoothly to afford the intermediate **D**, which finally underwent facile β-H elimination to give the macrocycle product.

The intramolecular alkene–alkene ring cyclization reaction using the rhodium catalyst led to various macrocycles containing a versatile dienoate moiety. The conjugated moiety present in the resulting macrocycles is highly versatile and can be converted into many different functional groups, thereby allowing easy access to macrolides containing different functionalities (Figure 5). For example, the hydrogenation of the macrocyclization product in the presence of the Pd/C catalyst afforded the saturated alkane product. The treatment of the product with a copper catalyst with PhMe$_2$Si-BPin furnished the 1,4-hydrosilylation product. On the other hand, reacting the product with the bis(pinacolato)diboron reagent led to the formation of the internal alkene product. Subjecting the product to nitromethane in the presence of 1,8-diazabicyclo[5.4.0]undec-7-ene (DBU) led to the 1,4-Michael addition product. Overall, the versatility of the dienoate products allowed easy access to a wide variety of useful macrocycles that were not easily accessible using the reported methods.

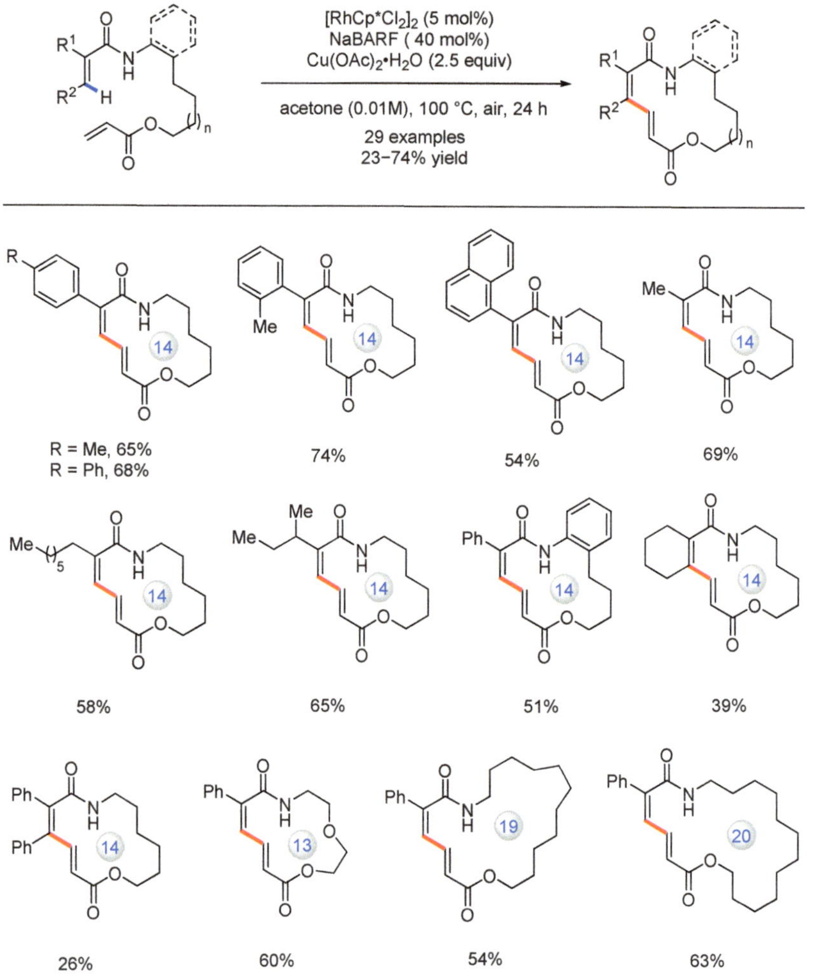

Figure 3. Macrolide synthesis via Cp*Rh(III)-catalyzed intramolecular oxidative cross-coupling of alkenes [56].

In 2020, the Loh group further expanded their intramolecular ring-closing alkene–alkene coupling strategy to α,β-unsaturated ketone substrates (Figure 6) [57]. Under the same conditions, a wide variety of macrolactams of different ring sizes were synthesized in moderate to good yields (33−72%). Of note, the intramolecular coupling of the alkene substrate with no substitution at the α-position also proceeded smoothly to furnish the corresponding macrolactam at a 44% yield. Moreover, this protocol could afford more strained 12- and 13-membered ring macrolactams in satisfactory yields. The authors proposed an analogous catalytic mechanism to elucidate this Cp*Rh(III)-catalyzed alkenyl C(sp^2)–H activation–macrocyclization pathway.

Figure 4. Proposed mechanism for the Cp*Rh(III)-catalyzed intramolecular alkenyl sp^2 C–H activation–macrocyclization [56].

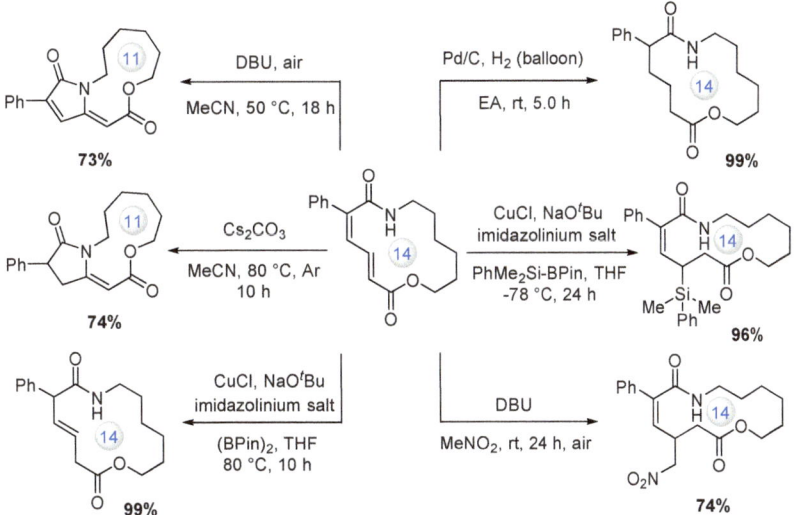

Figure 5. Synthetic derivatization of the cross-coupling macrocycle product [56].

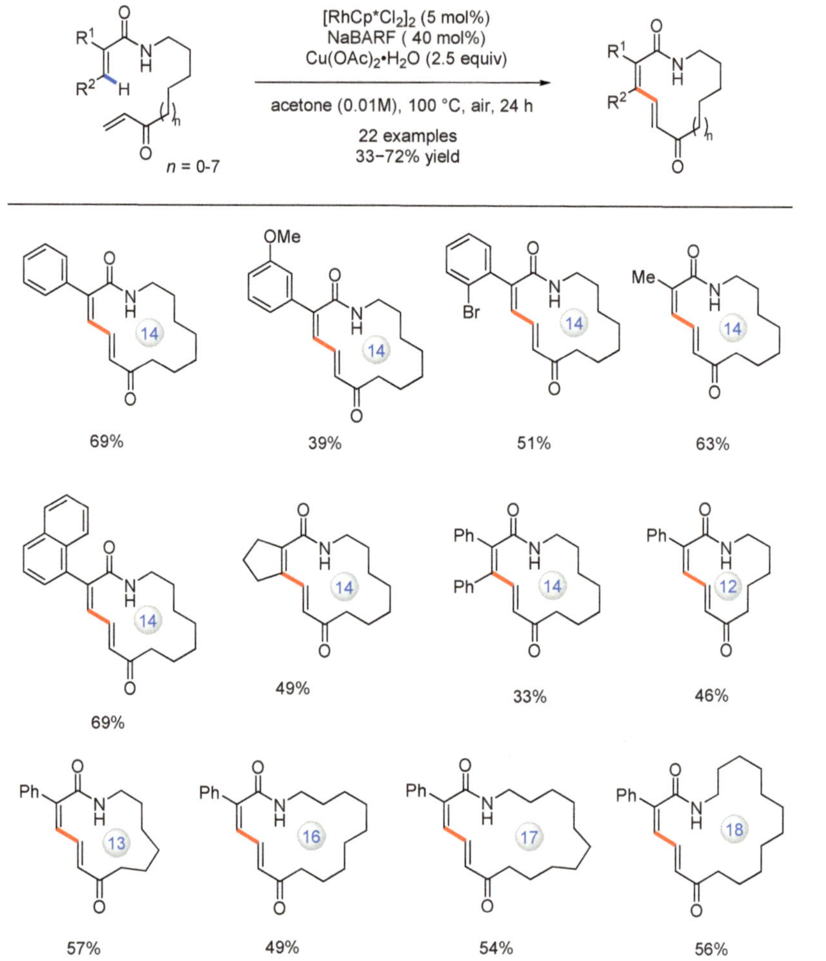

Figure 6. Macrolactam synthesis via Cp*Rh(III)-catalyzed intramolecular ring-closing alkene–alkene cross-couplings [57].

3. Polypeptide Macrocycles

Peptide-based macrocycles have been shown to have a wide variety of biological activities [58]. Some important drugs such as the antibiotic vancomycin [59], anti-cancer agent octreotide [60], and immunosuppressant cyclosporine [61] contain a peptide macrocycle structure. Furthermore, in contrast to linear peptides, peptide macrocycles have improved properties in terms of cell penetration, stability, and selectivity. In recent years, cyclic peptides have also been used as the delivery systems in peptide–drug conjugates such as ^{177}Lu-dotatate [62] and edotreotide gallium Ga-68 [63]. In addition to the amide bond formation strategy, the use of C–H bond functionalization strategies has gained tremendous attention in recent years due to their ability to construct many different stapled cyclopeptides. In this part of the review, we will divide them into 3 different categories: (1) in the first part, we will focus mainly on the sp^2 C–H bond functionalization of indole C–H bond macrocyclization; (2) in the second part, we will focus mainly on the C–H bond functionalization of the aryl C(sp^2)–H bond macrocyclization strategy; (3) in the final part,

In recent years, transition-metal-catalyzed C–H activation reactions have attracted increasing attention [24–31]. In contrast to the classical cross-coupling reactions, such as Suzuki coupling, Negishi coupling, and Stille coupling, the C–H bond functionalization coupling reactions do not need to use organometallic or organohalide reagents. Due to these remarkable advantages, the C–H bond functionalization reactions have been utilized widely in organic synthesis, and a large number of excellent reviews on this topic have also been reported [32–43]. While the intermolecular versions of metal-catalyzed, group-controlled $C(sp^2)$–H for aryl groups and vinyl groups and $C(sp^3)$–H for alkyl groups have been well studied, the application of the intramolecular version of macrocyclization to construct various macrocycles has remained relatively unexplored [44–47]. The lengthy synthesis of the precursors for the risky macrocyclization step may be one of the main reasons. In recent years, we have witnessed more activities in the polypeptide area as compared to polyketide synthesis. In this mini-review, we will mainly focus on the direct synthesis of polyketide and polypeptide macrocycles involving carbon–carbon-based cyclization reactions via intramolecular C–H bond macrocyclization reactions (Figure 2) [48].

2. Polyketide Macrocycles

Macrolides containing a diene moiety are found in many bioactive natural products such as vicenistatin [49–51], geldanamycin [52,53], and cyclamenol A [54]. Therefore, the development of robust methods to construct structurally diverse macrocycles via C–C bond macrocyclization containing a fixed configuration of the diene moiety can be challenging. Extending the reactions they developed for the intermolecular version [55], Loh's group in 2018 first reported an efficient method to construct macrolactams containing diene functionality with high Z and E selectivity via an alkenyl sp^2 C–H bond olefination using a Cp*Rh(III) catalyst (Figure 3) [56]. This strategy provides easy access to a wide variety of macrocodes with different ring sizes in an atom-economical manner. Mechanistically, this ring-closing alkene–alkene coupling reaction was proposed to proceed via a Z-olefinic vinylrohodium(III) intermediate of the acrylamide derivative (Figure 4). Initially, the anion exchange of the [RhCp*Cl$_2$]$_2$ catalyst with the NaBARF additive and Cu(OAc)$_2$•H$_2$O produced the highly reactive species **A**, which then selectively activated the Z-olefinic $C(sp^2)$–H bond of acrylamides to generate the vinylrhodium(III) intermediate **B**. Subsequent coordination and further migratory insertion occurred smoothly to afford the intermediate **D**, which finally underwent facile β-H elimination to give the macrocycle product.

The intramolecular alkene–alkene ring cyclization reaction using the rhodium catalyst led to various macrocycles containing a versatile dienoate moiety. The conjugated moiety present in the resulting macrocycles is highly versatile and can be converted into many different functional groups, thereby allowing easy access to macrolides containing different functionalities (Figure 5). For example, the hydrogenation of the macrocyclization product in the presence of the Pd/C catalyst afforded the saturated alkane product. The treatment of the product with a copper catalyst with PhMe$_2$Si-BPin furnished the 1,4-hydrosilylation product. On the other hand, reacting the product with the bis(pinacolato)diboron reagent led to the formation of the internal alkene product. Subjecting the product to nitromethane in the presence of 1,8-diazabicyclo[5.4.0]undec-7-ene (DBU) led to the 1,4-Michael addition product. Overall, the versatility of the dienoate products allowed easy access to a wide variety of useful macrocycles that were not easily accessible using the reported methods.

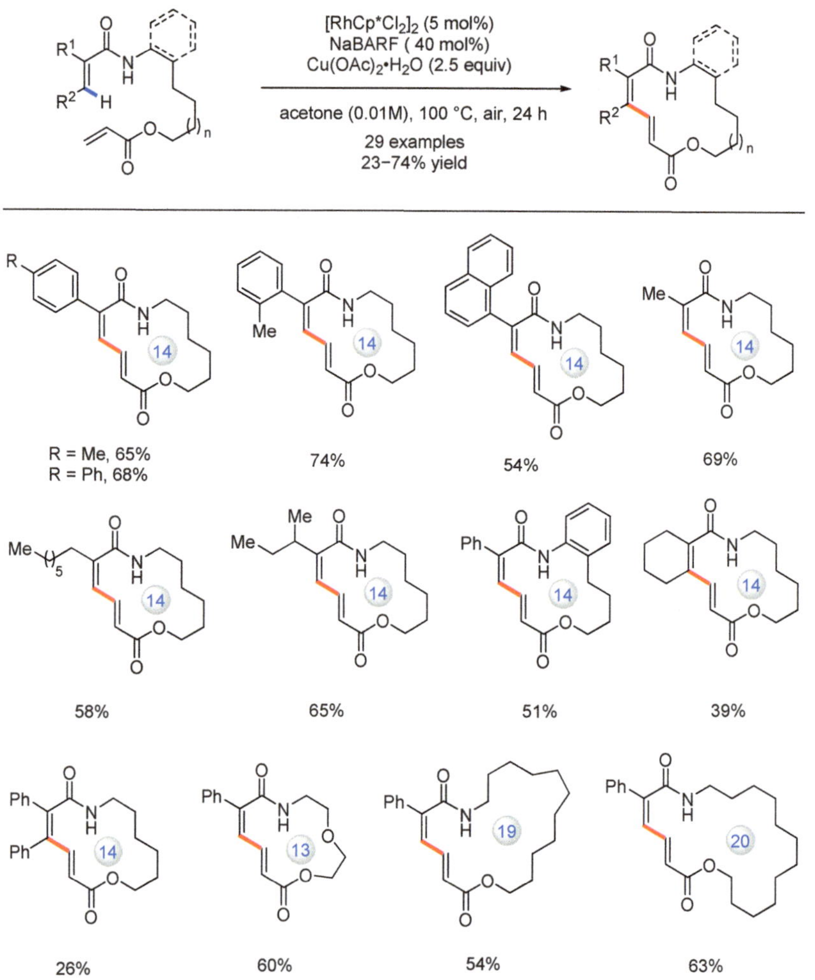

Figure 3. Macrolide synthesis via Cp*Rh(III)-catalyzed intramolecular oxidative cross-coupling of alkenes [56].

In 2020, the Loh group further expanded their intramolecular ring-closing alkene–alkene coupling strategy to α,β-unsaturated ketone substrates (Figure 6) [57]. Under the same conditions, a wide variety of macrolactams of different ring sizes were synthesized in moderate to good yields (33−72%). Of note, the intramolecular coupling of the alkene substrate with no substitution at the α-position also proceeded smoothly to furnish the corresponding macrolactam at a 44% yield. Moreover, this protocol could afford more strained 12- and 13-membered ring macrolactams in satisfactory yields. The authors proposed an analogous catalytic mechanism to elucidate this Cp*Rh(III)-catalyzed alkenyl C(sp^2)–H activation–macrocyclization pathway.

we will discuss the recent progress in the inert C(sp^3)–H bond (both with and without a directing group) functionalization macrocyclization strategy.

3.1. C(sp^2)–H Bond Functionalization of Indoles

The macrocyclization reactions involving C(sp^2)–H bond activation can be grouped into 2 categories. The first group involves the sp^2 C–H bond activation of the indole of Trp residues to couple with different coupling partners. With the assistance of microwave irradiation, James and co-workers in 2012 first developed a general palladium-catalyzed method to activate indole C–H bonds of tryptophan derivatives bearing aromatic iodide side chains for peptide macrocyclization through a Pd(0)/Pd(II) redox manifold, enabling the formation of diverse indole–aryl-bridged macrocycles (Figure 7) [64]. In the presence of 2-NO$_2$-BzOH and AgBF$_4$ additives, the macrocyclization reaction was typically finished within 30 min at 130 °C in dimethylacetamide (DMA), which could occur at a remarkably moderate concentration (30 mM). Both *para*- and *meta*-substituted aromatic iodides could be tolerated with the current protocol, generating a variety of 15- to 25-membered rings macrocyclization products in yields ranging from 40% to 75%. Of note, unprotected Tyr was also tolerated in this strategy.

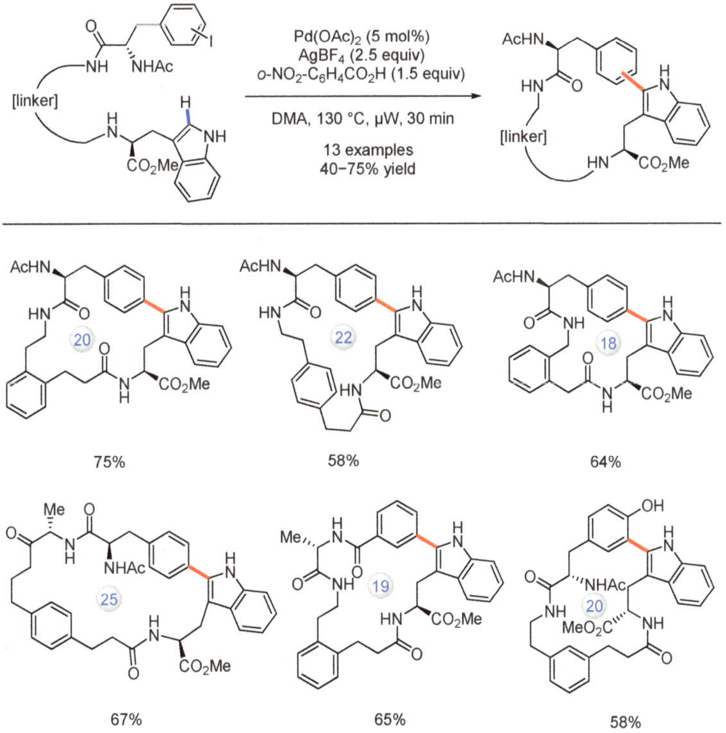

Figure 7. Synthesis of cyclic peptides via palladium-catalyzed intramolecular indole C-2 C–H bond arylation [64].

Later, in 2015, Albericio, Lavilla, and their co-workers extended their Pd-catalyzed C–H activation–macrocyclization protocol, and further demonstrated the intramolecular C–H bond arylation of indoles of Trp residues at the C2 position with different aryl halides to construct structurally more complex and constrained cyclopeptides containing unique Trp(C4)–arene cross-links (Figure 8a) [65] This transformation was compatible with a series of unprotected amino acid units such as Arg, Gln, Asp, and Ser, which could be applied to both solution-phase and solid-phase peptide synthesis processes. Using this strategy, the authors were able to synthesize a double stapled bicyclic peptide in a 25% HPLC conversion process, which proved extremely difficult to prepare using previously reported methods. It should be noted that the competing intermolecular cyclodimerization associated with this strategy selectively generated cyclodimeric peptides. In order to obtain the determining factors in controlling the fate of the transformation, Lavilla et al. further carried out a systematic study of the intermolecular C–H bond arylation of Trp residues bearing iodinated aromatic side chains, and the results showed that linear peptides containing a *meta-* or *para-*iodophenylalanine unit at adjacent positions selectively afforded cyclic dimers instead of stapled peptides. Increasing the chain length of the residues remarkably resulted in the generation of peptide-based macrocycles (Figure 8b) [66].

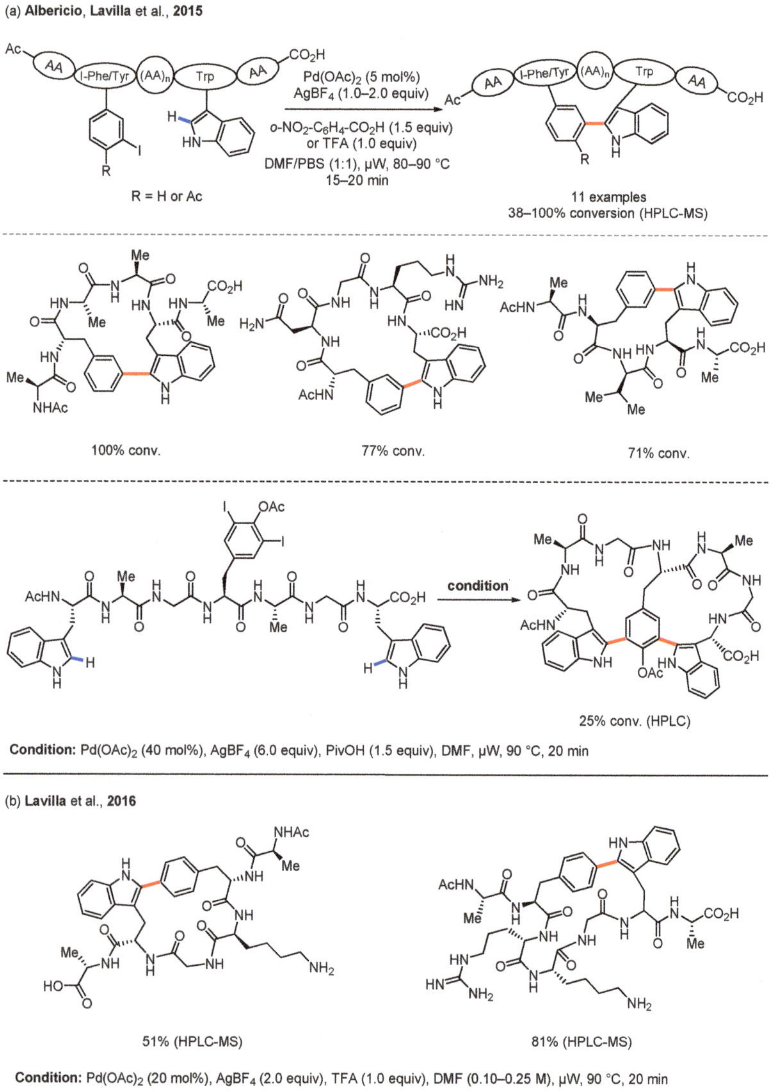

Figure 8. Synthesis of stapled peptides via palladium-catalyzed intramolecular C(sp^2)–H bond arylation [65,66].

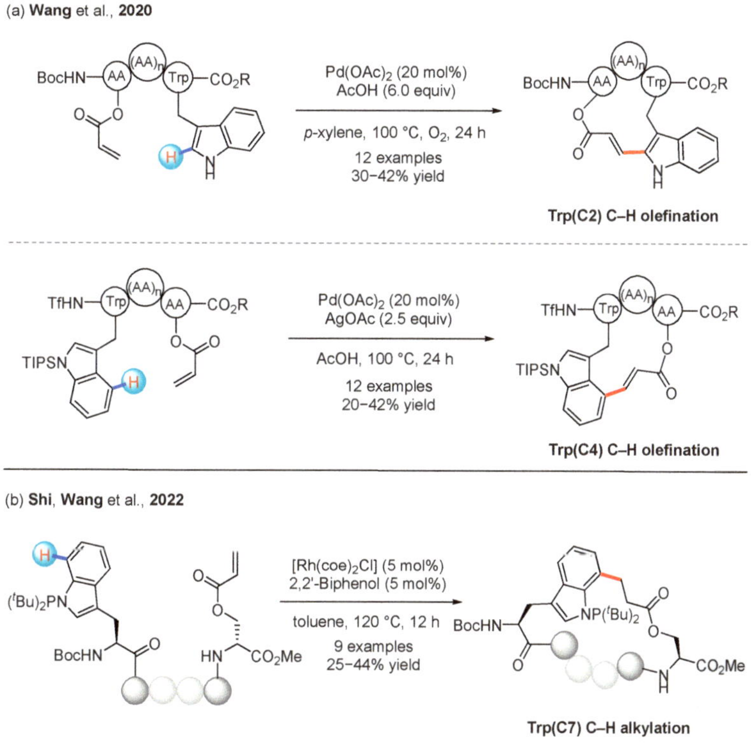

Figure 10. Peptide macrocyclizations via site-selective intramolecular C(sp^2)–H bond functionalization of tryptophan [68,69].

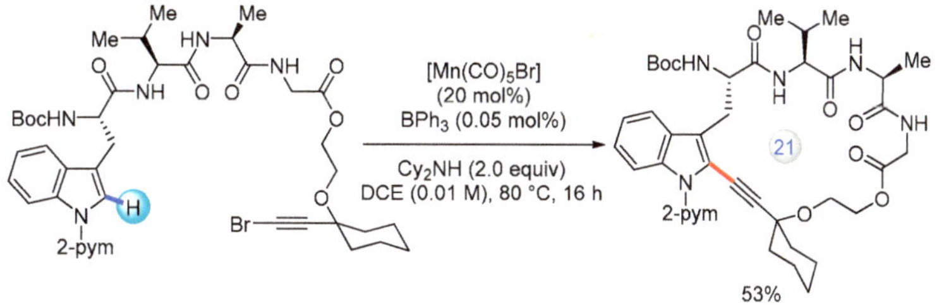

Figure 11. Peptide macrocyclization via manganese-catalyzed intramolecular C(sp^2)–H alkynylation [70].

3.2. C(sp^2)–H Bond Functionalization of Arenes

Besides the above-mentioned examples of C(sp^2)–H bond functionalization of indole of tryptophan, the intramolecular macrocyclization reactions via transition-metal-catalyzed C(sp^2)–H bond functionalization of simple arenes have also been investigated in recent years. In 2019, Wang and colleagues elaborated a general backbone-directed approach to fabricate macrocycle peptides with biaryl cross-links via the palladium-catalyzed intramolecular *ortho*-C(sp^2)–H arylation of short peptides with tethered aromatic iodide side chains (Figure 12) [71]. The N-terminal benzamides of the peptide backbone act as efficient directing groups in this strategy, which significantly facilitate this site-specific arylation process. Under the reaction conditions, biaryl-bridged cyclic peptides with different ring sizes were synthesized in reasonable yields (20–40%). Moreover, this protocol was also applicable to the intermolecular *ortho*-C(sp^2)–H arylation of various oligopeptides, furnishing a series of biaryl-linked products in good yields.

In 2020, Liu and co-workers reported the activation of the C(sp^2)–H bond of the N-protected indole at the C2 position with the assistance of the 2-pyridyl directing group enabled by Cp*Rh(III) catalysis. Under the typical reaction conditions of [RhCp*Cl$_2$]$_2$ (10 mol%), Ag$_2$O (20 mol%), and AgOAc (1.5 equivalent) in CH$_3$CN at 80 °C, the intramolecular C(sp^2)–H bond coupling with a tethered maleimide moiety proceeded uneventfully to provide 3 examples of maleimide-decorated, tryptophan-based macrocyclic peptides in moderate yields (Figure 9) [67].

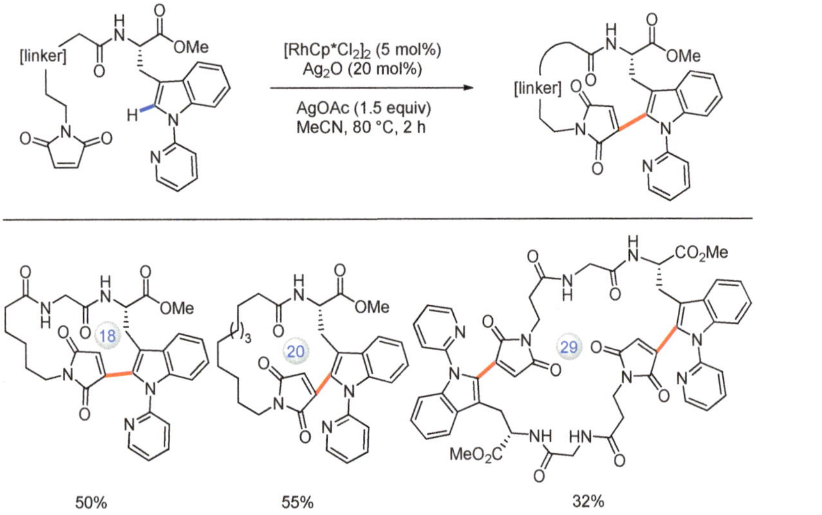

Figure 9. Synthesis of maleimide-decorated peptide macrocycles via the Cp*Rh(III)-catalyzed intramolecular C(sp^2)–H alkenylation of tryptophan [67].

In an effort to achieve highly regioselective late-stage peptide macrocyclization, Wang et al. in 2020 utilized the peptide backbone as endogenous directing groups to promote site-selective peptide macrocyclization at Trp(C2) (Figure 10a) [68]. Using the combination of a Pd(OAc)$_2$ (10 mol%) catalyst and AcOH (6.0 equivalent) additive under an O$_2$ (1.0 atm) atmosphere in *para*-xylene at 100 °C, 12 examples of peptide-based macrocycles with unique Trp–alkene cross-links were documented in reasonable yields (20–42%). Under modified conditions, the authors were able to synthesize macrocyclic peptides containing unique Trp(C4)–alkene cross-links enabled by the palladium-catalyzed C(sp^2)–H olefination at the C4 position of Trp residues bearing a TfNH-directing group (Figure 9a) [65]. More recently, the same group further extended this method to realize the rhodium(I)-catalyzed regioselective intramolecular C(sp^2)–H bond alkylation of Trp residues bearing a *N*-P^{t}Bu$_2$ directing group, providing efficient access to diverse peptide macrocycles with Trp(C7)-alkyl cross-links (Figure 10b) [69].

In 2017, The Ackerman group successfully utilized alkynyl bromide as an internal coupling partner for the cost-effective manganese(I)-catalyzed late-stage macrocyclization of 2-pym-protected indole of indole Trp residue, giving rise to a 21-membered, peptide-based macrocycle with aryl-alkyne cross-links at a 53% yield (Figure 11) [70].

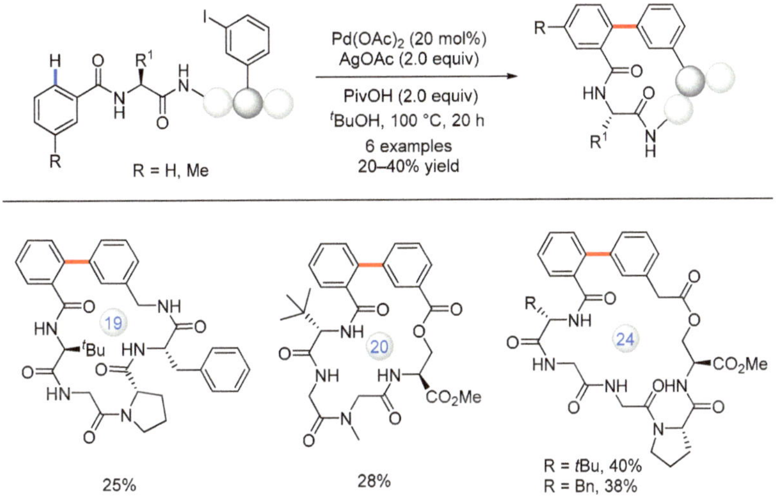

Figure 12. Macrocyclization of biaryl-bridged cyclic peptides via palladium-catalyzed intramolecular C(sp^2)–H bond activation [71].

In a subsequent study, Chen and co-workers disclosed a versatile approach for the synthesis of diverse cyclophane-braced macrocycle peptides through the efficient picolinamide-directed intramolecular aromatic γ- and δ-C(sp^2)–H arylation of readily accessible linear peptide precursors with aryl iodides (Figure 13) [72]. The judicious choice of silver additive was found to be critical for high efficiency. The intramolecular aromatic C(sp^2)–H arylation reactions occurred smoothly in the presence of the Pd(CH$_3$CN)$_4$(BF$_4$)$_2$ (10 mol%) catalyst and 1.5 equivalent of AgOBz additive at 130 °C. Under modified conditions of Pd(OAc)$_2$ and AgOAc in hexafluoroisopropanol (HFIP), the alkenyl counterparts were also compatible with the C–H sources for this approach, giving rise to the corresponding aryl–vinyl C(sp^2)-linked peptide-based macrocycles in moderate yields. However, a remarkable 1:1 mixture of γ- and δ-arylated products was observed for the alkenyl C(sp^2)–H arylation of a smaller ring size. Mechanistically, the reaction was suggested to occur via Pd(II)-catalyzed C(sp^2)–H activation to produce a putative palladacycle species. The subsequent oxidative addition (OA) with tethered aryl iodides generated the key Pd(IV) intermediate, which finally underwent reductive elimination (RE) to afford the C–H activation–macrocyclization product.

The Wang group in 2018 reported the straightforward synthesis of macrocycle peptides with unique aryl–vinyl cross-links via the late-stage palladium-catalyzed δ-C(sp^2)–H olefination of phenylalanine (Phe) residues (Figure 14) [73]. The peptide backbone amides were readily employed as internal directing groups to facilitate the macrocyclization of peptides in the N-to-C direction. This macrocyclization protocol tolerated a broad range of short peptides, generating a variety of peptide-based macrocycles of different sizes and shapes. Both activated and unbaised aliphatic alkenes were viable coupling partners for this macrocyclization. Moreover, the authors also demonstrated the utility of this protocol via the synthesis of a structurally constrained bicyclic peptide via the one-pot macrocyclization of β-C(sp^2)–H arylation and δ-C(sp^2)–H olefination reactions.

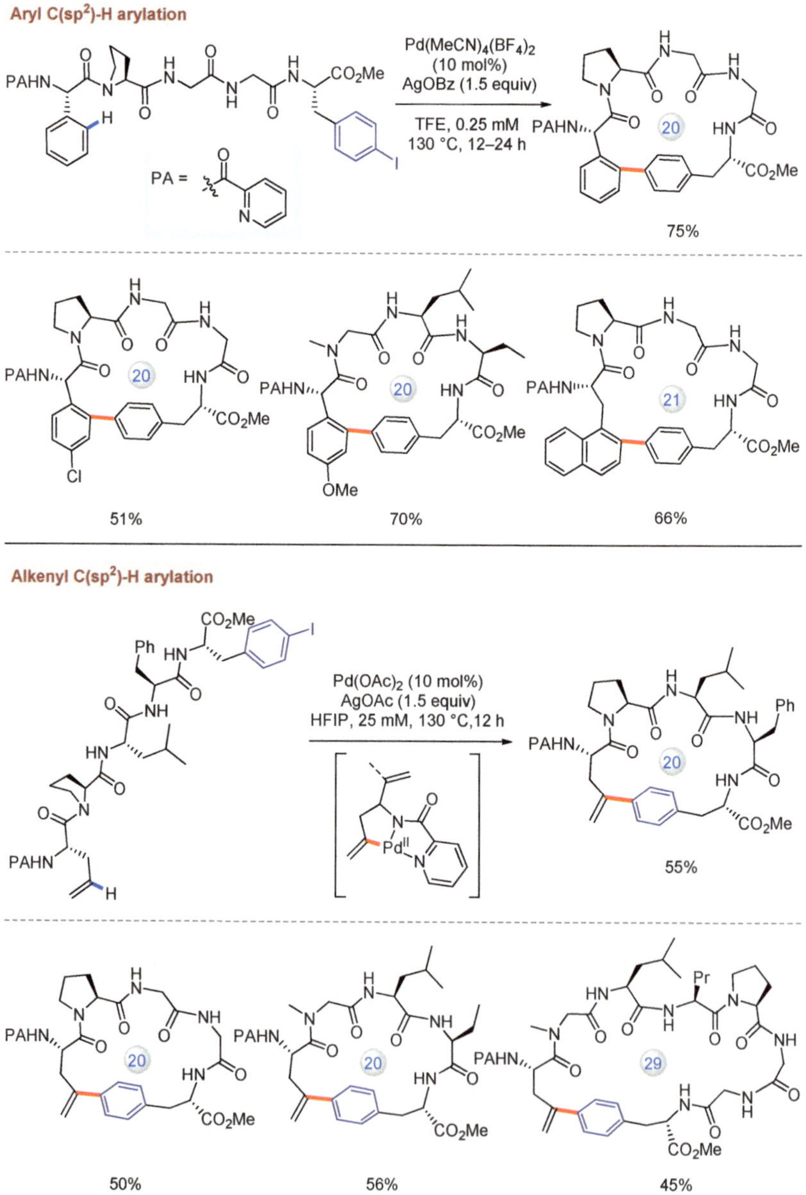

Figure 13. Synthesis of cyclophane-braced peptide macrocycles through palladium-catalyzed intramolecular vinyl and aryl C(sp^2)–H arylation [72].

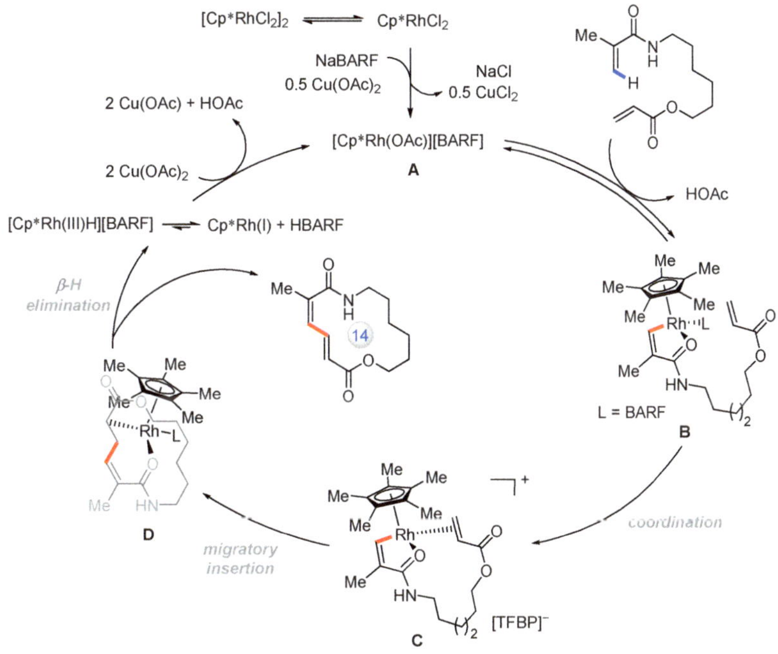

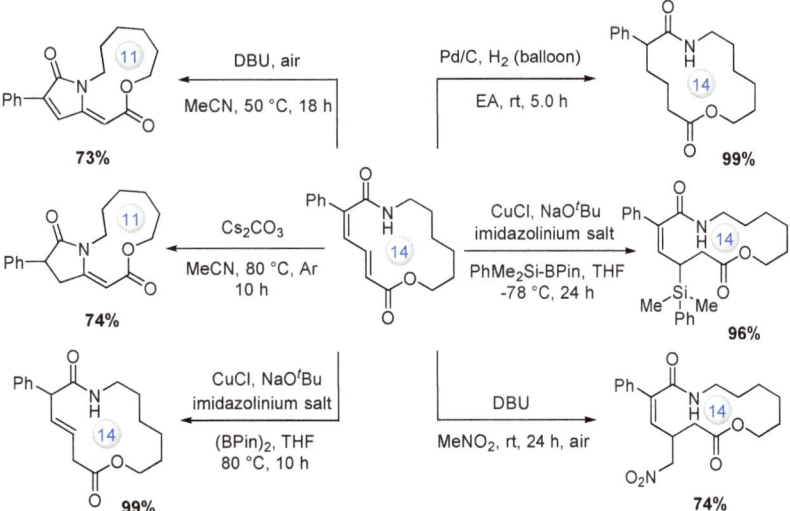

Figure 4. Proposed mechanism for the Cp*Rh(III)-catalyzed intramolecular alkenyl sp^2 C–H activation–macrocyclization [56].

Figure 5. Synthetic derivatization of the cross-coupling macrocycle product [56].

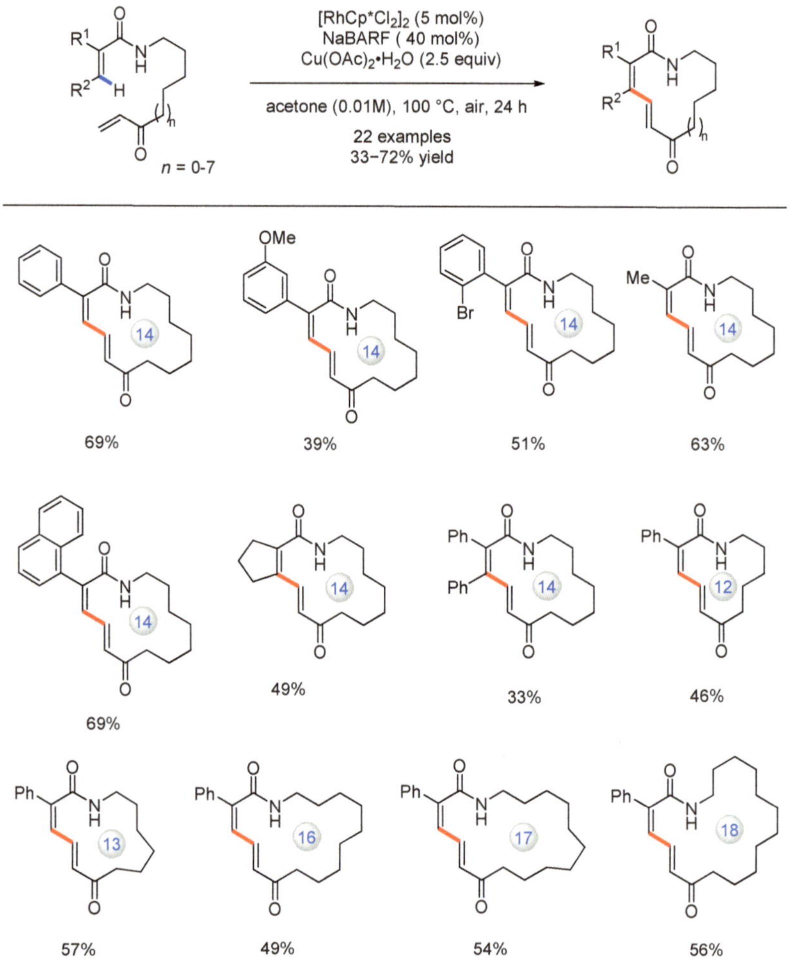

Figure 6. Macrolactam synthesis via Cp*Rh(III)-catalyzed intramolecular ring-closing alkene−alkene cross-couplings [57].

3. Polypeptide Macrocycles

Peptide-based macrocycles have been shown to have a wide variety of biological activities [58]. Some important drugs such as the antibiotic vancomycin [59], anti-cancer agent octreotide [60], and immunosuppressant cyclosporine [61] contain a peptide macrocycle structure. Furthermore, in contrast to linear peptides, peptide macrocycles have improved properties in terms of cell penetration, stability, and selectivity. In recent years, cyclic peptides have also been used as the delivery systems in peptide–drug conjugates such as ^{177}Lu-dotatate [62] and edotreotide gallium Ga-68 [63]. In addition to the amide bond formation strategy, the use of C–H bond functionalization strategies has gained tremendous attention in recent years due to their ability to construct many different stapled cyclopeptides. In this part of the review, we will divide them into 3 different categories: (1) in the first part, we will focus mainly on the sp^2 C–H bond functionalization of indole C–H bond macrocyclization; (2) in the second part, we will focus mainly on the C–H bond functionalization of the aryl C(sp^2)–H bond macrocyclization strategy; (3) in the final part,

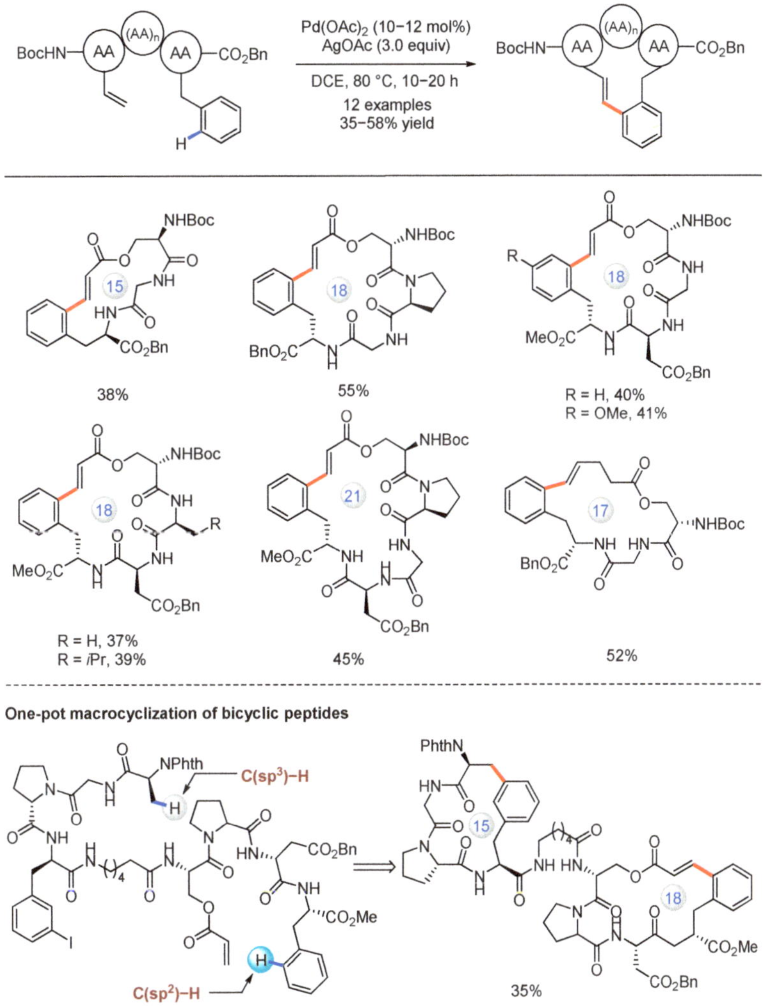

Figure 14. Peptide macrocyclization via Pd(II)-catalyzed intramolecular δ-C(sp^2)–H olefination of phenylalanines [73].

In the same year, Wang and co-workers also reported a peptide-guided C(sp^2)–H activation method for the late-stage macrocyclization of various sulfonamide-containing peptides via intramolecular *ortho*-directed C(sp^2)–H alkenylation, which was significantly facilitated by the internal peptide backbone (Figure 15) [74]. Using the *N*-sulfonated peptides as endogenous directing groups, this macrocyclization protocol featured a broad substrate scope and tolerated both activated acrylates and unbaised aliphatic alkenes, giving rise to a series of bioactive peptidosulfonamide macrocycles with aryl–alkene cross-links over a 34–72% yield range. As an extension of this approach, the same group in 2019 further achieved the macrocyclization reaction of peptidoarylacetamides under identical conditions (Figure 16) [75].

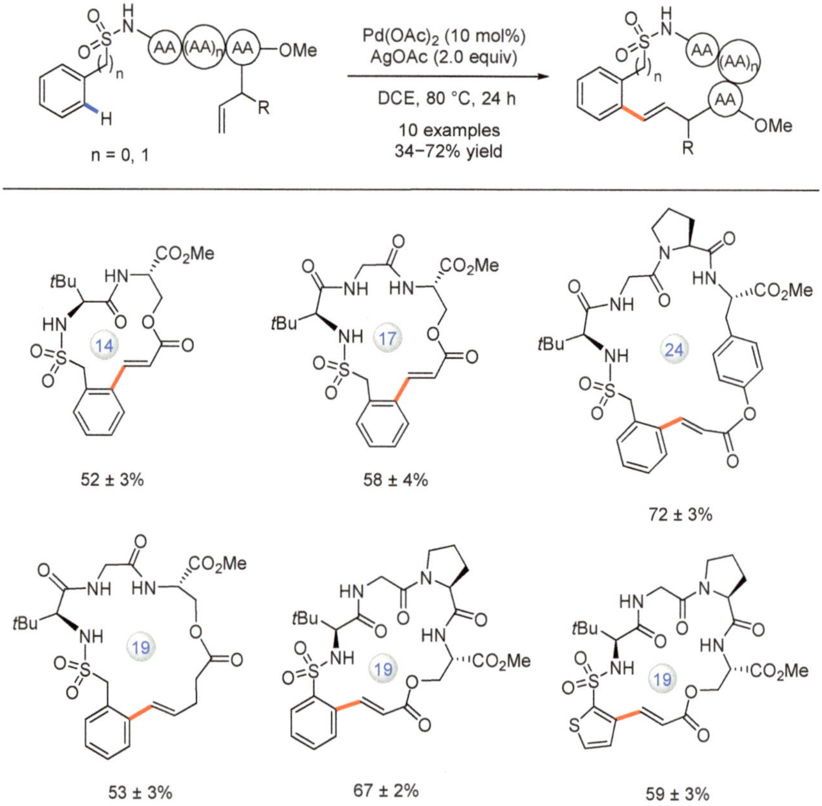

Figure 15. Macrocyclization of peptidosulfonamides via palladium(II)-catalyzed intramolecular C(sp^2)–H bond olefination [74].

Subsequently, Wang and colleagues continued their versatile macrocyclization strategy [76], and further accomplished an efficient late-stage macrocyclization reaction of various bioactive oxazole-containing peptides through the intramolecular palladium-catalyzed δ-C(sp^2)–H olefination (Figure 17a) [77]. In this report, the oxazole motifs in the peptide backbones acted as endogenous directing groups to promote this *ortho*-C(sp^2)–H olefination reactions. The resulting oxazole-containing cyclic peptides bearing aryl–alkene cross-links showed strong cytotoxicity toward cancer cells. Quite recently, Wang et al. further expanded to report on the late-stage modification and macrocyclization of diverse thiazole-containing peptides, generating a series of 21–25 membered bioactive peptide-based macrocycles in 35–59% yields (Figure 17b) [78].

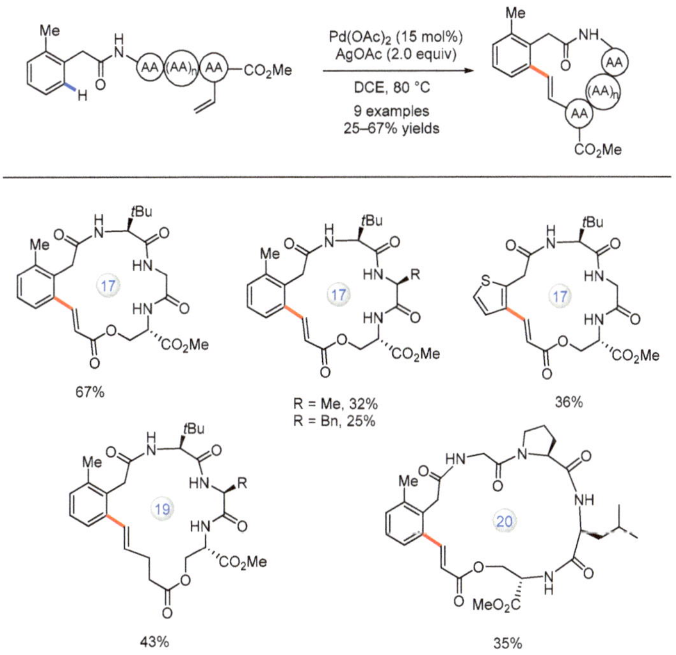

Figure 16. Macrocyclization via Pd(II)-catalyzed intramolecular C(sp²)–H olefination of peptidoarylacetamides [75].

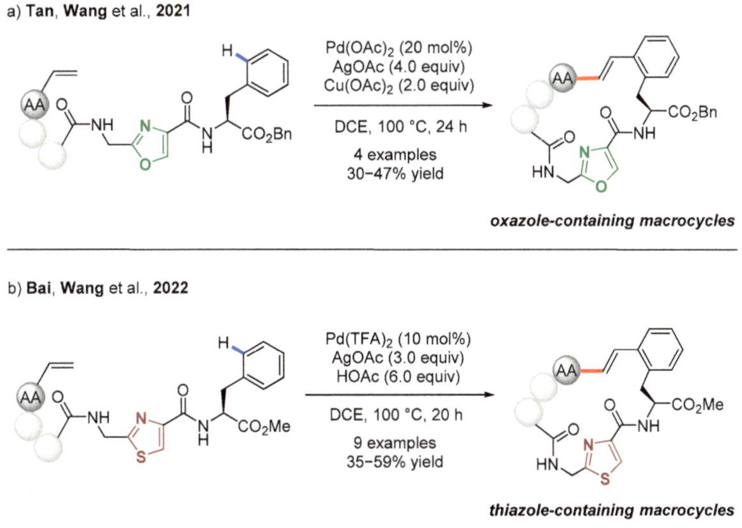

a) **Tan, Wang** et al., 2021

oxazole-containing macrocycles

b) **Bai, Wang** et al., 2022

thiazole-containing macrocycles

Figure 17. Late-stage macrocyclization via intramolecular δ-C(sp²)–H bond olefination of oxazole-and thiazole-containing peptides [77,78].

3.3. C(sp³)–H Bond Functionalization

In recent years, there has been much interest in the development of stapled peptides, including polypeptide macrocycles, for drug discovery. Simultaneously, the construction of C–C cross-linked stapled peptides using the C–H activation coupling strategies has also emerged as an attractive strategy for peptide-based macrocycles synthesis. Although there are many reports on the intermolecular C(sp^3)–H bond functionalization of amino acids and peptides [79–81], the intramolecular version of C(sp^3)–H bond functionalization for polypeptide macrocycles synthesis is still very rare. Inspired by Yu's backbone-assisted inert C(sp^3)–H activation strategy for the late-stage derivatization of various short peptides [82], Albericio, Noisier, and co-workers in 2017 elegantly established an efficient palladium-catalyzed C(sp^3)–H activation method without relying on any external directing group for peptide stapling (Figure 18) [83]. Unprecedentedly, this process allowed the linkage between N-terminal alanine (Ala) and C-terminal phenyl alanine (Phe) residues. The role of the solvent was found to be crucial for this peptide macrocyclization process, and using t-BuOH as the co-solvent they significantly suppressed the unwanted side reactions. Under the optimized reaction conditions, the intramolecular C(sp^3)–H arylation of N-terminal Ala with meta-iodinated Phe proceeded smoothly with good to excellent conversion rates, and a wide variety of cyclic hydrocarbon cross-linked polypeptides bearing different ring sizes macrocycles were synthesized in modest yields. Moreover, this strategy is also compatible with solid-phase peptide synthesis, enabling the rapid synthesis of new peptide motifs.

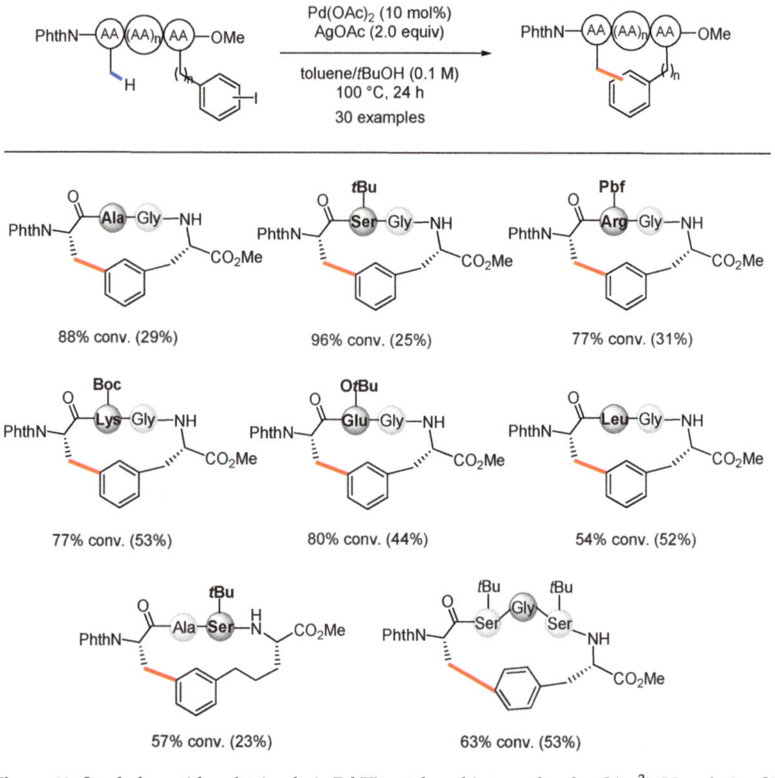

Figure 18. Stapled peptides obtained via Pd(II)-catalyzed intramolecular C(sp^3)–H arylation [83].

Almost at the same time, Wang and co-workers also demonstrated an analogous macrocyclization via the intramolecular β-C(sp^3)–H arylation of amino acids at the N-terminus of the peptides, which could proceed smoothly in both solution-phase and solid-phase peptide synthesis approaches (Figure 19) [84]. In this protocol, 1,2-dichloroethane (DCE) was used as the solvent to achieve remarkable efficiency. This approach tolerated a broad scope of peptide substrates, including tetra- and pentapeptides, delivering a diverse variety of polypeptide macrocycles bearing Cβ–Ar cross-links between the β-carbon of amino acids and the aromatic ring of Phe/Trp, with satisfactory diastereoselectivity. Successfully, the authors applied this macrocyclization methodology to the total synthesis of the key fragment in the bioactive natural product celogentin C.

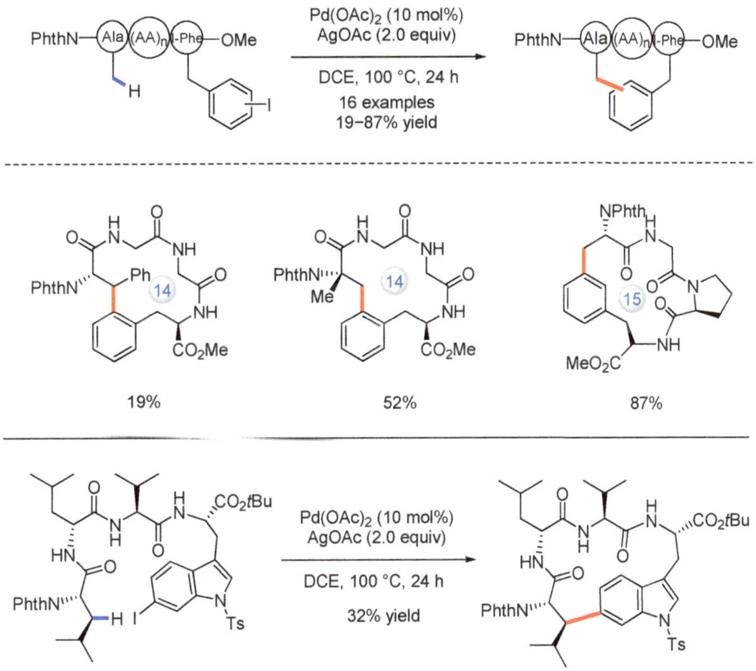

Figure 19. Synthesis of peptide macrocycles via palladium(II)-catalyzed intramolecular C(sp^3)–H arylation [84].

Through the assistance of 8-aminoquinoline (AQ) as the bidentate directing group [85], Chen et al. in 2018 developed a broadly applicable strategy for the efficient synthesis of cyclophane-braced peptide macrocycles from readily accessible linear peptide precursors via the *exo*-type intramolecular β-C(sp^3)–H macrocyclization of diverse alkyl appendants bearing iodinated aromatic side chains or tethered aryl iodides (Figure 20) [86]. The choice of *ortho*-phenyl benzoic acid (oPBA) as the efficacious additive significantly promoted this AQ-directed intramolecular β methylene C(sp^3)–H arylation reaction. Both protected and unprotected amino acid units were compatible with this macrocyclization strategy. Various cyclophane-braced peptide macrocycles of different sizes and shapes were synthesized in moderate to good yields. However, the amide-linked AQ auxiliary was found to be difficult to remove from the peptide macrocycles in this case. Mechanistically, this AQ-directed intramolecular macrocyclization process was proposed to occur through a concerted metalation–deprotonation (CMD) catalytic mechanism. With the assistance of the carboxylate ligand, the reversible β-C(sp^3)–H palladation produced a chelation-stabilized five-membered alkylpalladium(II) species. Subsequently, this putative palladacycle intermediate underwent intramolecular oxidative addition and reductive elimination to afford the expected peptide macrocycles.

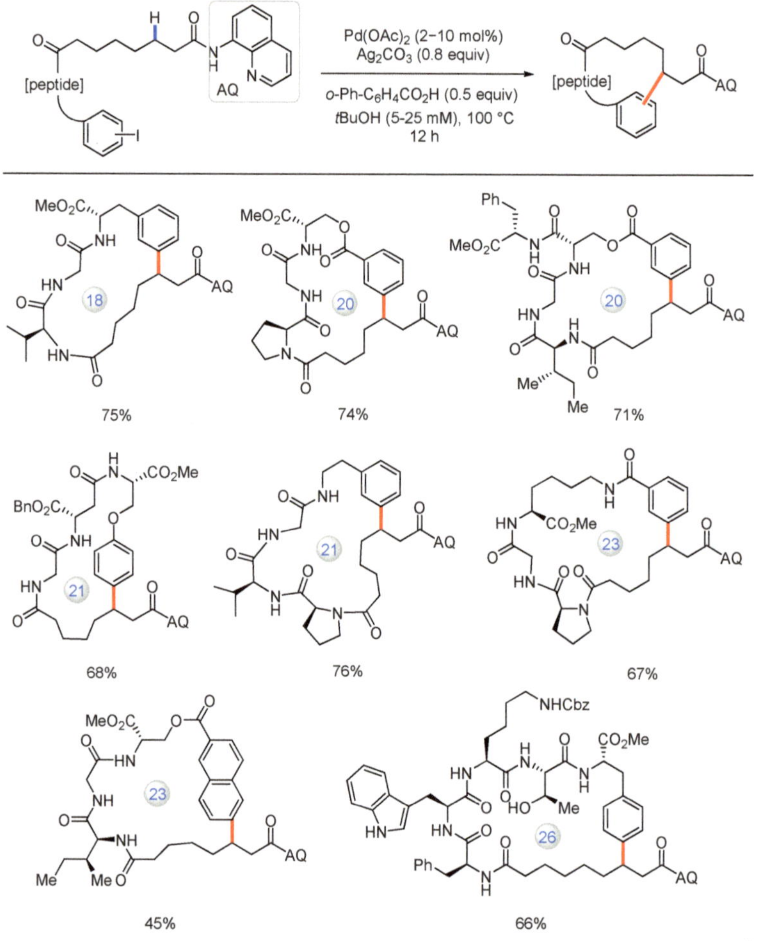

Figure 20. Synthesis of cyclophane-braced peptide macrocycles via palladium-catalyzed intramolecular C(sp³)–H arylation [86].

By taking advantage of readily accessible and removable N,N-bidentate picolinamide (PA) auxiliaries, Chen and co-workers in 2019 also described the efficient Pd-catalyzed intramolecular remote γ-C(sp^3)–H arylation of diverse N-terminal aliphatic amino acid units via Pd(II)/Pd(IV) catalysis (Figure 21a) [87]. This reaction tolerates γ-C(sp^3)–H bonds at both methyl and methylene positions of peptide substrates. Unprotected peptides bearing diverse free polar side chains were proven to be compatible substrates for this PA-directed C(sp^3)–H macrocyclization protocol, and the cyclization reaction occurred smoothly with high efficiency. Through the typical combination of a Pd(MeCN)$_4$(BF$_4$)$_2$ catalyst and AgOAc additive, 30 examples of this transformation were documented with reasonable yields up to 80%. Encouraged by this success, the same group further elaborated on the efficient synthesis of cyclophane-braced peptide macrocycles via the palladium(II)-catalyzed, AQ-directed *endo*-type intramolecular β methyl C(sp^3)–H arylation reaction with tethered aryl iodides, giving rise to a broad range of cyclic peptides featuring different ring sizes in satisfactory yields (Figure 21b) [88].

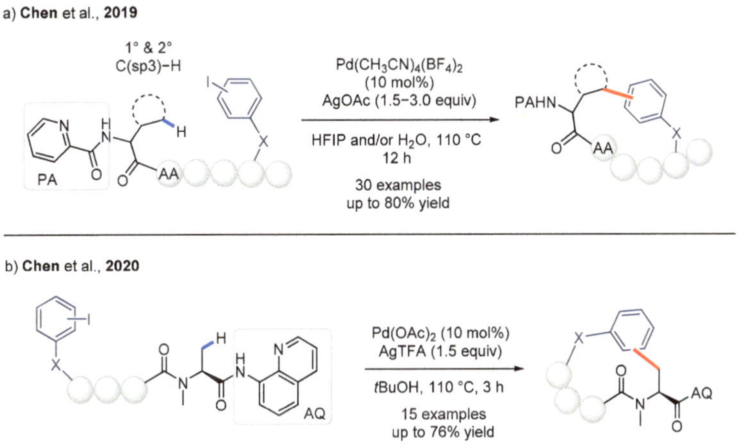

Figure 21. Synthesis of cyclophane-braced peptide macrocycles via intramolecular C(sp^3)–H arylation [87,88].

4. Conclusions

In summary, we have witnessed a surge in the use of intramolecular C–H bond activation–macrocyclization reactions to construct polypeptide macrocycles, and to a lesser extent polyketide macrocycles. Despite the remarkable advantages of using C–H activation strategies for the straightforward synthesis of diverse polyketide and polypeptide macrocycles, it is not yet possible to completely avoid the use of substrates bearing alkene or aryl iodide functionalities for the C–H bond activation–macrocyclization reactions. The application of the C–H bond macrocyclization strategy is still an untapped territory, and strategies utilizing transition-metal-free electrocatalytic photoredox C–H bond activation remain to be tested. Most of the reported strategies utilize the aryl C(sp^2)–H activation methods to construct macrocycles. Sporadic examples utilizing alkenyl C(sp^2)–H bond and alkyl C(sp^3)–H bond activation methods have also emerged in recent years. With the surge in activities related to C–H bond activation, more reports utilizing C–H activation strategies for macrocyclization can be envisaged in this fascinating field.

Author Contributions: X.W.: Literature collection and manuscript writing. M.-Z.L.: Literature collection, manuscript writing. and funding acquisition; T.-P.L.: Content design, manuscript revision, project administration, and funding acquisition. All authors have read and agreed to the published version of the manuscript.

Funding: We gratefully acknowledge the financial support from the National Natural Science Foundation of China (no. 22101095) and the "Innovation and Entrepreneurship Talents Plan" of Jiangsu Province (M.-Z. Lu). We also acknowledge the financial support from the Distinguished University Professor grant (Nanyang Technological University); the AcRF Tier 1 grants from the Ministry of Education of Singapore (RG11/20 and RT14/20); and the Agency for Science, Technology, and Research (A*STAR) under its MTC Individual Research Grants (M21K2c0114).

Data Availability Statement: Not applicable.

Conflicts of Interest: The authors declare no conflict of interest.

References

1. Gibson, S.E.; Lecci, C. Amino acid derived macrocycles—An area driven by synthesis or application? *Angew. Chem. Int. Ed.* **2006**, *45*, 1364–1377. [CrossRef]
2. Yudin, A.K. Macrocycles: Lessons from the distant past, recent developments, and future directions. *Chem. Sci.* **2015**, *6*, 30–49. [CrossRef]
3. Wessjohann, L.; Rivera, D.G.; Vercillo, O.E. Multiple multicomponent macrocyclizations (MiBs): A strategic development toward macrocycle diversity. *Chem. Rev.* **2009**, *109*, 796–814. [CrossRef]
4. White, C.J.; Yudin, A.K. Contemporary strategies for peptide macrocyclization. *Nat. Chem.* **2011**, *3*, 509–524. [CrossRef]
5. Martí-Centelles, V.; Pandey, M.D.; Burguete, M.I.; Luis, S.V. Macrocyclization reactions: The importance of conformational, configurational, and template-induced preorganization. *Chem. Rev.* **2015**, *115*, 8736–8834. [CrossRef] [PubMed]
6. Wu, J.; Tang, J.; Chen, H.; He, Y.; Wang, H.; Yao, H. Recent developments in peptide macrocyclization. *Tetrahedron Lett.* **2018**, *59*, 325–333. [CrossRef]
7. Mortensen, K.T.; Osberger, T.J.; King, T.A.; Sore, H.F.; Spring, D.R. Strategies for the diversity-oriented synthesis of macrocycles. *Chem. Rev.* **2019**, *119*, 10288–10317. [CrossRef]
8. Reguera, L.; Rivera, D.G. Multicomponent reaction toolbox for peptide macrocyclization and stapling. *Chem. Rev.* **2019**, *119*, 9836–9860. [CrossRef] [PubMed]
9. Zheng, K.; Hong, R. Stereoconfining macrocyclizations in the total synthesis of natural products. *Nat. Prod. Rep.* **2019**, *36*, 1546–1575. [CrossRef]
10. Rivera, D.G.; Ojeda-Carralero, G.M.; Reguera, L.; Van der Eycken, E.V. Peptide macrocyclization by transition metal catalysis. *Chem. Soc. Rev.* **2020**, *49*, 2039–2059. [CrossRef]
11. Parenty, A.; Moreau, X.; Campagne, J.-M. Macrolactonizations in the total synthesis of natural products. *Chem. Rev.* **2006**, *106*, 911–939. [CrossRef]
12. Li, Y.; Yin, X.; Dai, M. Catalytic macrolactonizations for natural product synthesis. *Nat. Prod. Rep.* **2017**, *34*, 1185–1192. [CrossRef] [PubMed]
13. Yang, M.; Wang, X.; Zhao, J. Ynamide-mediated macrolactonization. *ACS Catal.* **2020**, *10*, 5230–5235. [CrossRef]
14. Ishihara, K.; Kuroki, Y.; Hanaki, N.; Ohara, S.; Yamamoto, H. Antimony-templated macrolactamization of tetraamino esters. facile synthesis of macrocyclic spermine alkaloids, (±)-buchnerine, (±)-verbacine, (±)-verbaskine, and (±)-verbascenine. *J. Am. Chem. Soc.* **1996**, *118*, 1569–1570. [CrossRef]
15. Doi, T.; Kamioka, S.; Shimazu, S.; Takahashi, T. A synthesis of RGD model cyclic peptide by palladium-catalyzed carbonylative macrolactamization. *Org. Lett.* **2008**, *10*, 817–819. [CrossRef] [PubMed]
16. Lim, N.-K.; Linghu, X.; Wong, N.; Zhang, H.; Sowell, C.G.; Gosselin, F. Macrolactamization approaches to arylomycin antibiotics core. *Org. Lett.* **2019**, *21*, 147–151. [CrossRef] [PubMed]
17. Hayward, C.M.; Yohannes, D.; Danishefsky, S.M. Total synthesis of rapamycin via a novel titanium-mediated aldol macrocyclization reaction. *J. Am. Chem. Soc.* **1993**, *115*, 9345–9346. [CrossRef]
18. Wessjohann, L.A.; Scheid, G.O.; Eichelberger, U.; Umbreen, S. Total synthesis of epothilone D: The nerol/macroaldolization approach. *J. Org. Chem.* **2013**, *78*, 10588–10595. [CrossRef]
19. Abramite, J.A.; Sammakia, T. Application of the intramolecular Yamamoto vinylogous aldol reaction to the synthesis of macrolides. *Org. Lett.* **2007**, *9*, 2103–2106. [CrossRef]
20. Gradillas, A.; Pérez-Castells, J. Macrocyclization by ring-closing metathesis in the total synthesis of natural products: Reaction conditions and limitations. *Angew. Chem. Int. Ed.* **2006**, *45*, 6086–6101. [CrossRef]
21. Denmark, S.E.; Muhuhi, J.M. Development of a general, sequential, ring-closing metathesis/intramolecular cross-coupling reaction for the synthesis of polyunsaturated macrolactones. *J. Am. Chem. Soc.* **2010**, *132*, 11768–11778. [CrossRef] [PubMed]
22. Yu, M.; Wang, C.; Kyle, A.F.; Jakubec, P.; Dixon, D.; Schrock, R.R.; Hoveyda, A.H. Synthesis of macrocyclic natural products by catalyst-controlled stereoselective ring-closing metathesis. *Nature* **2011**, *479*, 88–93. [CrossRef] [PubMed]
23. Lecourt, C.; Dhambri, S.; Allievi, L.; Sanogo, Y.; Zeghbib, N.; Othman, R.B.; Lannou, M.-I.; Sorin, G.; Ardisson, J. Natural products and ring-closing metathesis: Synthesis of sterically congested olefins. *Nat. Prod. Rep.* **2018**, *35*, 105–124. [CrossRef]
24. Gensch, T.; Hopkinson, M.N.; Glorius, F.; Wencel-Delord, J. Mild metal-catalyzed C–H activation: Examples and concepts. *Chem. Soc. Rev.* **2016**, *45*, 2900–2936. [CrossRef] [PubMed]
25. He, J.; Wasa, M.; Chan, K.S.L.; Shao, Q.; Yu, J.-Q. Palladium-catalyzed transformations of alkyl C–H bonds. *Chem. Rev.* **2017**, *117*, 8754–8786. [CrossRef]
26. Loup, J.; Dhawa, U.; Pesciaioli, F.; Wencel-Delord, J.; Ackermann, L. Enantioselective C–H activation with earth-abundant 3d transition metals. *Angew. Chem. Int. Ed.* **2019**, *58*, 12803–12818. [CrossRef] [PubMed]
27. Woźniak, Ł.; Tan, J.-F.; Nguyen, Q.-H.; Madron du Vigne, A.; Smal, V.; Cao, Y.-X.; Cramer, N. Catalytic enantioselective functionalizations of C–H bonds by chiral iridium complexes. *Chem. Rev.* **2020**, *120*, 10516–10543. [CrossRef]
28. Liu, B.; Romine, A.M.; Rubel, C.Z.; Engle, K.M.; Shi, B.-F. Transition-metal-catalyzed, coordination-assisted functionalization of nonactivated C(sp^3)–H bonds. *Chem. Rev.* **2021**, *121*, 14957–15074. [CrossRef]
29. Murali, K.; Machado, L.A.; Carvalho, R.L.; Pedrosa, L.F.; Mukherjee, R.; da Silva Júnior, E.N.; Maiti, D. Decoding directing groups and their pivotal role in C–H activation. *Chem. Eur. J.* **2021**, *27*, 12453–12508. [CrossRef]

30. Carvalho, R.L.; Almeida, R.G.; Murali, K.; Machado, L.A.; Pedrosa, L.F.; Dolui, P.; Maiti, D.; da Silva Júnior, E.N. Removal and modification of directing groups used in metal-catalyzed C–H functionalization: The magical step of conversion into 'conventional' functional groups. *Org. Biomol. Chem.* **2021**, *19*, 525–547. [CrossRef]
31. Sinha, S.K.; Guin, S.; Maiti, S.; Biswas, J.P.; Porey, S.; Maiti, D. Toolbox for distal C–H bond functionalizations in organic molecules. *Chem. Rev.* **2022**, *122*, 5682–5841. [CrossRef] [PubMed]
32. Sambiagio, C.; Schönbauer, D.; Blieck, R.; Dao-Huy, T.; Pototschnig, G.; Schaaf, P.; Wiesinger, T.; Zia, M.F.; Wencel-Delord, J.; Besset, T.; et al. A comprehensive overview of directing groups applied in metal-catalysed C–H functionalisation chemistry. *Chem. Soc. Rev.* **2018**, *47*, 6603–6743. [CrossRef] [PubMed]
33. Sauermann, N.; Meyer, T.H.; Qiu, Y.; Ackermann, L. Electrocatalytic C–H activation. *ACS Catal.* **2018**, *8*, 7086–7103. [CrossRef]
34. Gandeepan, P.; Müller, T.; Zell, D.; Cera, G.; Warratz, S.; Ackermann, L. 3d Transition metals for C–H activation. *Chem. Rev.* **2019**, *119*, 2192–2452. [CrossRef] [PubMed]
35. Stepek, I.A.; Itami, K. Recent advances in C–H activation for the synthesis of π-extended materials. *ACS Mater. Lett.* **2020**, *2*, 951–974. [CrossRef]
36. Achar, T.K.; Maiti, S.; Jana, S.; Maiti, D. Transition metal catalyzed enantioselective C(sp^3)–H bond functionalization. *ACS Catal.* **2020**, *10*, 13748–13793. [CrossRef]
37. Meng, G.; Lam NY, S.; Lucas, E.L.; Saint-Denis, T.G.; Verma, P.; Chekshin, N.; Yu, J.-Q. Achieving site-selectivity for C–H activation processes based on distance and geometry: A carpenter's approach. *J. Am. Chem. Soc.* **2020**, *142*, 10571–10591. [CrossRef]
38. Zhang, Q.; Shi, B.-F. 2-(Pyridin-2-yl)isopropyl (PIP) amine: An enabling directing group for divergent and asymmetric functionalization of unactivated methylene C(sp^3)–H bonds. *Acc. Chem. Res.* **2021**, *54*, 2750–2763. [CrossRef]
39. Lam NY, S.; Wu, K.; Yu, J.-Q. Advancing the logic of chemical synthesis: C–H activation as strategic and tactical disconnections for C–C bond construction. *Angew. Chem. Int. Ed.* **2021**, *60*, 15767–15790. [CrossRef]
40. Kharitonov, V.B.; Muratov, D.V.; Loginov, D.A. Cyclopentadienyl complexes of group 9 metals in the total synthesis of natural products. *Coord. Chem. Rev.* **2022**, *471*, 214744. [CrossRef]
41. Mandal, D.; Roychowdhury, S.; Biswas, J.P.; Maiti, S.; Maiti, D. Transition-metal-catalyzed C–H bond alkylation using olefins: Recent advances and mechanistic aspects. *Chem. Soc. Rev.* **2022**, *51*, 7358–7426. [CrossRef] [PubMed]
42. Lucas, E.L.; Lam NY, S.; Zhuang, Z.; Chan HS, S.; Strassfeld, D.A.; Yu, J.-Q. Palladium-catalyzed enantioselective β-C(sp^3)–H activation reactions of aliphatic acids: A retrosynthetic surrogate for enolate alkylation and conjugate addition. *Acc. Chem. Res.* **2022**, *55*, 537–550. [CrossRef] [PubMed]
43. Lu, M.-Z.; Goh, J.; Maraswami, M.; Jia, Z.; Tian, J.-S.; Loh, T.-P. Recent advances in alkenyl sp^2 C–H and C–F bond functionalizations: Scope, mechanism, and applications. *Chem. Rev.* **2022**, *122*, 17479–17646. [CrossRef]
44. Sengupta, S.; Mehta, G. Late stage modification of peptides via C–H activation reactions. *Tetrahedron Lett.* **2017**, *58*, 1357–1372. [CrossRef]
45. Lu, X.; He, S.-J.; Cheng, W.-M.; Shi, J. Transition-metal-catalyzed C–H functionalization for late-stage modification of peptides and proteins. *Chin Chem Lett.* **2018**, *29*, 1001–1008. [CrossRef]
46. Wang, W.; Lorion, M.M.; Shah, J.; Kapdi, A.R.; Ackermann, L. Late-stage peptide diversification by position-selective C–H activation. *Angew. Chem. Int. Ed.* **2018**, *57*, 14700–14717. [CrossRef]
47. Tong, H.-R.; Li, B.; Li, G.; He, G.; Chen, G. Postassembly modifications of peptides via metal-catalyzed C–H functionalization. *CCS Chem.* **2020**, *2*, 1797–1820. [CrossRef]
48. Sengupta, S.; Mehta, G. Macrocyclization via C–H functionalization: A new paradigm in macrocycle synthesis. *Org. Biomol. Chem.* **2020**, *18*, 1851–1876. [CrossRef]
49. Minami, A.; Eguchi, T. Substrate flexibility of vicenisaminyltransferase vinC involved in the biosynthesis of vicenistatin. *J. Am. Chem. Soc.* **2007**, *129*, 5102–5107. [CrossRef]
50. Shinohara, Y.; Kudo, F.; Eguchi, T. A natural protecting group strategy to carry an amino acid starter unit in the biosynthesis of macrolactam polyketide antibiotics. *J. Am. Chem. Soc.* **2011**, *133*, 18134–18137. [CrossRef]
51. Miyanaga, A.; Iwasawa, S.; Shinohara, Y.; Kudo, F.; Eguchi, T. Structure-based analysis of the molecular interactions between acyltransferase and acyl carrier protein in vicenistatin biosynthesis. *PNAS* **2016**, *113*, 1802–1807. [CrossRef] [PubMed]
52. Qin, H.-L.; Panek, J.S. Total synthesis of the Hsp90 inhibitor geldanamycin. *Org. Lett.* **2008**, *10*, 2477–2479. [CrossRef] [PubMed]
53. Hilton, M.J.; Brackett, C.M.; Mercado, B.Q.; Blagg BS, J.; Miller, S.J. Catalysis-enabled access to cryptic geldanamycin oxides. *ACS Cent. Sci.* **2020**, *6*, 426–435. [CrossRef]
54. Nazare, M.; Waldmann, H. Synthesis of the (9S,18R) diastereomer of cyclamenol A. *Angew. Chem. Int. Ed.* **2000**, *39*, 1125–1128. [CrossRef]
55. Zhang, J.; Loh, T.-P. Ruthenium- and rhodium-catalyzed cross-coupling reaction of acrylamides with alkenes: Efficient access to (Z,E)-dienamides. *Chem. Commun.* **2012**, *48*, 11232–11234. [CrossRef] [PubMed]
56. Jiang, B.; Zhao, M.; Li, S.-S.; Xu, Y.-H.; Loh, T.-P. Macrolide synthesis through intramolecular oxidative cross-coupling of alkenes. *Angew. Chem. Int. Ed.* **2018**, *57*, 555–559. [CrossRef]
57. Maraswami, M.; Goh, J.; Loh, T.-P. Macrolactam synthesis via ring-closing alkene-alkene cross-coupling reactions. *Org. Lett.* **2020**, *22*, 9724–9728. [CrossRef]
58. Driggers, E.M.; Hale, S.P.; Lee, J.; Terrett, N.K. The exploration of macrocycles for drug discovery—An underexploited structural class. *Nat. Rev. Drug Discovery* **2008**, *7*, 608–624. [CrossRef]

59. Shchelik, I.S.; Gademann, K. Thiol- and disulfide-containing vancomycin derivatives against bacterial resistance and biofilm formation. *ACS Med. Chem. Lett.* **2021**, *12*, 1898–1904. [CrossRef]
60. Lelle, M.; Kaloyanova, S.; Freidel, C.; Theodoropoulou, M.; Musheev, M.; Niehrs, C.; Stalla, G.; Peneva, K. Octreotide-mediated tumor-targeted drug delivery via a cleavable doxorubicin–peptide conjugate. *Mol. Pharm.* **2015**, *12*, 4290–4300. [CrossRef]
61. Peruzzi, M.T.; Gallou, F.; Lee, S.J.; Gagné, M.R. Site selective amide reduction of cyclosporine A enables diverse derivation of an important cyclic peptide. *Org. Lett.* **2019**, *21*, 3451–3455. [CrossRef]
62. Banerjee, S.; Pillai, M.R.; Knapp, F.F. Lutetium-177 therapeutic radiopharmaceuticals: Linking chemistry, radiochemistry, and practical applications. *Chem. Rev.* **2015**, *115*, 2934–2974. [CrossRef] [PubMed]
63. Patel, T.K.; Adhikari, N.; Amin, S.A.; Biswas, S.; Jha, T.; Ghosh, B. Small molecule drug conjugates (SMDCs): An emerging strategy for anticancer drug design and discovery. *N. J. Chem.* **2021**, *45*, 5291–5321. [CrossRef]
64. Dong, H.; Limberakis, C.; Liras, S.; Price, D.; James, K. Peptidic macrocyclization via palladium-catalyzed chemoselective indole C-2 arylation. *Chem. Commun.* **2012**, *48*, 11644–11646. [CrossRef]
65. Mendive-Tapia, L.; Preciado, S.; García, J.; Ramón, R.; Kielland, N.; Albericio, F.; Lavilla, R. New peptide architectures through C–H activation stapling between tryptophan-phenylalanine/tyrosine residues. *Nat. Commun.* **2015**, *6*, 7160–7169. [CrossRef] [PubMed]
66. Mendive-Tapia, L.; Bertran, A.; García, J.; Acosta, G.; Albericio, F.; Lavilla, R. Constrained cyclopeptides: Biaryl formation through Pd-catalyzed C–H activation in peptides-structural control of the cyclization vs. cyclodimerization outcome. *Chem.–Eur. J.* **2016**, *22*, 13114–13119. [CrossRef] [PubMed]
67. Peng, J.; Li, C.; Khamrakulov, M.; Wang, J.; Liu, H. Rhodium(III)-catalyzed C–H alkenylation: Access to maleimide-decorated tryptophan and tryptophan-containing peptides. *Org. Lett.* **2020**, *22*, 1535–1541. [CrossRef]
68. Bai, Z.; Cai, C.; Sheng, W.; Ren, Y.; Wang, H. Late-stage peptide macrocyclization by palladium-catalyzed site-selective C–H olefination of tryptophan. *Angew. Chem. Int. Ed.* **2020**, *59*, 14686–14692. [CrossRef]
69. Liu, L.; Fan, X.; Wang, B.; Deng, H.; Wang, T.; Zheng, J.; Chen, J.; Shi, Z.; Wang, H. P^{III}-Directed late-stage ligation and macrocyclization of peptides with olefins by rhodium catalysis. *Angew. Chem. Int. Ed.* **2022**, *61*, e202206177.
70. Ruan, Z.; Sauermann, N.; Manoni, E.; Ackermann, L. Manganese-catalyzed C–H alkynylation: Expedient peptide synthesis and modification. *Angew. Chem. Int. Ed.* **2017**, *56*, 3172–3176. [CrossRef]
71. Bai, Q.; Bai, Z.; Wang, H. Macrocyclization of biaryl-bridged peptides through late-stage palladium-catalyzed $C(sp^2)$–H arylation. *Org. Lett.* **2019**, *21*, 8225–8228. [CrossRef] [PubMed]
72. Han, B.; Li, B.; Qi, L.; Yang, P.; He, G.; Chen, G. Construction of cyclophane-braced peptide macrocycles via palladium-catalyzed picolinamide-directed intramolecular $C(sp^2)$–H arylation. *Org. Lett.* **2020**, *22*, 6879–6883. [CrossRef] [PubMed]
73. Bai, Z.; Cai, C.; Yu, Z.; Wang, H. Backbone-enabled directional peptide macrocyclization through late-stage palladium-catalyzed δ-$C(sp^2)$–H olefination. *Angew. Chem. Int. Ed.* **2018**, *57*, 13912–13916. [CrossRef] [PubMed]
74. Tang, J.; Chen, H.; He, Y.; Sheng, W.; Bai, Q.; Wang, H. Peptide-guided functionalization and macrocyclization of bioactive peptidosulfonamides by Pd(II)-catalyzed late-stage C–H activation. *Nat. Commun.* **2018**, *9*, 3383–3390. [CrossRef] [PubMed]
75. Tang, J.; Wu, J.; Liu, S.; Yao, H.; Wang, H. Macrocyclization of peptidoarylacetamides with self-assembly properties through late-stage palladium-catalyzed $C(sp^2)$ –H olefination. *Sci. Adv.* **2019**, *5*, eaaw0323.
76. Bai, Z.; Wang, H. Backbone-enabled peptide macrocyclization through late-stage palladium-catalyzed C–H activation. *Synlett* **2020**, *31*, 199–204. [CrossRef]
77. Liu, S.; Cai, C.; Bai, Z.; Sheng, W.; Tan, J.; Wang, H. Late-stage macrocyclization of bioactive peptides with internal oxazole motifs via palladium-catalyzed C–H olefination. *Org. Lett.* **2021**, *23*, 2933–2937. [CrossRef]
78. Cai, C.; Wang, F.; Xiao, X.; Sheng, W.; Liu, S.; Chen, J.; Zheng, J.; Xie, R.; Bai, Z.; Wang, H. Macrocyclization of bioactive peptides with internal thiazole motifs via palladium-catalyzed C–H olefination. *Chem. Commun.* **2022**, *58*, 4861–4864. [CrossRef]
79. Noisier, A.F.M.; Brimble, M.A. C–H functionalization in the synthesis of amino acids and peptides. *Chem. Rev.* **2014**, *114*, 8775–8806. [CrossRef]
80. He, G.; Wang, B.; Nack, W.A.; Chen, G. Syntheses and transformations of α-amino acids via palladium-catalyzed auxiliary-directed sp^3 C–H functionalization. *Acc. Chem. Res.* **2016**, *49*, 635–645. [CrossRef]
81. Brandhofer, T.; Mancheño, O.G. Site-selective C–H bond activation/functionalization of alpha-amino acids and peptide-like derivatives. *Eur. J. Org. Chem.* **2018**, *2018*, 6050–6067. [CrossRef]
82. Gong, W.; Zhang, G.; Liu, T.; Giri, R.; Yu, J.-Q. Site-selective $C(sp^3)$–H functionalization of di-, tri-, and tetrapeptides at the N-terminus. *J. Am. Chem. Soc.* **2014**, *136*, 16940–16946. [CrossRef] [PubMed]
83. Noisier AF, M.; García, J.; Ionuț, I.A.; Albericio, F. Stapled peptides by late-stage $C(sp^3)$–H activation. *Angew. Chem. Int. Ed.* **2017**, *56*, 314–318. [CrossRef] [PubMed]
84. Tang, J.; He, Y.; Chen, H.; Sheng, W.; Wang, H. Synthesis of bioactive and stabilized cyclic peptides by macrocyclization using $C(sp^3)$–H activation. *Chem. Sci.* **2017**, *8*, 4565–4570. [CrossRef]
85. Daugulis, O.; Roane, J.; Tran, L.D. Bidentate, monoanionic auxiliary-directed functionalization of carbon–Hydrogen bonds. *Acc. Chem. Res.* **2015**, *48*, 1053–1064. [CrossRef]
86. Zhang, X.; Lu, G.; Sun, M.; Mahankali, M.; Ma, Y.; Zhang, M.; Hua, W.; Hu, Y.; Wang, Q.; Chen, J.; et al. A general strategy for synthesis of cyclophane-braced peptide macrocycles via palladium-catalysed intramolecular sp^3 C–H arylation. *Nat. Chem.* **2018**, *10*, 540–548. [CrossRef]

87. Li, B.; Li, X.; Han, B.; Chen, Z.; Zhang, X.; He, G.; Chen, G. Construction of natural-product-like cyclophane-braced peptide macrocycles via sp^3 C–H arylation. *J. Am. Chem. Soc.* **2019**, *141*, 9401–9407. [CrossRef]
88. Li, X.; Qi, L.; Li, B.; Zhao, Z.; He, G.; Chen, G. Synthesis of cyclophane-braced peptide macrocycles via palladium-catalyzed intramolecular C(sp^3)–H arylation of *N*-methyl alanine at C-termini. *Org. Lett.* **2020**, *22*, 6209–6213. [CrossRef]

Disclaimer/Publisher's Note: The statements, opinions and data contained in all publications are solely those of the individual author(s) and contributor(s) and not of MDPI and/or the editor(s). MDPI and/or the editor(s) disclaim responsibility for any injury to people or property resulting from any ideas, methods, instructions or products referred to in the content.

Review

Construction of Benzo-Fused Polycyclic Heteroaromatic Compounds through Palladium-Catalyzed Intramolecular C-H/C-H Biaryl Coupling

Yuji Nishii [1,*] and Masahiro Miura [2,*]

[1] Department of Applied Chemistry, Graduate School of Engineering, Osaka University, Osaka 565-0871, Japan
[2] Innovative Catalysis Science Division, Institute for Open and Transdisciplinary Research Initiatives (ICS-OTRI), Osaka University, Osaka 565-0871, Japan
* Correspondence: y_nishii@chem.eng.osaka-u.ac.jp (Y.N.); miura@chem.eng.osaka-u.ac.jp (M.M.)

Abstract: Dibenzo-fused five-membered heteroaromatic compounds, including dibenzofuran, carbazole, and dibenzothiophene, are fundamental structural units in various important polycyclic heteroaromatic compounds. The intramolecular C-H/C-H biaryl coupling of diaryl (thio)ethers and amines based on palladium(II) catalysis under oxidative conditions is known to be one of the most effective, step-economic methods for their construction. Representative examples for the construction of structurally intriguing π-extended polycyclic heteroaromatics through catalytic coupling reactions are briefly summarized in this mini-review.

Keywords: C-H/C-H coupling; dibenzofuran; carbazole; dibenzothiophene; palladium

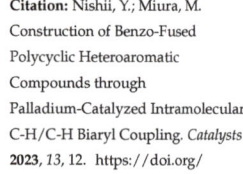

Citation: Nishii, Y.; Miura, M. Construction of Benzo-Fused Polycyclic Heteroaromatic Compounds through Palladium-Catalyzed Intramolecular C-H/C-H Biaryl Coupling. *Catalysts* **2023**, *13*, 12. https://doi.org/10.3390/catal13010012

Academic Editors: Ewa Kowalska and Yuichi Kobayashi

Received: 6 December 2022
Revised: 15 December 2022
Accepted: 17 December 2022
Published: 22 December 2022

Copyright: © 2022 by the authors. Licensee MDPI, Basel, Switzerland. This article is an open access article distributed under the terms and conditions of the Creative Commons Attribution (CC BY) license (https:// creativecommons.org/licenses/by/ 4.0/).

1. Introduction

Polycyclic heteroaromatic compounds often exhibit interesting biological and physical properties, and thus organic chemists have devoted tremendous effort to developing their effective synthetic methods. Recently, those involving C-H bond activation have attracted much attention as they can provide short-step synthetic sequences leading to target molecules [1–4]. Among the polycyclic heteroaromatics, dibenzofuran, carbazole, and dibenzothiophene are well recognized to be fundamental structural units as their derivatives have many applications in pharmaceutical and materials chemistry areas [5]. Various palladium-catalyzed methods are available for their construction [3,6]. The direct intramolecular aromatic coupling reactions involving C-H bond cleavage are illustrated in Scheme 1 [3]: (a) C-H/C-X coupling, (b) C-H/C-H coupling, and (c) C-H/E-H coupling (E = O, NR, S etc.). Of these, the C-H/C-H coupling reactions (Scheme 1b) have recently been extensively studied as versatile bond-formation methods [3]. This type of reaction is considered to proceed via double C-H bond cleavages by Pd(II) species to provide the key six-membered palladacycle intermediates and successive reductive elimination leading to the desired products (Scheme 2). The formed Pd(0) species after reductive elimination should be re-oxidized to Pd(II) species for the catalytic conversion of the substrates. As the oxidants, molecular oxygen and/or metallic species such as Cu(II) and Ag(I) salts are usually employed. Thus, proper choices of Pd(II) species and oxidants as well as solvents and additional promoters are essential for an efficient coupling. It is noted that Ag(I) species are now known to be especially versatile oxidants, and they may act as such not only in the re-oxidation step but also in the initial aromatic C-H activation step [7–9].

In this mini-review, we briefly summarize recent examples, including those from our group, for the construction of the dibenzofuran, carbazole, and dibenzothiophene derivatives, especially focusing on structurally intriguing π-extended polycyclic heteroaromatics through multiple catalytic C-H/C-H couplings, along with citing seminal reactions.

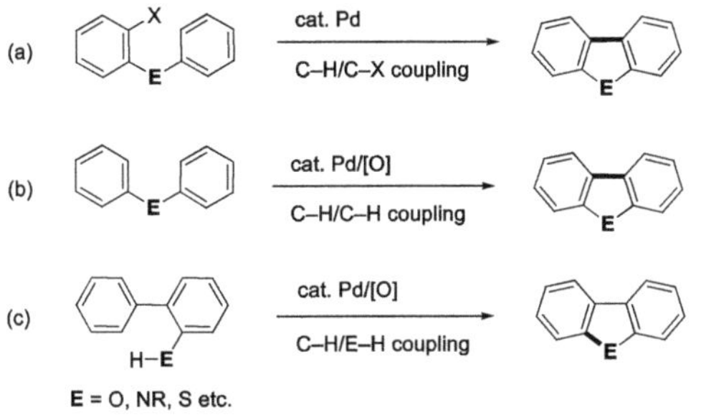

Scheme 1. Synthesis of dibenzo-fused five-membered heterocycles through C–H couplings: (**a**) direct C–H/C–X coupling, (**b**) oxidative C–H/C–H coupling, and (**c**) oxidative C–H/E–H coupling.

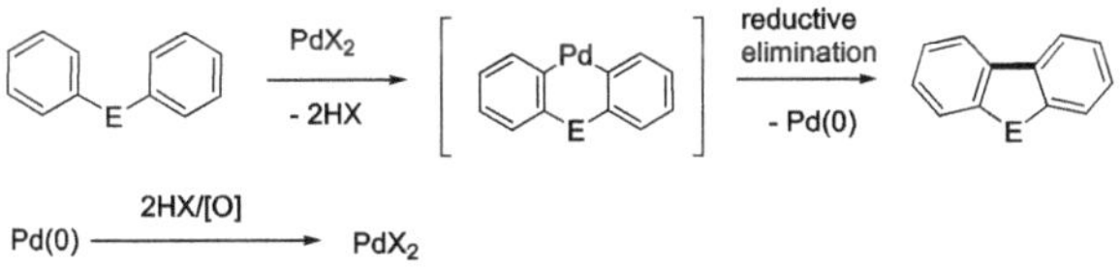

Scheme 2. Schematic concept of Pd(II)-catalyzed intramolecular biaryl coupling.

2. Dibenzofuran Derivatives and Related Compounds

The first example of intramolecular oxidative coupling of diphenyl ether to produce dibenzofuran (Scheme 2, E = O) was reported by Yoshimoto and Itatani in 1973 [10]. The reaction using the ether as a substrate and a solvent was carried out using Pd(OAc)$_2$ and acetylacetone as a palladium source and an additive, respectively, under a pressurized oxygen/nitrogen mixture at 150 °C. It was found that dibenzofuran was formed together with the dimers of diphenyl ether with several catalyst turnovers [10,11]. Subsequently, Åkermark and coworkers reported that a stoichiometric use of Pd(OAc)$_2$ in acetic acid resulted in the selective formation of dibenzofuran [12], and then the reaction occurred catalytically in the presence of Pd(TFA)$_2$ (TFA = trifluoroacetate) and Sn(OAc)$_2$ as a catalyst and an additive, respectively, under a normal pressure of oxygen (Scheme 3a) [13]. Fagnou and coworkers discovered that pivalic acid (PivOH) and a carbonate base such as K$_2$CO$_3$ acted effectively as a solvent and an additive, respectively, and the product was obtained in a good yield from the reaction under air at ambient pressure (Scheme 3b) [14].

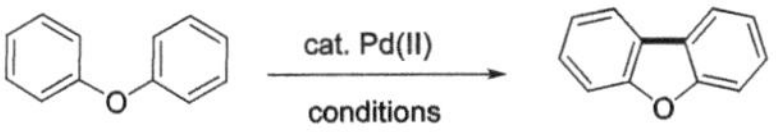

(a) Pd(TFA)$_2$ (5 mol%), Sn(OAc)$_2$ (10 mol%), in AcOH under O$_2$, 116 °C, 48 h, 60% yield
(b) Pd(OAc)$_2$ (5 mol%), K$_2$CO$_3$ (10 mol%), in PivOH under air, 120 °C, 42 h, 75% yield

Scheme 3. Pd(II)-catalyzed cyclization of diphenyl ether to dibenzofuran.

Double [15] and triple [16] cyclizations of 1,4-diphenoxy- and 1,3,5-triphenoxy-benzenes were previously undertaken using stoichiometric Pd(OAc)$_2$, and the corresponding products were found to be formed albeit with moderate to low yields. We recently disclosed that the double cyclization of 1,2-diaryloxybenzenes catalytically occurred using Pd(TFA)$_2$ and AgOAc as a catalyst and an oxidant, respectively, in PivOH (Scheme 4) [17]. Similarly, 1,4-di(4-*tert*-butylphenoxy)-benzene and -naphthalene underwent double cyclization to form the corresponding five- and six-ring compounds. The former product was found to show ultraviolet fluorescence in both solution (CHCl$_3$, 347 nm) and solid state (369 nm). The five-ring compound also exhibited green phosphorescence in a crystalline solid matrix at room temperature [18]. The triple cyclization of 1,3,5-triaryloxybenzenes leading to trioxatruxenes as products were efficiently proceeded under similar conditions. The trioxatruxenes showed blue phosphorescence in solution with relatively long lifetimes at a low temperature of 77 K [19,20].

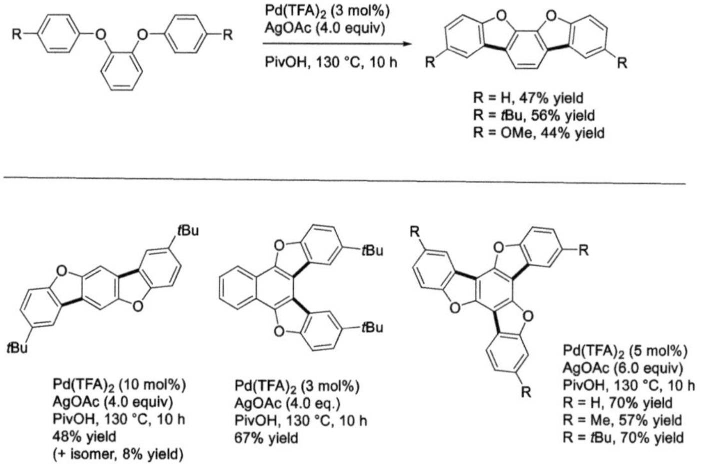
Pd(TFA)2 (3 mol%)
AgOAc (4.0 equiv)
PivOH, 130 °C, 10 h
R = H, 47% yield
R = tBu, 56% yield
R = OMe, 44% yield
Pd(TFA)2 (10 mol%)
AgOAc (4.0 equiv)
PivOH, 130 °C, 10 h
48% yield
(+ isomer, 8% yield)
Pd(TFA)2 (3 mol%)
AgOAc (4.0 eq.)
PivOH, 130 °C, 10 h
67% yield
Pd(TFA)2 (5 mol%)
AgOAc (6.0 equiv)
PivOH, 130 °C, 10 h
R = H, 70% yield
R = Me, 57% yield
R = tBu, 70% yield

Scheme 4. Double and triple cyclization reactions of diaryloxy- and triaryloxy-benzenes.

The dehydrogenative cyclization of aryl hereroaryl ethers is also possible. For example, Kanai, Kuninobu, and coworkers reported the reaction of 3-phenoxybenzo[*b*]thiophene in the presence of $Pd(OPiv)_2$ and AgOPiv in DMF to provide the cyclized four-ring compound (Scheme 5a) [21]. We independently developed the same reaction under different conditions (Scheme 5b) [22]. Kanai and Kuninobu extended their method to the double cyclization reactions to produce the π-extended compounds illustrated in Scheme 6 [21].

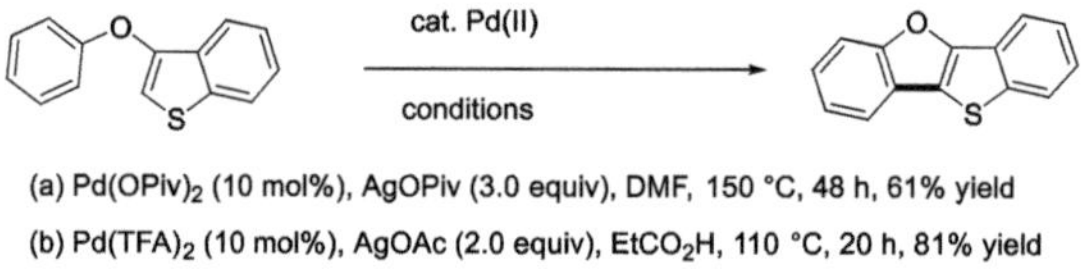
cat. Pd(II)
conditions
(a) Pd(OPiv)2 (10 mol%), AgOPiv (3.0 equiv), DMF, 150 °C, 48 h, 61% yield
(b) Pd(TFA)2 (10 mol%), AgOAc (2.0 equiv), EtCO2H, 110 °C, 20 h, 81% yield

Scheme 5. Cyclization of 3-phenoxybenzo[*b*]thiophene.

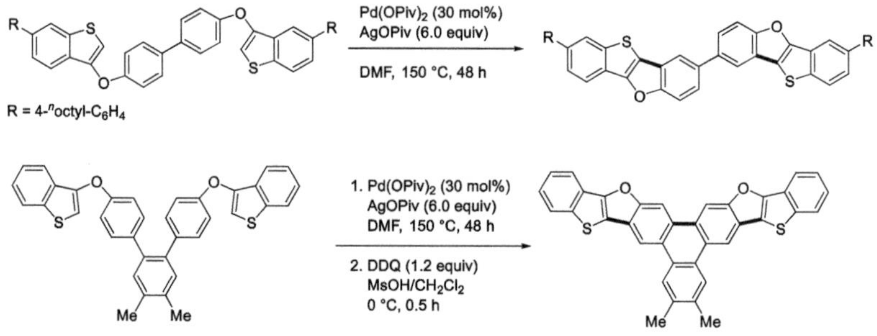
Pd(OPiv)2 (30 mol%)
AgOPiv (6.0 equiv)
DMF, 150 °C, 48 h
R = 4-noctyl-C6H4
1. Pd(OPiv)2 (30 mol%)
AgOPiv (6.0 equiv)
DMF, 150 °C, 48 h
2. DDQ (1.2 equiv)
MsOH/CH2Cl2
0 °C, 0.5 h

Scheme 6. Double cyclization reactions of 3-phenoxybenzo[*b*]thiophene derivatives.

We next reported the double dehydrogenative cyclization of 2,6-diaryloxypyridines, which are readily prepared from 2,6-dihalopyridines and phenols, in the presence of $Pd(TFA)_2$ and AgOAc in PivOH to afford the corresponding five-ring systems (Scheme 7) [23]. Both electron-donating and electron-withdrawing groups on the aryl groups were tolerable. The chloro-function on a five-ring product was effectively substituted by a carbazolyl group [24] by an elaborated Buchwald–Hartwig reaction developed by Suzuki and coworkers at Takasago (Scheme 8) [25]. The product was found to show interesting photophysical properties, including a small ΔE_{ST} value, which is required for TADF (thermally activated delayed fluorescence).

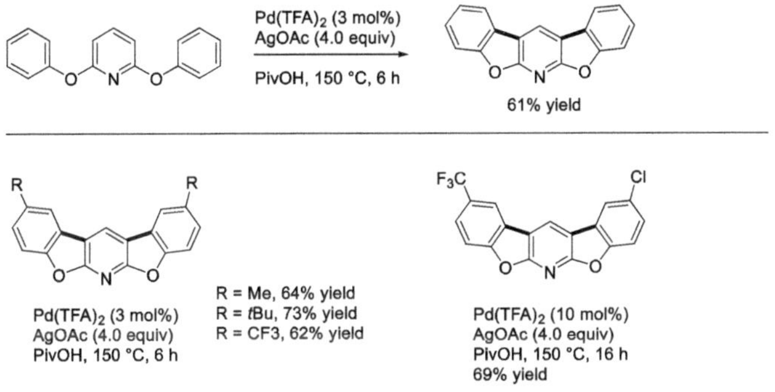

Scheme 7. Double cyclization of 2,6-diaryloxypyridines.

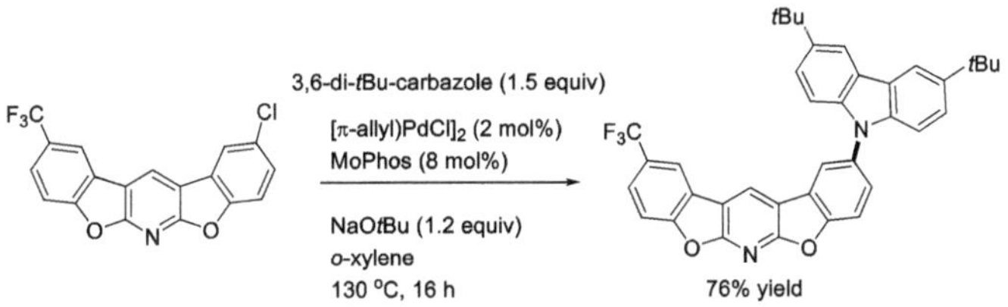

Scheme 8. Introduction of a carbazoyl function to a chloro-bisbenzofuropyridine.

We then developed a twofold double cyclization to form four C-C bonds in a BINOL derived bis(naphthyloxypyridyl aryl ether), as shown in Scheme 9 [26]. The chiral products showed CPL (circularly polarized luminescence) properties in both solution and solid states. Interestingly, a considerably enhanced CPL was observed in the solid state compared to that in solution.

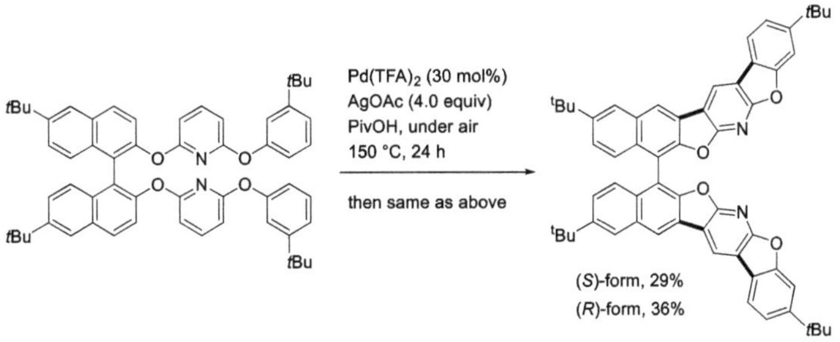

Scheme 9. Twofold double cyclization of a chiral bis(naphthyloxypyridyl aryl ether).

3. Carbazole Derivatives and Related Compounds

In 1994, Knölker and O'Sullivan reported one of the early examples for the synthesis of carbazoles by palladium-catalyzed dehydrogenative coupling using $Cu(OAc)_2$ as an oxidant [27]. Thus, they developed a method for the construction of a benzo[b]carbazoloquinone derivative aiming at synthesizing natural antibiotics (Scheme 10). Subsequently, Åkermark and coworkers described a similar reaction under oxygen [13]. Carbazole itself was shown to be effectively obtained from diphenylamine under the Åkermark's conditions [13]. Fagnou and coworkers showed a more general method for the synthesis of carbazole derivatives employing PivOH and K_2CO_3 as a solvent and an additive under air [14]. The Fagnou's conditions allowed one to construct carbazoles bearing both strongly electron-withdrawing and electron-donating substituents. It is worth noting that a carbazole synthesis by the palladium-catalyzed arylamination of aryl triflates followed by direct biaryl coupling in a one-pot manner was also reported by Fujii and Ohno [28].

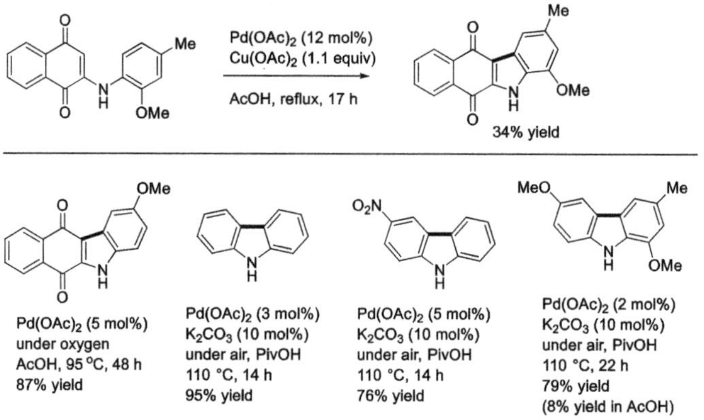

Scheme 10. Synthesis of carbazoles by intramolecular cyclization.

Hellwinkel and Kistenmacher described that the double cyclization of 1,4-di(phenylamino)benzene did not occur even with a stoichiometric amount of $Pd(OAc)_2$ (Scheme 11), while the reaction of triphenylamine gave N-phenylcarbazole in a good yield (see below) [15]. In contrast, Kober and Knölker reported that the treatment of 1,4-diphenylaminobenzene bearing two ethoxycarbonyl groups at the central benzene ring effectively afforded the corresponding doubly cyclized product [29]. The five-ring compound was then utilized to synthesize a natural product, malasseziazole C. The catalytic triple cyclization of a 1,3,5-triarylaminobenzene (aryl = 4-ethoxycarbonylphenyl) leading to a triazatruxene derivative was successfully achieved using a $Pd(TFA)_2$ catalyst under air (Scheme 12) [30]. After hydrolysis to the corresponding tricarboxylic acid, it was utilized to construct a bor-network for the encapsulation of a ruthenium catalyst by Zhou and coworkers [30].

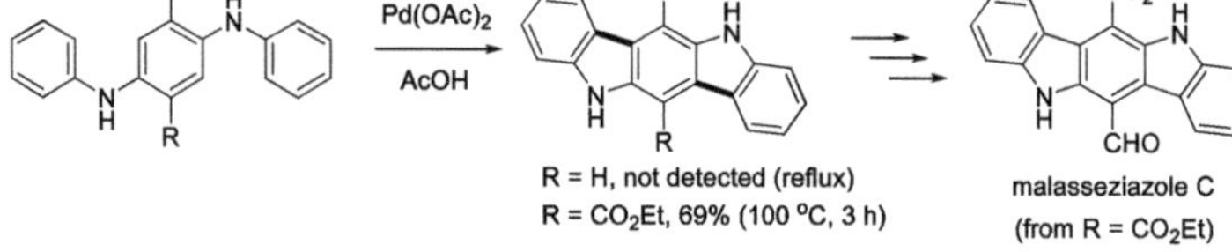

Scheme 11. Synthesis of malasseziazole C through double cyclization.

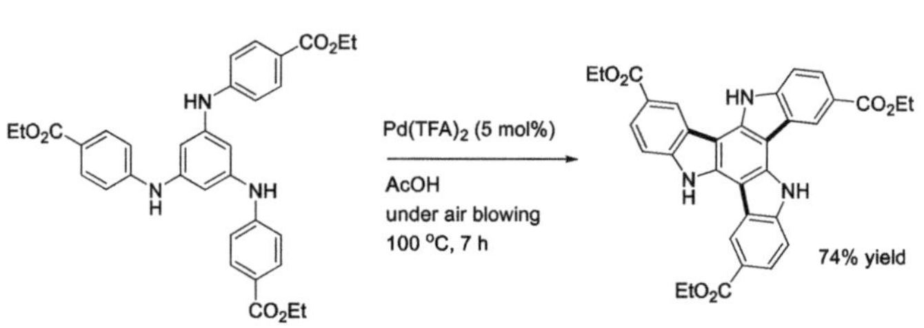

Scheme 12. Triple cyclization of a 1,3,5-tri(arylamino)benzene.

As described above, the stoichiometric reaction of triphenylamine could provide N-phenylcarbazole, but no further cyclization was observed (Scheme 13, left) [15]. Recently,

Patureau and coworkers realized the palladium-catalyzed dehydrogenative cyclization leading to the strained compound, indolo[3,2,1-*jk*]carbazole (ICz), using an oxidant mixture of Ag$_2$O and CuO in PivOH under air (Scheme 13, right) [31].

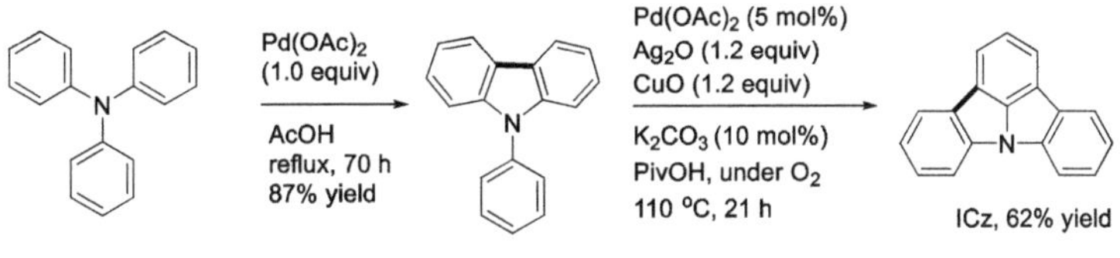

Scheme 13. Synthesis of indole[3,2,1-*jk*]carbazole (ICz).

Since ICz and its derivatives are known to have useful applications as electron donor materials, we undertook to perform the double cyclization of bis(carbazol-9-yl)benzenes [32]. To gain the tractable solubility of the products, 3,6-di-*tert*-butyl-9*H*-carbazole was used as the carbazolyl group. The treatment of the 1,3-bis(carbazolyl)benzene with an oxidizing system of Pd(TFA)$_2$ and AgOAc in PivOH gave the expected double cyclization product (Scheme 14, X = CH). The reaction of the 2,6-bis(carbazolyl)pyridine proceeded similarly (Scheme 14, X = N). Two possible doubly cyclized products were obtained from the reaction of the 1,4-bis(carbazolyl)benzene (Scheme 15). In contrast, a singly cyclized compound in a different manner was isolated from the reaction mixture using the 1,2-bis(carbazolyl)benzene as substrate (Scheme 16). This is probably due to steric reasons. The cardinal optoelectronic properties of the products in Schemes 14–16 were also estimated. Among the compounds, the linear product in Scheme 15 showed the highest HOMO level, whereas it was the lowest with the pyridine analog in Scheme 14.

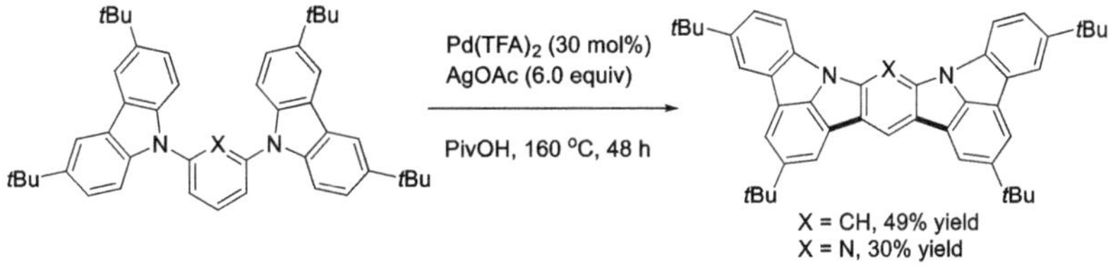

Scheme 14. Double cyclization of a 1,3-bis(carbazolyl)benzene and a 2,6-bis(carbazolyl)pyridine.

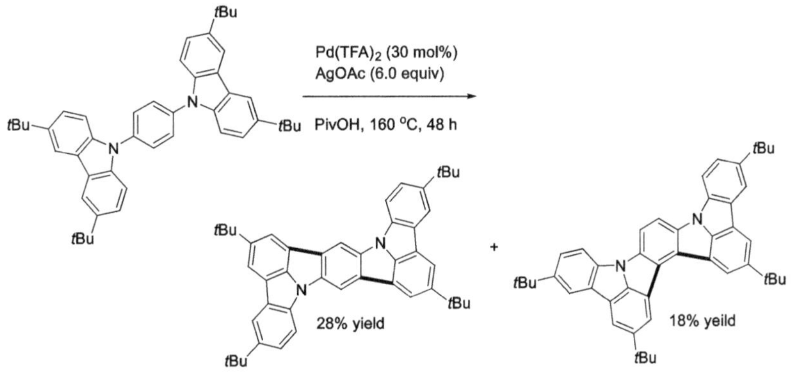

Scheme 15. Double cyclization of a 1,4-bis(carbazoryl)benzene.

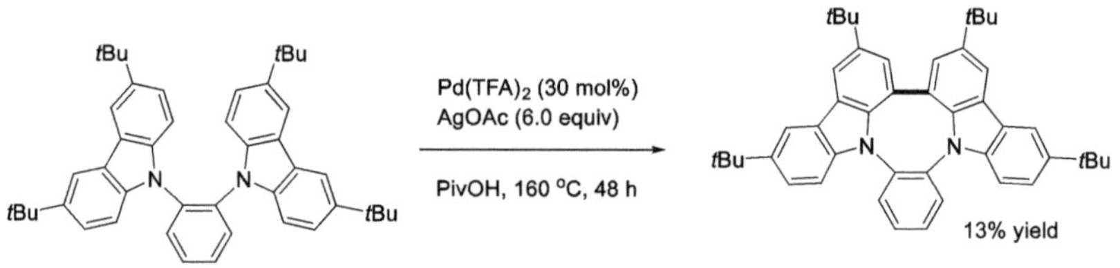

Scheme 16. Reaction of a 1,2-bis(carbazoryl)benzene.

We then undertook to develop a method for the synthesis of some azahelicenes involving carbazole moieties [33]. We designed two 1,1′-biphenyl-3-yl (X = CH)- and 2-phenypyridin-6-yl (X = N)-substituted indolo[2,3-a]carbazole as the cyclization platforms. The treatment of the indolocarbazoles with Pd(TFA)$_2$ and AgOAc in PivOH allowed one to produce the desired azahelicenes (Scheme 17). The racemic helicenes were then subjected to preparative HPLC equipped with conventional stationary phases. The chiral compounds thus obtained showed typical CPL properties in CHCl$_3$ solutions. It is worth noting that the addition of trifluoroacetic acid to the solution of the pyridinyl compound exhibited a red-shifted CPL with an enhanced anisotropy factor.

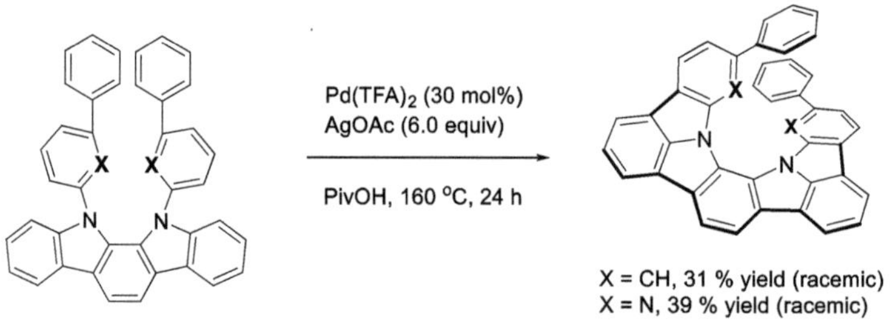

Scheme 17. Synthesis of azahelicenes.

4. Dibenzothiophene Derivatives and Related Compounds

The synthesis of dibenzothiophenes by the catalytic dehydrogenative cyclization of diaryl sulfides has a relatively short history compared to that of dibenzofurans and carbazoles. This might probably be due to the formation of relatively stable complexes that are catalytically less active under common conditions [34]. In 2014, Zhou and coworkers reported the synthesis of dibenzothiophenes from diphenyl sulfides [35]. They found that the reaction occurred effectively under modified Fagnou's conditions using Pd(TFA)$_2$, AgOAc, and K$_2$CO$_3$ in PivOH (Scheme 18). Not only 4,4′-disubstututid diphenyl sulfides but also 2,2′-disubstituted ones could provide the desired, sterically hindered dibenzothiophenes (Scheme 18, down-left). We undertook to develop a method for the synthesis of a linear dithienothiophene of interest in materials chemistry from di(3-thienyl) sulfide (Scheme 18, down-middle) [22]. The reaction occurred with the use of Pd(TFA)$_2$ and AgOAc in EtCO$_2$H, albeit the product yield was moderate. Oechsle and Paradies also reported the synthesis of dithienothiophenes from dithienyl sulfides [36] under modified Mori's thiophene homocoupling conditions with the use of a combination of Pd(II)/AgNO$_3$/KF in DMSO (Scheme 18, down-right) [37].

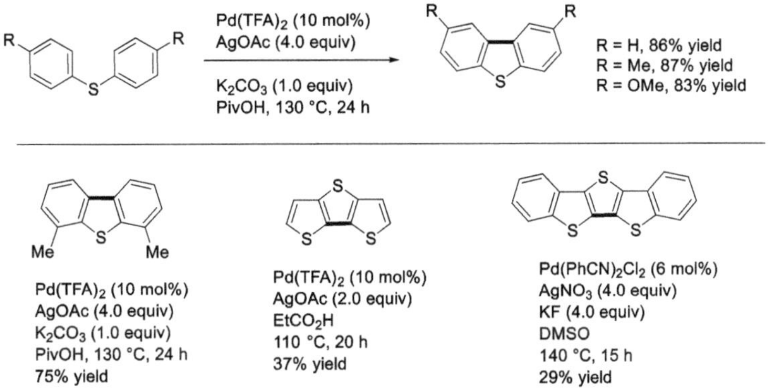

Scheme 18. Synthesis of dibenzothiophenes and related compounds.

We examined the double and triple cyclization reactions [19,38]. Shown in Scheme 19 are those of 1,2- and 1,4-di(4-*tert*-butylphenylthio)benzenes and 1,3,5-tri(3- or 4-*tert*-butylphenylthio)benzene. Thus, we could obtain the corresponding benzothienobenzothiopenes and trithiatruxecenes. We also estimated their cardinal optoelectronic properties. The π-extended products exhibited blue to green phosphorescence by dispersing them in a PMMA film [19,38]. The trithiatruxenes showed semiconducting properties in thin film states [19].

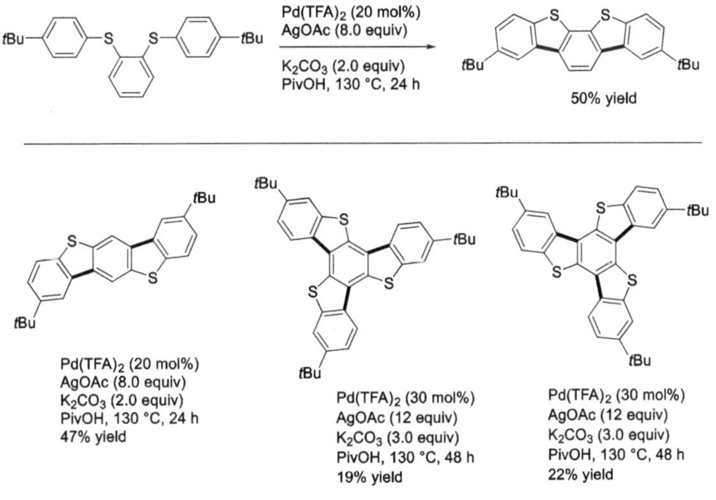

Scheme 19. Synthesis of benzothienobenzothiophenes and trithiatruxenes.

5. Concluding Remarks

We herein summarized the synthetic methods of dibenzofurans, carbazoles, and dibenzothiophenes by the palladium-catalyzed dehydrogenative cyclization reactions of diaryl (thio)ethers and diaryl amines along with the brief history. It was emphasized that the multiple cyclization methods can allow one to prepare π-extended polycyclic heteroaromatic compounds by short-step sequences. Not only the heterocycles focused herein but also various other dibenzo heterocycles can be constructed with the dehydrogenative coupling strategy [1–4]. While potentially useful for constructing the compounds of substantial importance in pharmaceutical and materials chemistry, the reported methods to date usually require relatively high loading of Pd, often together with stoichiometric amounts of metallic salts. Thus, to utilize the dehydrogenative strategy in industrial synthesis, new catalytic systems of high efficiency should be developed. We wish further substantial advances of this research area, so that various useful chemical products can be produced in practical amounts.

Author Contributions: Y.N. and M.M. jointly performed the part of our work with coworkers and wrote this mini-review. All authors have read and agreed to the published version of the manuscript.

Funding: Our work was supported by the Japan Society for the Promotion of Science (JSPS) KAKENHI Grants JP 19K15586 and 21K14627 (Grant-in-Aid for Young Scientists) to Y.N. and JP 17H06092 (Grant-in-Aid for Specially Promoted Research) to M.M.

Acknowledgments: In contributing this article to the memorial issue for the late Jiro Tsuji, M.M. is deeply grateful for his continuous encouragement to our research group, which allowed us to perform our work with confidence. M.M. would also like to note that he learned a lot while Tsuji was writing the book "Palladium Reagents and Catalysts" (2004). M.M. read the crude manuscript of every chapter and learned not only palladium chemistry but also how to write an excellent review. Y.N. and M.M. thank the coworkers whose names appear in the publications from our group.

Conflicts of Interest: The authors declare no conflict of interest.

References

1. Jin, T.; Zhao, J.; Asao, N.; Yamamoto, Y. Metal-Catalyzed Annulation Reactions for π-Conjugated Polycycles. *Chem. Eur. J.* **2014**, *20*, 3554–3576. [CrossRef]
2. Segawa, Y.; Maekawa, T.; Itami, K. Synthesis of Extended π-Systems through C–H Activation. *Angew. Chem. Int. Ed.* **2015**, *54*, 66–81. [CrossRef]
3. Yang, Y.; Lan, J.; You, J. Oxidative C-H/C-H Coupling Reactions between Two (Hetero)arenes. *Chem. Rev.* **2017**, *117*, 8787–8863. [CrossRef]
4. Grzybowski, M.; Sadowski, B.; Butenschön, H.; Gryko, D.T. Synthetic Applications of Oxidative Aromatic Coupling—From Biphenols to Nanographenes. *Angew. Chem. Int. Ed.* **2020**, *59*, 2998–3027. [CrossRef]
5. Schatz, J.; Brendgen, T.; Schühle, D. *Comprehensive Heterocyclic Chemistry IV, Vol. 3, Five-Membered Rings with One Heteroatom Together with Their Benzo- and other Carbocyclic-Fused Derivatives*; Wong, N.H.C., Ed.; Elsevier: Amsterdam, The Netherlands, 2022.
6. Tsuji, J. *Palladium Reagents and Catalysts*; Wiley: Chichester, UK, 2004.
7. Colletto, C.; Panigrahi, A.; Fernández-Casado, J.; Larrosa, I. Ag(I)–C–H Activation Enables Near-Room-Temperature Direct α-Arylation of Benzo[b]thiophenes. *J. Am. Chem. Soc.* **2018**, *140*, 9638–9643. [CrossRef]
8. Tlahuext-Aca, A.; Lee, S.Y.; Sakamoto, S.; Hartwig, J.F. Direct Arylation of Simple Arenes with Aryl Bromides by Synergistic Silver and Palladium Catalysis. *ACS Catal.* **2021**, *11*, 1430–1434. [CrossRef]
9. Mudarra, A.L.; de Salinas, S.M.; Pérez-Temprano, M.H. Beyond the traditional roles of Ag in catalysis: The transmetalating ability of organosilver(I) species in Pd-catalysed reactions. *Org. Biomol. Chem.* **2019**, *17*, 1655–1667. [CrossRef]
10. Yoshimoto, H.; Itatani, H. Palladium-Catalyzed Coupling Reaction of Aromatic Compounds. *Bull. Chem. Soc. Jpn.* **1973**, *46*, 2490–2492. [CrossRef]
11. Shiotani, A.; Itatani, H. Dibenzofurans by Intramolecular Ring Closire Reactions. *Angew. Chem. Int. Ed. Engl.* **1974**, *13*, 471–472. [CrossRef]
12. Åkermark, B.; Eberson, L.; Jonsson, E.; Pettersson, E. Palladium-Promoted Cyclization of Diphenyl Ether, Diphenylamine, and Related Compounds. *J. Org. Chem.* **1975**, *40*, 1365–1367. [CrossRef]
13. Hagelin, H.; Oslob, D.J.; Åkermark, B. Oxygen as Oxidant in Palladium-Catalyzed Inter- and Intramolecular Coupling Reactions. *Chem. Eur. J.* **1999**, *5*, 2413–2416. [CrossRef]
14. Liégault, B.; Lee, D.; Huestis, M.P.; Stuart, D.R.; Fagnou, K. Intramolecular Pd(II)-Catalyzed Oxidative Biaryl Synthesis Under Air: Reaction Development and Scope. *J. Org. Chem.* **2008**, *73*, 5022–5028. [CrossRef]
15. Hellwinkel, D.; Kistenmacher, T. Palladium Acetate-Mediated Cyclizations of Di- and Trifunctional Triarylamines, Diaryl Ethers, and Diaryl Ketones. *Liebigs Ann. Chem.* **1989**, *1989*, 945–949. [CrossRef]
16. Bergman, J.; Egestad, B. Cyclocondensation of 3(2H)-benzofuranone. *Tetrahedron Lett.* **1978**, *19*, 3143–3146. [CrossRef]
17. Kaida, H.; Satoh, T.; Nishii, Y.; Hirano, K.; Miura, M. Synthesis of Benzo-bis- and Benzo-tris-benzofurans by Palladium-Catalyzed Multiple Intramolecular C–H/C–H Coupling. *Chem. Lett.* **2016**, *45*, 1069–1071. [CrossRef]
18. Nakamura, S.; Tsuboi, M.; Taniguchi, T.; Nishii, Y.; Tohnai, N.; Miura, M. Room Temperature Phosphorescent Crystals Consisting of Cyclized Guests and Their Uncyclized Mother Host Molecules. *Chem. Lett.* **2020**, *49*, 921–924. [CrossRef]
19. Nakamura, S.; Okamoto, M.; Tohnai, M.; Nakayama, K.; Nishii, Y.; Miura, M. Synthesis and Properties of Tri-*tert*-Butylated Trioxa and Trithia Analogues of Truxene. *Bull. Chem. Soc. Jpn.* **2020**, *93*, 99–108. [CrossRef]
20. Ogaki, T.; Ohta, E.; Oda, Y.; Sato, H.; Matsui, Y.; Komeda, M.; Ikeda, H. Intramolecular Triple Cyclization Strategy for Sila- and Oxa-Analogues of Truxene with Long-Lived Phosphorescence. *Asian J. Org. Chem.* **2017**, *6*, 290–296. [CrossRef]

21. Saito, K.; Chikkade, P.K.; Kanai, M.; Kuninobu, Y. Palladium-Catalyzed Construction of Heteroatom-Containing π-Conjugated Systems by Intramolecular Oxidative C-H/C-H Coupling Reaction. *Chem. Eur. J.* **2015**, *21*, 8365–8368. [CrossRef]
22. Kaida, H.; Satoh, T.; Hirano, K.; Miura, M. Synthesis of Thieno[3,2-*b*]benzofurans by Palladium-Catalyzed Intramolecular C-H/C-H Coupling. *Chem. Lett.* **2015**, *44*, 1125–1127. [CrossRef]
23. Kaida, H.; Goya, T.; Nishii, Y.; Hirano, K.; Satoh, T.; Miura, M. Construction of Bisbenzofuro[2,3-*b*:3′,2′-*e*]pyridines by Palladium-Catalyzed Double Intramolecular Oxidative C-H/C-H Coupling. *Org. Lett.* **2017**, *19*, 1236–1239. [CrossRef] [PubMed]
24. Itai, Y.; Nishii, Y.; Stachelek, P.; Data, P.; Takeda, Y.; Minakata, S.; Miura, M. Syntheses of Diverse Donor-Substituted Bisbenzofuro[2,3-*b*:3′,2′-*e*]pyridines (BBZFPys) via Pd Catalysis, and Their Photophysical Properties. *J. Org. Chem.* **2018**, *83*, 10289–10302. [CrossRef] [PubMed]
25. Suzuki, K.; Hori, Y.; Kobayashi, T. A New Hybrid Phosphine Ligand for Palladium-Catalyzed Amination of Aryl Halides. *Adv. Synth. Catal.* **2008**, *350*, 652–656. [CrossRef]
26. Takishima, R.; Nishii, Y.; Hinoue, T.; Imai, Y.; Miura, M. Synthesis and circularly polarized luminescence properties of BINOL-derived bisbenzofuro[2,3-*b*:3′,2′-*e*]pyridines (BBZFPys). *Beilstein J. Org. Chem.* **2020**, *16*, 325–336. [CrossRef]
27. Knölker, H.-J.; O'Sullivan, N. Indoloquinones—3. Palladium-promoted synthesis of hydroxy-substituted 5-Cyano-5*H*-benzo[*b*]carbazole-6, 11-diones. *Tetrahedron* **1994**, *50*, 10893–10908. [CrossRef]
28. Watanabe, T.; Oishi, S.; Fujii, N.; Ohno, H. Palladium-Catalyzed Direct Synthesis of Carbazoles via One-Pot *N*-Arylation and Oxidative Biaryl Coupling: Synthesis and Mechanistic Study. *J. Org. Chem.* **2009**, *74*, 4720–4726. [CrossRef]
29. Kober, U.; Knölker, H.-J. Palladium-Catalyzed Approach to Malasseziazole A and Fiest Total Synthesis of Malasseziazole C. *Synlett* **2015**, *26*, 1549–1552.
30. Wang, X.; Lu, W.; Gu, Z.-Y.; Weia, Z.; Zhou, H.-C. Topology-guided design of an anionic bor-network for photocatalytic [Ru(bpy)$_3$]$_2$$^+$ encapsulation. *Chem. Commun.* **2016**, *52*, 1926–1929. [CrossRef]
31. Jones, A.W.; Louillat-Habermeyer, M.-L.; Patureau., F.W. Strained Dehydrogenative Ring Closure of Phenylcarbazoles. *Adv. Synth. Catal.* **2015**, *357*, 945–949. [CrossRef]
32. Taniguchi, T.; Itai, Y.; Nishii, Y.; Tohnai, N.; Miura, M. Construction of Nitrogen-Containing Polycyclic Aromatic Compounds by Intramolecular Oxidative C-H/C-H Coupling of Bis(9*H*-Carbazol-9-yl)benzenes and Their Properties. *Chem. Lett.* **2019**, *48*, 1160–1163. [CrossRef]
33. Taniguchi, T.; Nishii, Y.; Mori, T.; Nakayama, K.; Miura, M. Synthesis, Structure, and Chiroptical Properties of Indolo- and Pyridopyrrolo- Carbazole-Based C2-Symmetric Azahelicenes. *Chem. Eur. J.* **2021**, *27*, 7356–7361. [CrossRef]
34. Vicente, J.; Abad, J.A.; López-Nicolás, R.-M.; Jones, P.G. Palladated Oligophenylene Thioethers: Synthesis and Reactivity toward Isocyanides, Carbon Monoxide, and Alkynes. *Organometallics* **2011**, *30*, 4983–4998. [CrossRef]
35. Che, R.; Wu, Z.; Li, Z.; Xiang, H.; Zhou, X. Synthesis of Dibenzothiophenes by Pd-Catalyzed Dual C-H Activation from Diaryl Sulfides. *Chem. Eur. J.* **2014**, *20*, 7258–7261. [CrossRef]
36. Oechsle, P.; Paradies, J. Ambidextrous Catalytic Access to Dithieno[3,2-*b*:2′,3′-*d*]thiophene (DTT) Derivatives by Both Palladium-Catalyzed C-S and Oxidative Dehydro C-H Coupling. *Org. Lett.* **2014**, *16*, 4086–4089. [CrossRef]
37. Masui, K.; Ikegami, H.; Mori, A. Palladium-Catalyzed C−H Homocoupling of Thiophenes: Facile Construction of Bithiophene Structure. *J. Am. Chem. Soc.* **2004**, *126*, 5074–5075. [CrossRef]
38. Tsuboi, M.; Nakamura, S.; Nandi, S.; de Silva, P.; Takeda, Y.; Miura, M. Syntheses and Room Temperature Phosphorescence Properties of Dibenzobenzodithiophenes and Dibenzothiophenes. *Bull. Chem. Soc. Jpn.* **2021**, *94*, 2498–2504. [CrossRef]

Disclaimer/Publisher's Note: The statements, opinions and data contained in all publications are solely those of the individual author(s) and contributor(s) and not of MDPI and/or the editor(s). MDPI and/or the editor(s) disclaim responsibility for any injury to people or property resulting from any ideas, methods, instructions or products referred to in the content.

Review

Coupling Reactions on Secondary Allylic, Propargylic, and Alkyl Carbons Using Organoborates/Ni and RMgX/Cu Reagents

Yuichi Kobayashi

Organization for the Strategic Coordination of Research and Intellectual Properties, Meiji University, Kawasaki 214-8571, Japan; ykobayas@bio.titech.ac.jp

Abstract: In the first part of this review, secondary carbon-carbon bond formation by using allylic coupling reactions with aryl and alkenyl borates is presented. Early investigations have revealed the suitability of a nickel catalyst and [R^{T}B(OMe)$_3$]Li (R^T: transferable group). Due to their low reactivity, the borates were converted to more reactive congeners possessing an alkanediol ligand, such as 2,3-butanediol and 2,2-dimethyl-1,3-propanediol. Borates with such diol ligands were used to install aryl and alkenyl groups on the monoacetate of 4-cyclopentenyl-1,3-diol. Furthermore, alkenyl borates showed sufficient reactivity toward less reactive allylic alcohol derivatives with bromine atoms at the cis position, producing dieneyl alcohols. In the second part, copper-based and/or copper-catalyzed substitutions of secondary allylic picolinates, propargylic phosphonates, and alkyl (2-pyridine)sulfonates with RMgX are briefly summarized. The application of these reactions to the synthesis of biologically active compounds is also discussed.

Keywords: borate; boronate ester; nickel; secondary carbon; coupling; substitution; Grignard reagent; copper; organic synthesis

Citation: Kobayashi, Y. Coupling Reactions on Secondary Allylic, Propargylic, and Alkyl Carbons Using Organoborates/Ni and RMgX/Cu Reagents. *Catalysts* **2023**, *13*, 132. https://doi.org/10.3390/catal13010132

Academic Editor: Laura Antonella Aronica

Received: 2 December 2022
Revised: 29 December 2022
Accepted: 3 January 2023
Published: 6 January 2023

Copyright: © 2023 by the author. Licensee MDPI, Basel, Switzerland. This article is an open access article distributed under the terms and conditions of the Creative Commons Attribution (CC BY) license (https://creativecommons.org/licenses/by/4.0/).

1. Introduction

When I was a doctoral course student of Professor Tsuji (1978–1981), transition metal-catalyzed reactions were little recognized as tools for organic synthesis. The obvious reason for this is the cost of transition metal complexes for stoichiometric use. However, Professor Tsuji emphasized the advantages of using transition metal-catalyzed reactions that could not be attained by the classical reactions [1,2]. Furthermore, Professor Tsuji was interested in carbon-carbon bond-forming reactions and natural product synthesis. According to him, such inclinations were grown in the Stork laboratory at Columbia University in New York, where he pursued his doctoral research [3,4].

Among the studies undertaken in the Tsuji group at that time, the palladium-catalyzed allylic substitution reaction with soft nucleophiles was especially attractive because of the ability to form carbon-carbon bonds. Later, the allylic substitution prompted me to investigate the coupling reaction of secondary allylic substrates with *hard* nucleophiles to furnish chiral carbon-carbon bonds connecting various types of carbon groups.

Borates were selected as hard nucleophiles because of their expected high nucleophilicity based on the general consideration of ate complexes, although such reactivity was not observed for [BuCH=CHB(Sia)$_2$Me]$^-$Li$^+$ in a mechanistic study of the Suzuki–Miyaura coupling reaction [5]. The recent progress in borates and boronate esters for carbon-carbon bond formation has been well summarized in several reviews [6,7]. In contrast, the reactions of borates in combination with nickel catalysts have been less well documented [8,9]. Herein, we present a review of our studies on nickel-catalyzed allylic coupling reactions on secondary allylic carbons and C(sp^2)–C(sp^2) coupling reactions with various types of borates, with a hope of spurring future studies on borate chemistry and organic synthesis. In addition, we briefly present our investigation of secondary C–C bond formation using allylic and propargylic substitutions with copper-based Grignard reagents. The substitution of secondary *alkyl* carbons is also mentioned.

2. Design of Reactive Borates

In an early investigation, allylic carbonates **1** and acetates were used as substrates to identify effective borates (R^T: transferable group) and nickel catalysts (Scheme 1, Equation (1)). Trimethoxy borate **7** was studied first to find a nickel catalyst [10] and then advanced to borate **8**, which possesses the 2,2-butanediol and methyl ligands [11]. Subsequently, carbonyl group-friendly borates **9** were developed [12]. Allylic coupling with cyclopentenyl acetate **3** was established using borate **10**, which possesses an *n*-Bu ligand to produce higher reactivity (Equation (2)) [13]. Products **4** were used to synthesize prostaglandins and aristeromycin. Furthermore, the concept of using a diol ligand led us to develop new alkenyl borates **11** for the coupling reaction of cis-bromo olefins **5**, which are readily available but suffer from low reactivity (Equation (3)) [14]. The synthesis of biologically active dienes has been achieved. Due to their high reactivity, borates were successfully used in the coupling reaction with aryl mesylates and tosylates [15], which are less reactive than triflates in catalytic coupling reactions.

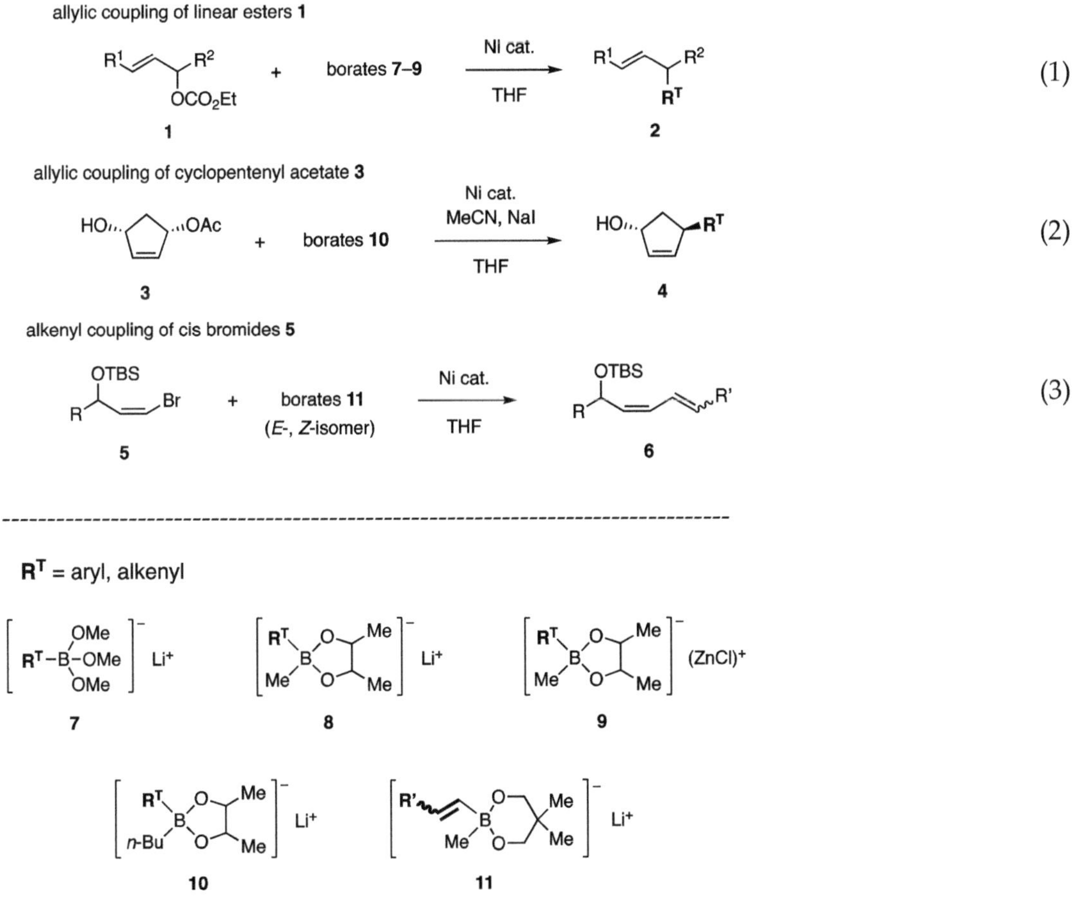

Scheme 1. Allylic and alkenyl coupling reactions with borates.

3. Preparation of Borates

The borates studied in this work were prepared using the methods described in Scheme 2 (R^T = aryl or alkenyl). Commercially available $B(OMe)_3$ was converted with R^T-Li to borates **7** (Method 1) [10,11], which was used for coupling without isolation. Borates **8–10** were obtained by the esterification of boronic acids $R^T B(OH)_2$ (**12**) with 2,3-butanediol, followed by the addition of MeLi [11], n-BuLi [11], or MeZnCl [12] to the resulting boronate esters **13** (Method 2). Some of the (E)-alkenyl borates **8–10** (R^T = (E)-alkenyl) and **11** were obtained via hydroboration of acetylenes **14** with $(Ipc)_2BH$, followed by the conversion of Ipc to EtO with MeCHO (Method 3). Hydroboration with $(c\text{-Hex})_2BH$, followed by oxidation with Me_3NO, has also been used to prepare boronic acids. A transformation consisting of lithiation of stereo-defined halo olefins **17** (X = Br, I) followed by reactions similar to Method 1 produced the (E)- and (Z)-isomers of **8–11** (Method 4). Boronate esters **13** and **16** are much less polar than the free diol and are chemically stable, allowing purification by chromatography on silica gel. Distillation of boronate esters is recommended before the reaction with MeLi, n-BuLi, or MeZnCl. Currently, various boronic acids, $R^T B(OH)_2$, and boronate esters are commercially available. However, sterically demanding diol ligands decreased their reactivity in our study, as discussed below. Pinacol, known as "pin", is one of such ligands.

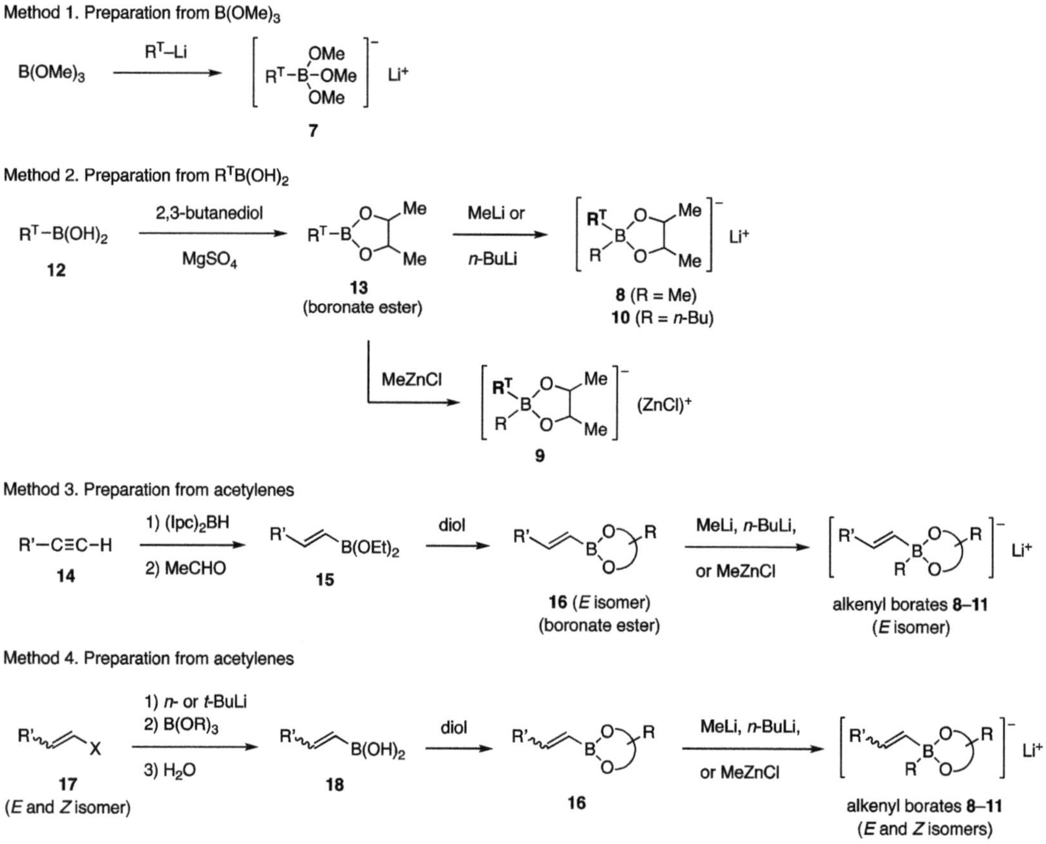

Scheme 2. Preparation of borates.

4. Allylic Coupling

4.1. Trimethoxyborates

The reaction of carbonate **1a** with phenylborate **7a** (prepared from B(OMe)$_3$ and PhLi) was first examined [10,11]. Unfortunately, palladium catalysts at 60–65 °C produced methyl ether **19** (Table 1, entries 1 and 2). In contrast, the Ni catalysts unexpectedly afforded coupling product **2a** in high yields (entries 3 and 4). Other aryl borates **7b–e** afforded **2b–e** in good yields (entries 5–8). The borate/Ni catalyst was successfully applied to the reaction of another substrate, **1b** ($R^1 = CO_2Et$, $R^2 = C_5H_{11}$).

Table 1. Allylic substitution with aryl borates.

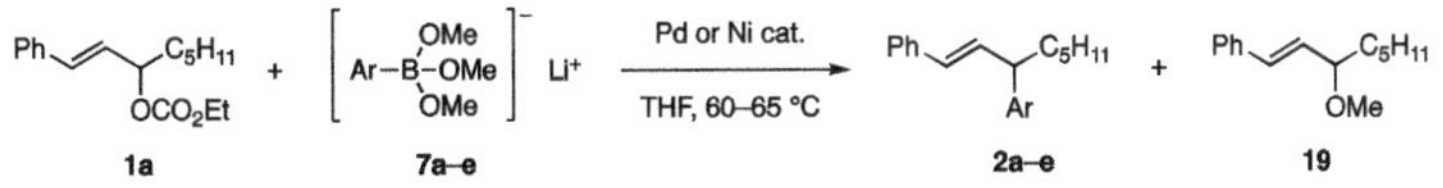

Entry	Cat.	2, Ar	2 (%)	19 (%)
1	Pd(PPh$_3$)$_2$	2a, Ph	<10	66
2	Pd(dppf)	2a, Ph	0	54
3	NiCl$_2$(PPh$_3$)$_2$	2a, Ph	97	0
4	NiCl$_2$(dppf)	2a, Ph	98	0
5	NiCl$_2$(dppf)	2b, 2-MeOC$_6$H$_4$	86	0
6	NiCl$_2$(dppf)	2c, 3-MeOC$_6$H$_4$	81	0
7	NiCl$_2$(dppf)	2d, 4-MeOC$_6$H$_4$	89	0
8	NiCl$_2$(dppf)	2e, 2-furyl	88	0

Before our investigation, the Pd-catalyzed reaction of primary allylic acetate **22a** with boronate ester **20** was reported by Suzuki and Miyaura in their mechanistic study of the "Suzuki–Miyaura coupling reaction" (Scheme 3, Equation (4)) [5]. Due to the reactivity toward secondary acetates has not been reported, acetate **24a**, and carbonate **1a** were subjected to Pd-catalyzed reactions in our study. However, undesired diene **25** or Et ether **26** was produced (Equations (5) and (6)) [11]. In contrast, the reaction of **1a** with borate **21** and a Ni catalyst proceeded smoothly to afford **2f** with a 70% yield (Equation (7)).

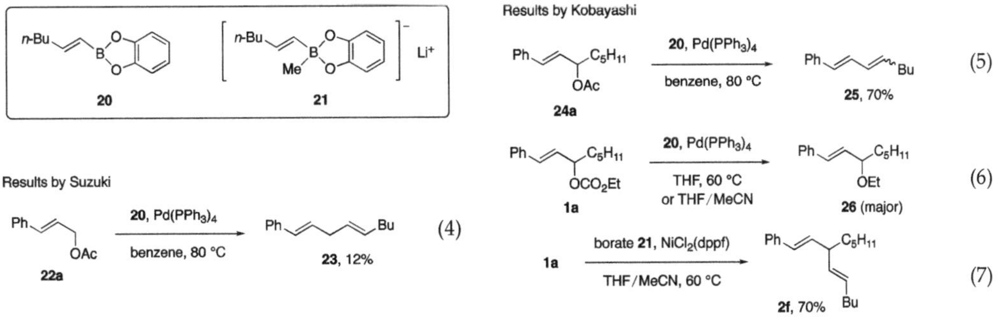

Scheme 3. Related reactions by Suzuki (Equation (4)) and by us (Equations (5)–(7)).

4.2. Borates with Alkanediol Ligands

Although the allylic coupling reactions with trimethoxyborates [R^T-B(OMe)$_3$]$^-$ Li$^+$ (**7**) proceeded in the presence of nickel catalysts, the required temperatures for the reaction were high (60–65 °C), and thus new borates with high reactivity at low temperatures were investigated. We postulated a low concentration of reactive species **7** in equilibrium with less reactive R^TB(OMe)$_2$ (**27**) and LiOMe (Scheme 4, Equation (8)), and envisaged that diol ligands in borates **28** would prevent the formation of the open form **29** by analogy with bidentate ligands that coordinate strongly with transition metals (Equation (9)).

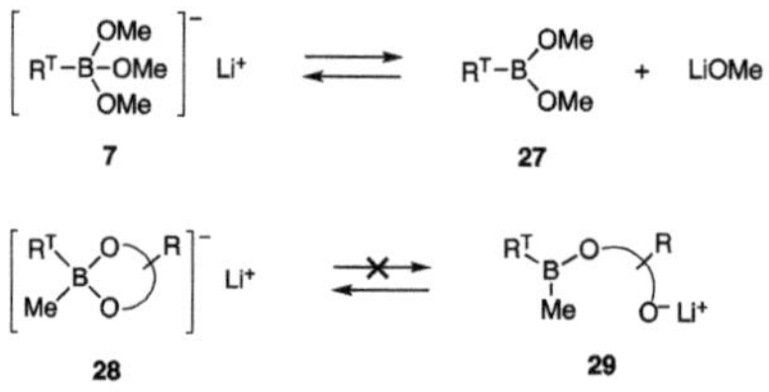

$$\begin{bmatrix} \text{OMe} \\ R^T-B-\text{OMe} \\ \text{OMe} \end{bmatrix} \text{Li}^+ \quad \rightleftharpoons \quad R^T-B\begin{matrix}\text{OMe}\\\text{OMe}\end{matrix} + \text{LiOMe} \qquad (8)$$

7   27

$$\begin{bmatrix} R^T & O & R \\ & B & \\ Me & O & \end{bmatrix} \text{Li}^+ \quad \stackrel{\times}{\rightleftharpoons} \quad R^T\!-\!B\!\begin{matrix}O\\\\O\end{matrix}\!\!\!\begin{matrix}R\\\\ \text{O}^-\text{Li}^+\end{matrix}\!\!\!\begin{matrix}\\\\Me\end{matrix} \qquad (9)$$

28   29

Scheme 4. Borates in likely equilibria.

Phenylborates **31A** (=**8a**) and **31B–F** possessing various diol ligands **A–F** were chosen as test borates (Scheme 5). Among these ligands, **C** is known as a pin. The reactivity order of **31A** (=**8a**) > **31B** > **31D** > **31C** >> **31F** > **31E** was determined for the coupling reaction shown in Scheme 6. This order is consistent with the working hypothesis proposed in Scheme 4. Unanticipatedly, the methyl group on the diol ligand increased the reactivity (**A** vs. **B**), whereas the congested methyl groups were affected reversely because of the steric reason (**A** vs. **C** (pin)). The most active ligand **A** allowed the reaction to proceed even at 5 °C. NiCl$_2$(PPh$_3$)$_2$ showed similar catalytic activity to NiCl$_2$(dppf). In contrast, **31A**/Pd catalysts were marginally productive.

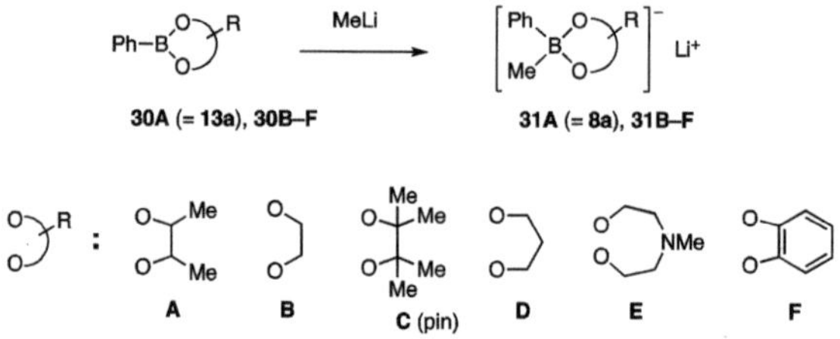

Scheme 5. Phenylborates with diol ligands for allylic coupling.

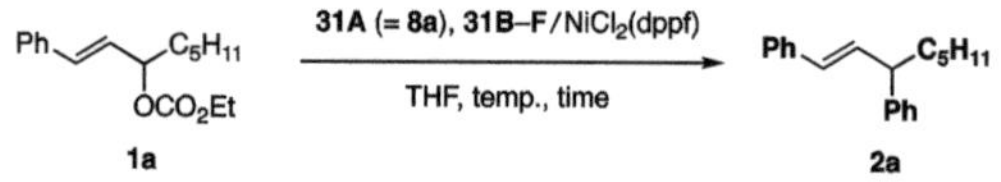

31A (= 8a)[1]	31A	31B	31C	31D	31E	31F
rt, 4 h	5 °C, 12 h	35 °C, 12 h	40 °C, 17 h	40 °C, 12 h	40 °C, 17 h	65 °C, 15 h
88%	97%	84%	60%	95%	NR	<10%

Scheme 6. Allylic coupling reaction of **1a** with phenylborates **31A–F**. [1] 2,3-Butanediol ligand consisted of *dl*- and *meso*-forms in a 4:1 ratio.

Various aryl borates with ligand **A** also showed high reactivity in allylic coupling to produce **2b–d** with 2-, 3-, and 4-MeOC$_6$H$_4$ and **2e** with 2-furyl in 85–99% yield (Figure 1). Products **2g–j** demonstrated the compatibility of the borates with the ester groups present in the allylic substrates. The reaction of **1c** with borates proceeded with inversion to afford **2i** and **2j**.

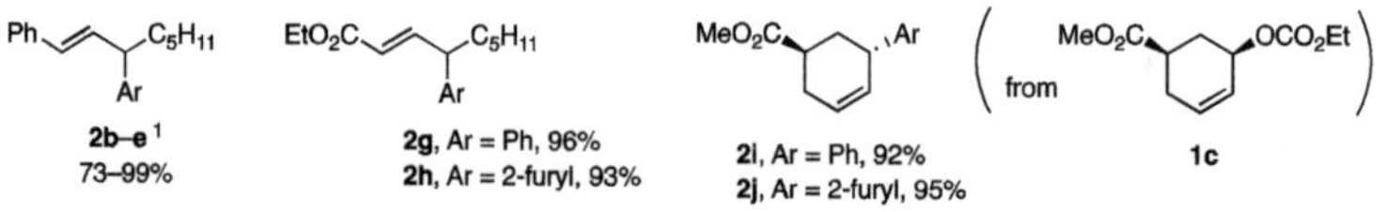

2b–e[1]
73–99%

2g, Ar = Ph, 96%
2h, Ar = 2-furyl, 93%

2i, Ar = Ph, 92%
2j, Ar = 2-furyl, 95%

(from **1c**)

Figure 1. Products **2b–j** from borates with ligand **A**. [1] **b, c, d**: 2-, 3-, and 4-MeOC$_6$H$_4$; **e** and 2-fury.

4.3. Carbonyl Group-Friendly Zinc Borates

Since the carbonyl group in **32** was incompatible with MeLi, which is used for the transformation of boronate ester to borate, other organometallics that are reactive and friendly with such functional groups were explored (Scheme 7). Among the candidates examined, MeZnCl, upon addition to **32**, provided zinc borate **9**, which exhibited sufficient nucleophilicity for the allylic coupling [12].

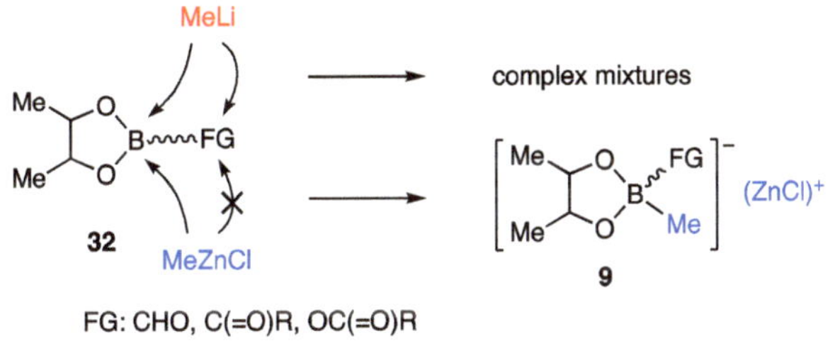

Scheme 7. Reasons for developing zinc borates **9**.

The reaction of acetate **20b** with phenylborate **9a** derived from **13a** and MeZnCl at 40–50 °C produced **2a** and byproducts (diene **25** and methyl coupling **33**) (Table 2, entry 1). Similar results were obtained for the borates derived from MeZnBr (entry 2) and MeZnI. The use of postulated borates derived from MeZnF, n-BuZnCl, and Et$_2$Zn was unsuccessful. Fortunately, the byproducts were eliminated by the addition of DMI or DMF (entries 3 and 4).

Table 2. Allylic coupling reaction with zinc phenylborates.

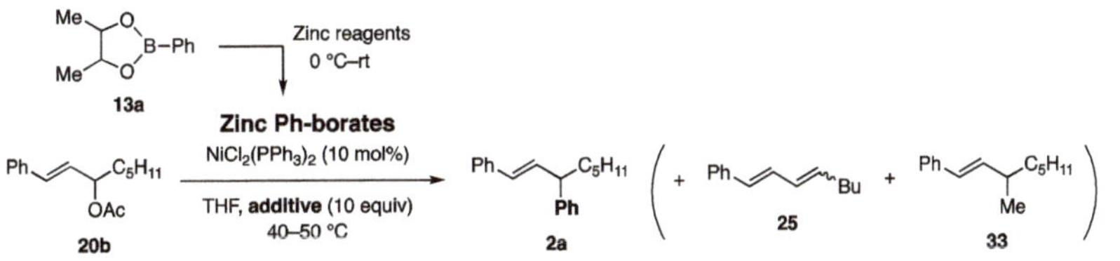

Entry	Zinc Reagent	Additive	2a, Yield (%)	Byproducts [1] (%)
1	MeZnCl	-	88	9
2	MeZnBr	-	85	6
3	MeZnCl	DMI	87	0
4	MeZnCl	DMF	80	0

[1] A mixture of **25** and **33**.

Under the reaction conditions used in entry 3 of Table 2, the zinc borates in Figure 2 underwent coupling reactions to give the products shown in Figure 3 in good to high yields. An anti-stereochemical course was established using cyclic substrates.

4.4. Allylic Coupling on Cyclopentenediol Monoacetate

The Ni-catalyzed allylic coupling reaction with borates was then applied to cyclopentenediol monoacetate **3** (Scheme 1, Equation (2)) [13]. As substrate **3** is available in racemic as well as enantioenriched forms by several methods, the reaction was expected to provide new access to the cyclopentanoids such as prostaglandins. Unfortunately, an attempted reaction with phenylborate **8a** (R^T = Ph) gave a mixture of enone **34** and Ph ketone **35** (Scheme 8, Equation (10)). The likely steps to these byproducts involve predominant β-H elimination of the π-allyl Ni intermediate (derived from **3** and Ni) over the allylic coupling (step 1), isomerization of the resulting olefin **36** to enone **34** (step 2), and 1,4-addition probably catalyzed by the Ni catalyst (step 3). Reactions similar to steps 1 and 3 have been reported [16,17].

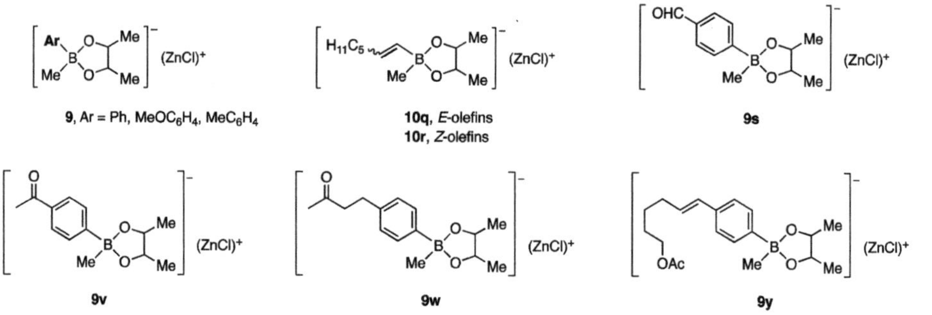

9, Ar = Ph, MeOC$_6$H$_4$, MeC$_6$H$_4$

10q, E-olefins
10r, Z-olefins

9s

9v

9w

9y

Figure 2. Zinc borates prepared for allylic coupling reaction.

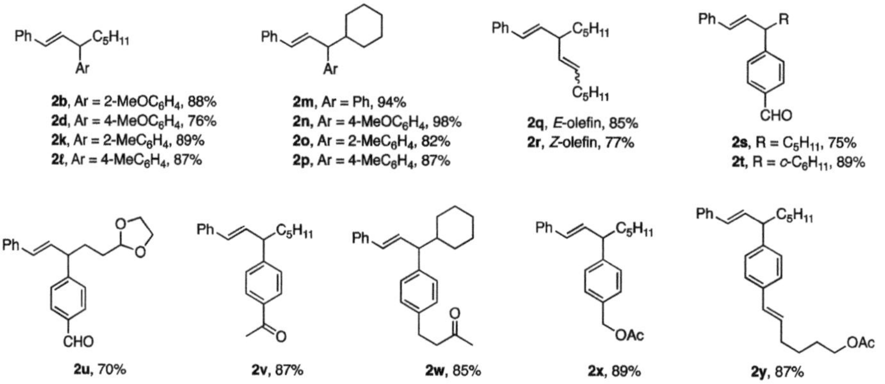

2b, Ar = 2-MeOC$_6$H$_4$, 88%
2d, Ar = 4-MeOC$_6$H$_4$, 76%
2k, Ar = 2-MeC$_6$H$_4$, 89%
2t, Ar = 4-MeC$_6$H$_4$, 87%

2m, Ar = Ph, 94%
2n, Ar = 4-MeOC$_6$H$_4$, 98%
2o, Ar = 2-MeC$_6$H$_4$, 82%
2p, Ar = 4-MeC$_6$H$_4$, 87%

2q, E-olefin, 85%
2r, Z-olefin, 77%

2s, R = C$_5$H$_{11}$, 75%
2t, R = o-C$_6$H$_{11}$, 89%

2u, 70%   **2v**, 87%   **2w**, 85%   **2x**, 89%   **2y**, 87%

Figure 3. Coupling products derived from zinc borates.

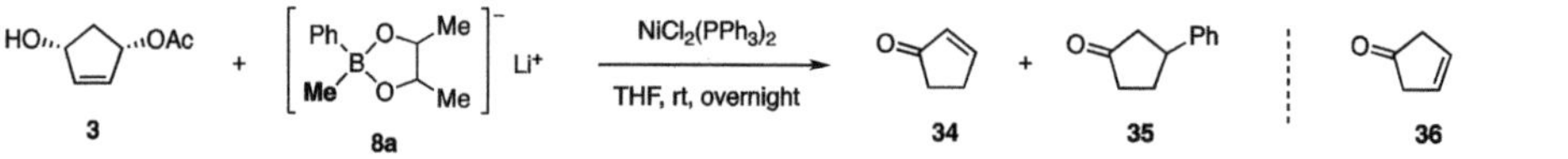

$$\text{3} + \text{8a} \xrightarrow[\text{THF, rt, overnight}]{\text{NiCl}_2(\text{PPh}_3)_2} \text{34} + \text{35} + \text{36} \quad (10)$$

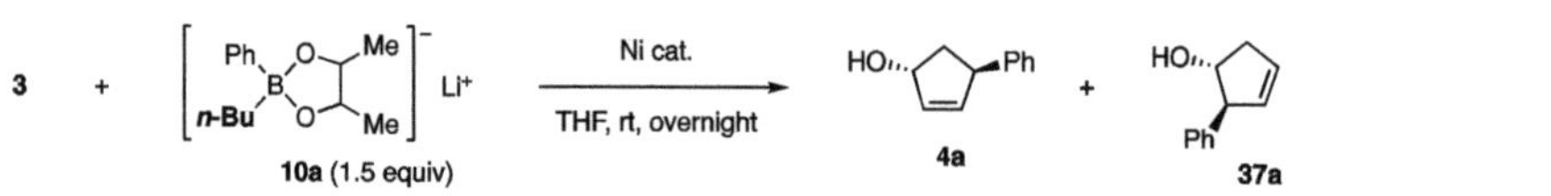

$$\text{3} + \text{10a (1.5 equiv)} \xrightarrow[\text{THF, rt, overnight}]{\text{Ni cat.}} \text{4a} + \text{37a} \quad (11)$$

Entry	Cat.	Additive	4a/37a	Yield
1	NiCl$_2$(PPh$_3$)$_2$	–	0.9:1	82% isolated yield
2	NiCl$_2$(dppf)	–	0.9:1	nd
3	NiCl$_2$(PPh$_3$)$_2$	t-BuCN (2 equiv)	6.6:1	nd
4	NiCl$_2$(PPh$_3$)$_2$	NaI (1 equiv)	5:1	nd
5	NiCl$_2$(PPh$_3$)$_2$	t-BuCN + NaI	13:1	84% isolated yield
6	cf. Pd(PPh$_3$)$_4$	–	35 (major)	nd

Scheme 8. Reaction of cyclopentenyl acetate **3** with borates **8a** and **10a** [1]. [1] Diol ligand consisted of *dl*- and *meso*-forms at a 4:1 ratio.

Based on the finding that the attachment of electron-donating Me groups to the diol ligand increased the nucleophilicity of borates (Scheme 6, **31A** (=**8a**) vs. **31B**), the Me ligand in **8a** was replaced by the more electron-donating *n*-Bu. Indeed, the new phenylborate **10a** transferred Ph to cyclopentenyl acetate **3** in the presence of NiCl$_2$(PPh$_3$)$_2$ (Scheme 8, Equation (11), entry 1). However, a mixture of regioisomers **4a** and **37a** was produced at a 0.9:1 ratio. To solve this problem, catalysts, polar solvents, and inorganic salts were examined without any promising guidelines. Although NiCl$_2$(dppf) did not change the product ratio (entry 2), *t*-BuCN and NaI favored the production of **4a** (entries 3 and 4). Furthermore, the cooperative action between both additives resulted in a higher regioisomeric ratio of 13:1 (entry 5). In contrast, a Pd catalyst produced ketone **35** (entry 6).

The new reagent type consisting of 2,3-butanediol and *n*-Bu ligands was successfully applied to aryl borates **10b–e**, which underwent a regioselective reaction in the presence of the two additives (*t*-BuCN and NaI) to produce **4b–e** in good yields (Table 3, entries 2–5). Alkenyl borates **10f–k** also afforded **4f–k** (entries 6–11). Among the entries, the regioselectivity for **4h–j** was especially high, whereas that for **4f** and **4k** was moderate. A hypothesis that accounts for this difference is the high nucleophilicity caused by the conjugation of the C–OTBS σ bond with the olefinic π-bond. In an additional entry, the MOM ether of **3** upon reaction with borate **10f** afforded the MOM ethers of **4f** and **37f** at a ratio of 6:1 (equation not shown). This ratio was similar to that recorded for **3**, indicating that the hydroxy group in **3** was not a regiocontroller.

Table 3. Coupling reaction of cyclopentenyl acetate **3** with borates **10**.

Entry	Suffix for R^T	R^T	Yield	4/37 Additive	4/37 No Additive
1 [1]	a	R = H	84	13:1	0.9:1
2	b	R = 3-Me	80	7:1	1.2:1
3	c	R = 4-Me	81	12:1	1.5:1
4	d	R = 3-MeO	84	6:1	0.7:1
5	e	R = 4-MeO	81	9:1	–
6	f	*n*-C$_5$H$_{11}$	89	6:1	3.0:1
7	g	*n*-C$_5$H$_{11}$	85	5:1	1.3:1
8	h	R = *n*-C$_5$H$_{11}$	67	15:1	5.4:1
9	i	R = *c*-C$_6$H$_{11}$	85	15:1	5.5:1
10	j	R = CH$_2$OPh	80	17:1	8.0:1
11	k	*n*-Bu, *n*-Bu	85	6:1	2.5:1

[1] Repost of Scheme 8, entry 5.

4.5. Determination of the Stereochemistry between R^T and OH in the Major Products 4

Substituents OH and R^T sterically affect the chemical shifts of Ha and Hb in **4** and Hc and Hd in cis isomer **38** (Figure 4) [18]. For example, Ha in **4** is cis to OH and trans to R^T, whereas Hb is arranged in the opposite direction. This relationship (one cis, one trans) between Ha and Hb results in a small chemical shift difference (Δδ) of ca. 0.3 ppm. On the other hand, Hc in **38** is trans to OH and R^T, whereas Hd is cis to these substituents, resulting in a large Δδ of 1 ppm. In our study, the stereochemistry of **4** was inverted to produce isomer **38** by the Mitsunobu inversion, and the vicinal protons (Ha/Hb, Hc/Hd) in the ^{1}H NMR spectra were easily found based on the large coupling constants of ca. 14 Hz.

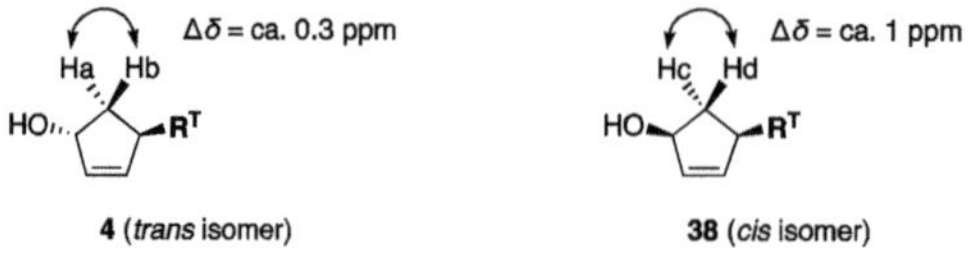

Figure 4. Chemical shift difference (Δδ).

4.6. Synthetic Application of Allylic Coupling

4.6.1. Prostaglandin A$_2$

The coupling conditions disclosed in Table 3 were then applied to enantiomerically enriched (1R,4S)-**3** and (S)-**10h** to afford (S,S,S)-**4h** in 79% yield with a regioisomeric ratio of 19:1 (Scheme 9) [13]. The product was transformed to iodolactone **39** through the Claisen rearrangement, followed by iodolactonization of the derived acid. Deiodination with AgOAc produced olefin **40**, the intermediate in the synthesis of PGA$_2$ (**41**) [19].

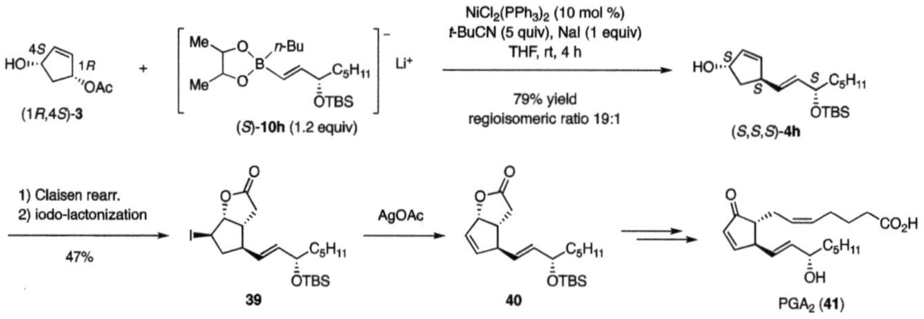

Scheme 9. Coupling reactions and further conversion to PGA$_2$.

4.6.2. Δ^7-PGA$_1$ Methyl Ester

The coupling reaction also produced (R,R,S)-**4h** regioselectively from (1S,4R)-**3** and (S)-**10h** (Scheme 10) [13]. This product was then transformed into mesylate **42** by oxidation to the ketone, followed by an aldol reaction under kinetic conditions and mesylation. The mesylate was eliminated using basic Al$_2$O$_3$, which afforded the E-olefin stereoselectively, irrespective of the aldol stereochemistry. Finally, desilylation to the target (**43**) [20] was achieved with NBS in aqueous DMSO, whereas Bu$_4$NF caused deprotonation of the C12-hydrogen in the dienone core.

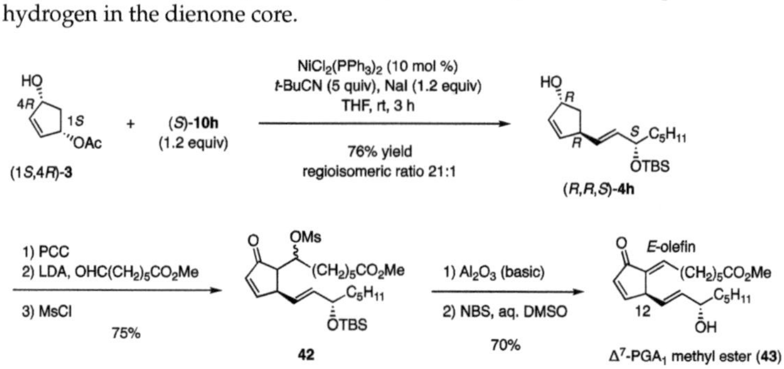

Scheme 10. Synthesis of Δ^7-PGA$_1$ methyl ester.

4.6.3. Aristeromycin

The furyl moiety as a precursor of CO_2Me was installed to (1R,4S)-**3** with furylborate **10ℓ** (Scheme 11) [21]. The hydroxy group in the product **4ℓ** was then replaced by N_3 with inversion using $(PhO)_2P(=O)N_3$, and the furyl group was oxidatively cleaved to afford **45**, which was transformed into the target **46** without difficulty.

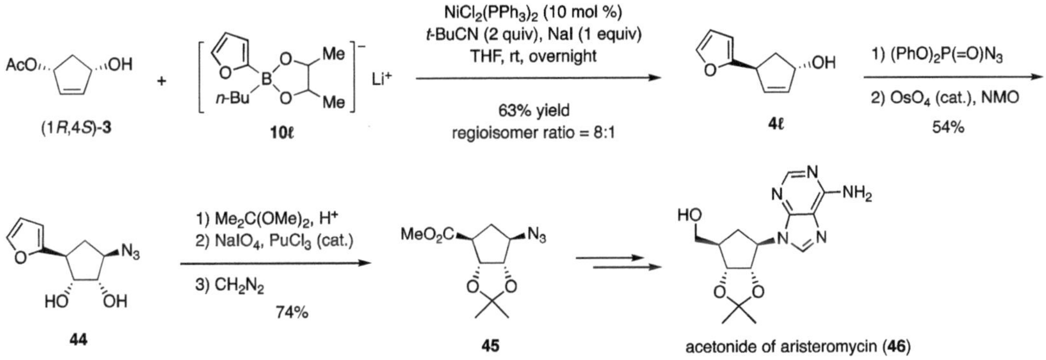

Scheme 11. Synthesis of aristeromycin.

5. Alkenyl-Alkenyl Coupling Reaction
5.1. Optimal Alkanediol Ligand for Alkenyl Borates

The ready availability of allylic alcohol derivatives possessing a bromine atom at the cis position [22] was one of the reasons to investigate coupling reactions with alkenyl borates (Scheme 1, Equation (3)). Other reasons include the possibility of using the reaction for the synthesis of biologically active dienes and the stereoselective functionalization of the dienes. Although the low reactivity of the cis-bromides was a concern due to steric congestion with the OR (R = TBS, etc.), we expected compensation from the high reactivity of borates.

Borate **47A**, with the most effective ligand **A** in the allylic coupling, was subjected to a test coupling with cis-bromide **5a**. However, the reaction was capricious at rt and required heating at 40–45 °C to produce diene **6a** (Table 4, entry 1), indicating that steric hindrance around the C–Br of **5a** for oxidative addition and/or C–Ni–Br thereof for transmetalation was an obstacle to accessing Ni or the borate. Borate **47C** with pin ligand **C** showed no reactivity (entry 2). Fortunately, ligand **G** exhibited the highest reactivity among the tested diols, affording **6a** in 90% isolated yield (entry 4) [14]. An attempted $PdCl_2(PPh_3)_2$-catalyzed coupling between **5a** and **44G** was unsuccessful.

The reactivity with ligand **G** was sufficient for the coupling of cis-bromides possessing bulkier TBDPS (t-BuPh$_2$Si) than TBS (t-BuMe$_2$Si) and a secondary alkyl group (c-C$_6$H$_{11}$), giving **6b** and **6c** in good yields. Furthermore, (2-Ph-vinyl)borate and a cis-borate afforded dienes **6d** and **6e** (Figure 5).

5.2. Synthetic Application of the Alkenyl-Alkenyl Coupling
5.2.1. Synthesis of 10,11-Dihydro-LTB$_4$

Based on the above findings for the synthesis of dienes, the necessary borate **11f** and cis-bromide **5f** were easily elucidated [23]. Among the existing methods for synthesizing borate **11f**, the following method was selected based on the potential regio- and stereoselectivity (Scheme 12). The hydroboration of acetylene **48** with (Ipc)$_2$BH (Ipc: diisopinocampheyl), followed by ligand exchange to EtO using MeCHO and transesterification, produced **16f**, which upon reaction with MeLi afforded borate **11f**. On the other hand, allylic alcohol **49** was prepared with high stereo- and enantioselectivity by the kinetic resolution using the Sharpless asymmetric epoxidation. Subsequently, **49** was converted to cis-bromide **5f** [24].

The coupling reaction proceeded smoothly to afford **6f** in 77% yield, and the subsequent desilylation of TBS with Bu$_4$NF conveniently induced "formal hydrolysis" of the Me ester group to produce the target **50**, which is an inflammatory mediator.

Table 4. Preliminary coupling reaction of (cis-bromo)olefin **5a** with alkenyl borates possessing diol ligands.

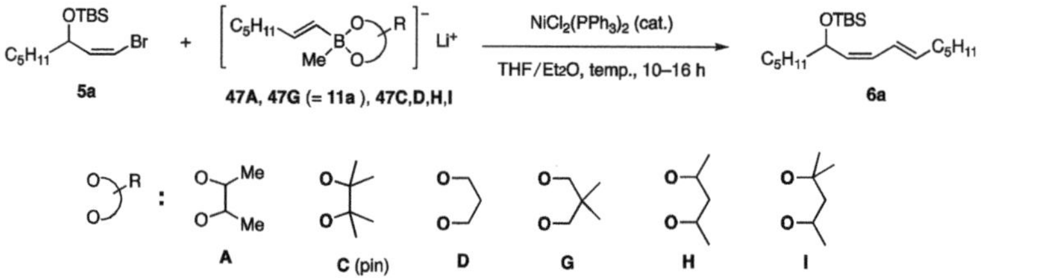

Entry	Borate	Diol Ligand	Temp.	6a, Yield (%)	Conversion (%)
1	47A	A	40–45 °C [1]	76	100
2	47C	C	rt	-	0
3	47D	D	rt	-	68
4	47G (=11a)	G	rt	90	100
5	47H	H	rt	-	79
6	47I	I	rt	-	0

[1] Capricious at rt.

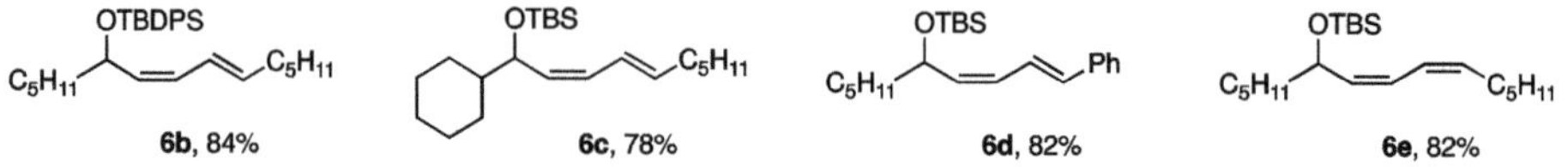

Figure 5. Other dienes produced by the coupling reaction.

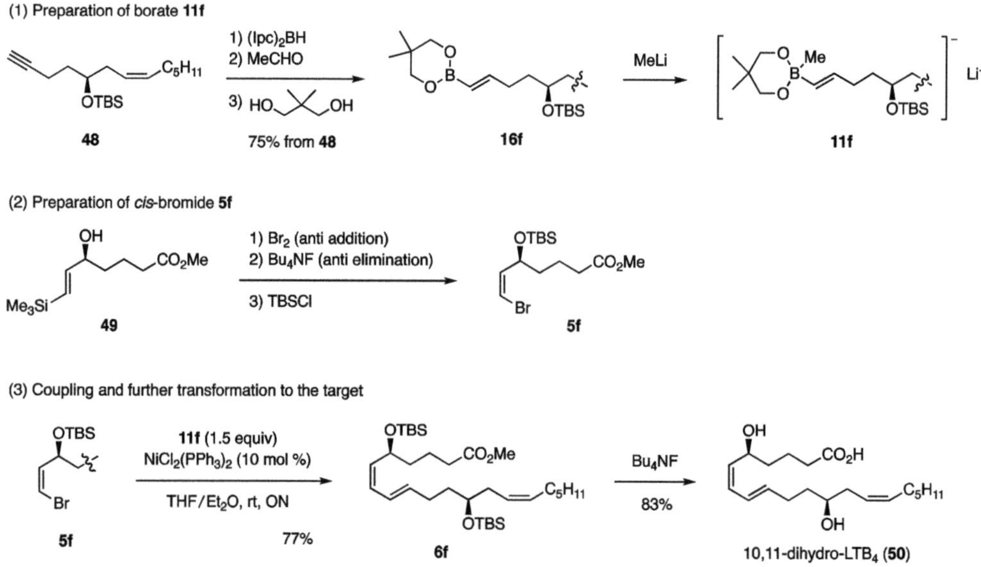

(1) Preparation of borate **11f**

(2) Preparation of *cis*-bromide **5f**

(3) Coupling and further transformation to the target

Scheme 12. Synthesis of 10,11-dihydro-LTB$_4$.

5.2.2. Synthesis of Korormicin

The title compound inhibits the growth of marine gram-negative bacteria. The planar structure of this compound was reported based on the NMR and mass spectral data. Consequently, four diastereomers were synthesized to determine the stereochemistry by comparing the $[\alpha]_D$ values of the diastereomers with those published for natural korormicin. The synthesis of the diastereomer, which showed a close $[\alpha]_D$ value to that of the natural one, is shown in Scheme 13 [25]. The necessary borate **11g** and *cis*-iodide **5g**, were synthesized through **51** and **52**, respectively. Due to the structural specificity of **52** possessing the COOH group, the corresponding bromide could not be obtained. Instead, iodide **52** was transformed to **53** via iodolactonization. Acid **53** was then reacted with amine **54** under somewhat specific reaction conditions to afford amide **5g**. The Ni-catalyzed coupling reaction of iodide **5g** with borate **11g** produced the TBS ether of korormicin, which upon desilylation afforded korormicin (**55**).

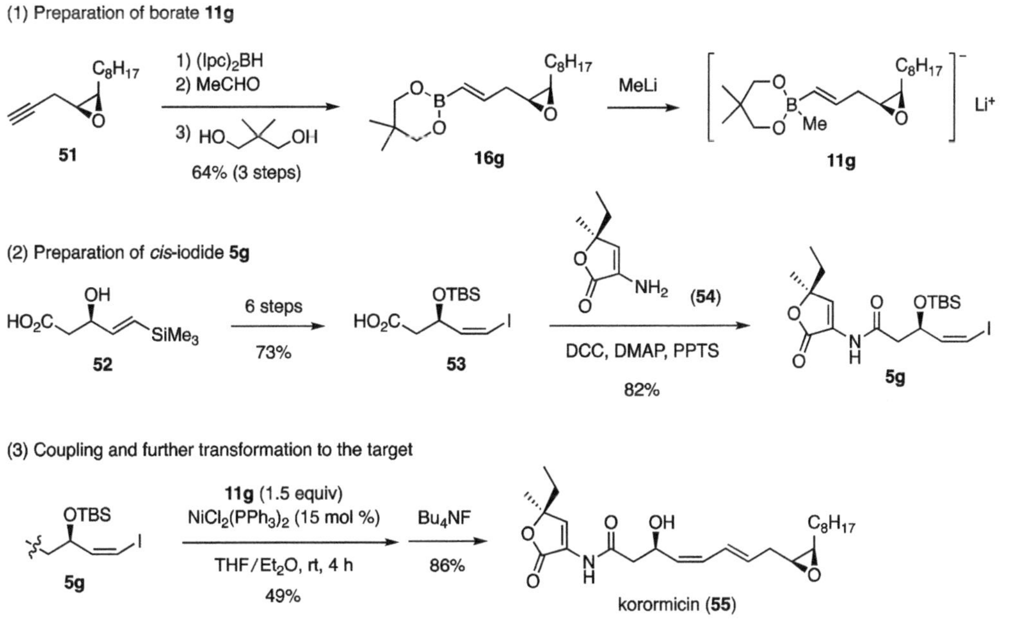

Scheme 13. Synthesis of korormicin for determination of its stereochemistry.

5.3. Functionalization of the Alkenyl-Alkenyl Coupling Products

Since the epoxidation of *cis* allylic alcohols with *m*-CPBA proceeds stereoselectively, the regio- and stereoselectivity of the subsequent Pd-catalyzed reaction with AcOH were investigated for the synthesis of polyhydroxy compounds.

In practice, the epoxidation of alcohol **56** derived from **6a** was stereoselective, and the subsequent Pd(PPh$_3$)$_4$-catalyzed reaction of the resulting epoxide **57** with AcOH produced dihydroxy acetate **58** with high regio- and stereoselectivity (Scheme 14, Equation (12)) [26]. Similarly, **59** was converted to **61** through **60** (Equation (13)). No *cis* olefins were produced. Several dienyl alcohols were also selectively converted (products not shown). A plausible mechanism involves the rapid isomerization of the *anti-syn*-π-allyl Pd intermediate formed from **60** to *syn-syn*-π-allyl Pd before the attack by AcOH.

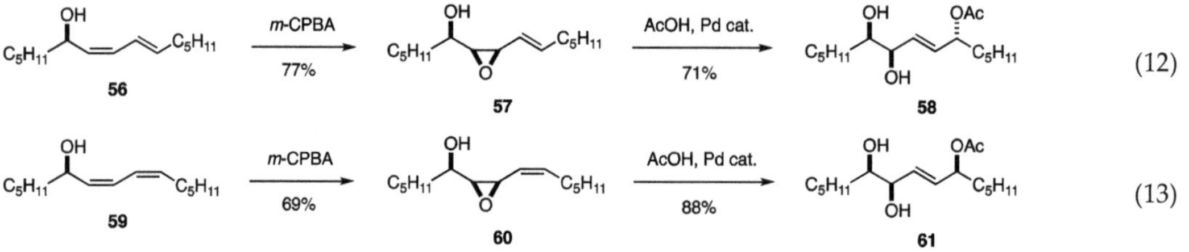

Scheme 14. Functionalization of dienyl alcohols.

The same structural pattern was extracted from decarestrictin D (**62**), which exhibits inhibitory activity against HMG-CoA reductase. Based on the stereochemical relationship summarized in Scheme 14, dienyl alcohol **6h** was designed as the precursor of the key intermediate **63** (Scheme 15) [26,27]. Coupling partners **5h** and **11h** were prepared using methods similar to those used for the synthesis of 10,11-dihydroxy-LTB$_4$ (**50**) (Scheme 12). The use of *cis*-bromide **5h**, which possesses a free hydroxy group, was a new case of the Ni-catalyzed coupling reactions. The reaction proceeded smoothly, requiring 1.5 equiv of **11h** to produce **6h** with a 76% yield. This result indicates that the borate was not affected by the free hydroxy group in **5h**. The epoxidation of **6h** with *m*-CPBA and subsequent Pd-catalyzed reaction with AcOH proceeded with high selectivity to afford the intermediate **63** with a 68% yield. Further transformation to decarestrictin D (**62**) was successful.

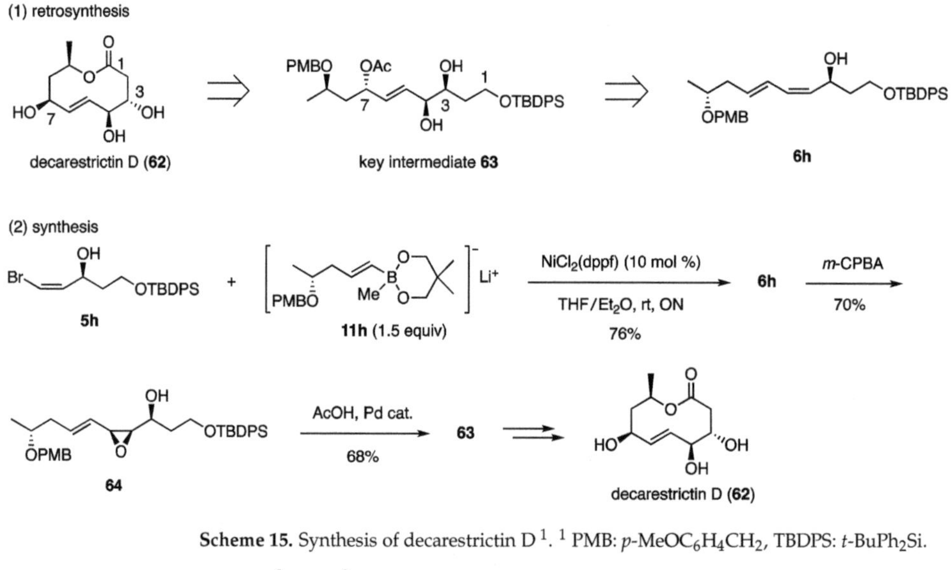

Scheme 15. Synthesis of decarestrictin D [1]. [1] PMB: *p*-MeOC$_6$H$_4$CH$_2$, TBDPS: *t*-BuPh$_2$Si.

5.4. C(sp^2)–C(sp^2) Coupling

Borates with diol ligands have been applied to the coupling of other types of substrates with low reactivity. First, mesylates and tosylates were subjected to a coupling reaction with phenylborates **8a** and **10a** (Table 5) [15]. The former insufficiently produced a mixture of the desired product **67a** and the Me derivative **68** (R = Me) in a ca. 1:1 ratio with 24–32% yields of **67a** (entries 1 and 2). In contrast, borate **10a** exhibited high reactivity, affording **67a** with a 95% yield (entry 3). Tosylate **66** also underwent coupling to produce **67a** with an 83% yield (entry 4).

Table 5. Coupling reactions of mesylates.

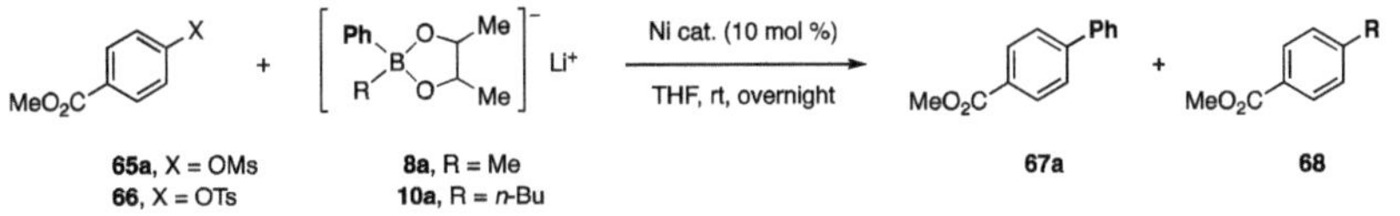

Entry	Substrate	Borate	Cat.	67a, Yield (%)	67a/68 [1]
1	65a	8a	$NiCl_2(PPh_3)_2$	24	57:43
2	65a	8a	$NiCl_2(dppf)$	32	55:45
3	65a	10a	$NiCl_2(PPh_3)_2$	95	>95:5
4	66	10a	$NiCl_2(PPh_3)_2$	83	>95:5

[1] R = Me or n-Bu.

Several mesylates and borates of type **10** were subjected to the coupling conditions to afford the products shown in Scheme 16. Mesylates possessing electron-withdrawing groups, such as CO_2Me, Ac, and CN, were good substrates. The CO_2Me group at the ortho position was not an obstacle to the reaction (**67d, 67e**). The fluorine atom and the additional aromatic moiety were positive for the coupling. On the contrary, mesylate bearing the p-MeO group afforded **67j** with a 58% yield. This result suggests that electron-donating groups play a somewhat negative role. In this regard, the high yields of **67k–m** are not surprising.

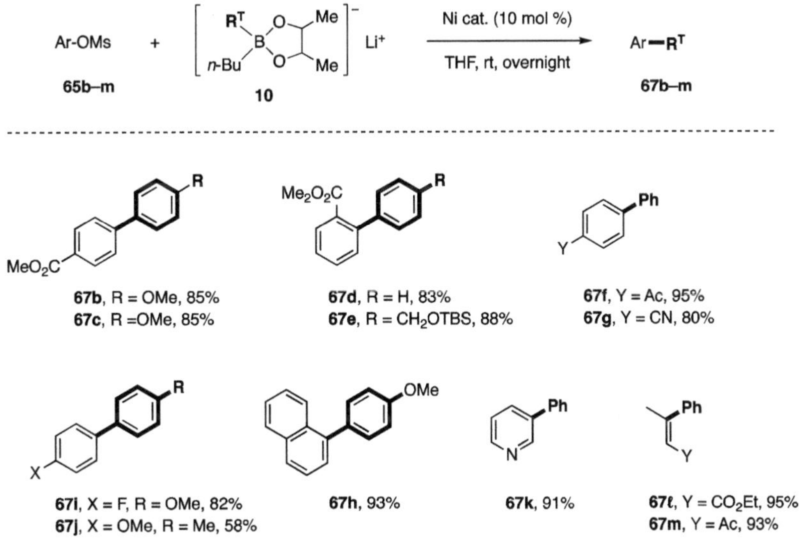

Scheme 16. Coupling reaction of mesylates.

Although the coupling of mesylates with borates was successfully explored, we found little synthetic advantage because the same products could be synthesized using triflates. Therefore, investigation of the mesylates was discontinued.

The construction of a carbon framework on small phosphonates was investigated starting with α-bromoalkenyl phosphonates [28,29], which were prepared by bromination of the corresponding olefins followed by dehydrobromination with Et_3N. The Suzuki-Miyaura coupling of (Z)- and (E)-**69** with $PhB(OH)_2$ at 90–95 °C afforded (E)- and (Z)-**70** in high yields (Scheme 17, Equations (14) and (17)), whereas a similar coupling with an alkenyl borane was unsuccessful (Equation (15)). In contrast, alkenyl borate **11a** was reactive toward (Z)- and (E)-**69** under the catalytic action of Ni to afford **71** as the major product from (Z)-**69** and **72** from (E)-**69**, respectively (Equations (16) and (18)). Although the mixture of **71** and **72** was separated by chromatography, a more selective method is still necessary. Several entries that use different bromides and alkenyl borates yielded similar results.

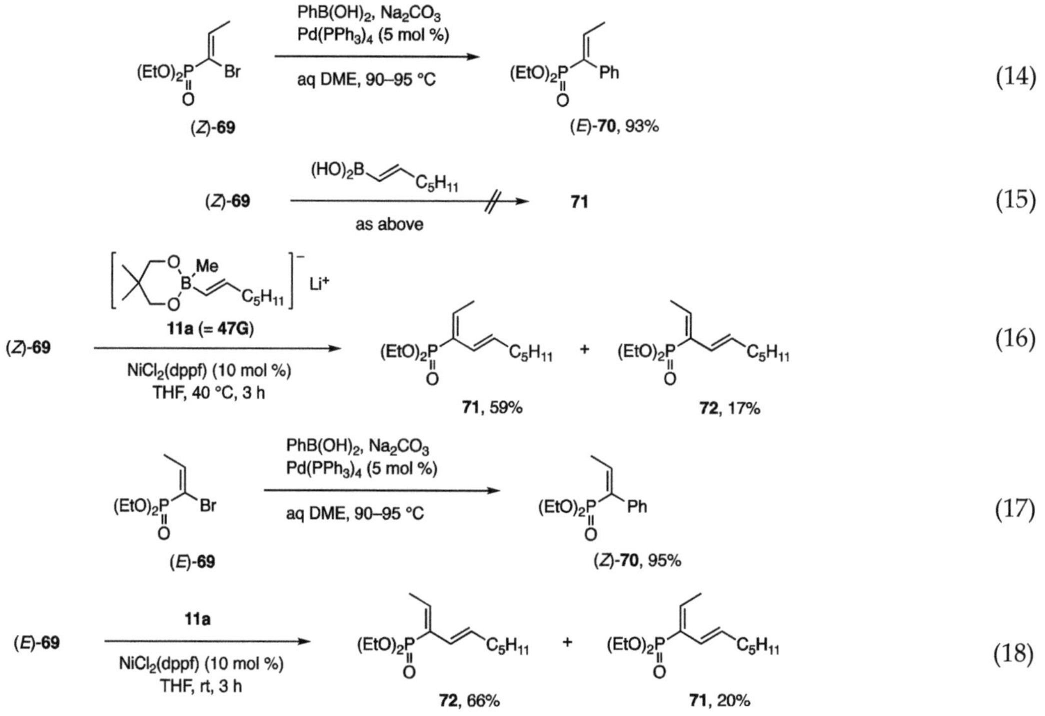

$$(EtO)_2P(O)\text{-}C(Br)\text{=}CH\text{-}CH_3 \;(Z)\text{-}69 \xrightarrow[\text{aq DME, 90–95 °C}]{\text{PhB(OH)}_2,\ Na_2CO_3,\ Pd(PPh_3)_4\ (5\ \text{mol\%})} (EtO)_2P(O)\text{-}C(Ph)\text{=}CH\text{-}CH_3 \;(E)\text{-}70,\ 93\%\quad(14)$$

$$(Z)\text{-}69 \xrightarrow[\text{as above}]{(HO)_2B\text{-}CH\text{=}CH\text{-}C_5H_{11}} \nrightarrow \mathbf{71} \quad(15)$$

$$(Z)\text{-}69 \xrightarrow[\substack{\text{NiCl}_2(\text{dppf})\ (10\ \text{mol\%}) \\ \text{THF, 40 °C, 3 h}}]{\mathbf{11a}\ (=47G)} \mathbf{71},\ 59\% + \mathbf{72},\ 17\% \quad(16)$$

$$(E)\text{-}69 \xrightarrow[\text{aq DME, 90–95 °C}]{\text{PhB(OH)}_2,\ Na_2CO_3,\ Pd(PPh_3)_4\ (5\ \text{mol\%})} (Z)\text{-}70,\ 95\% \quad(17)$$

$$(E)\text{-}69 \xrightarrow[\substack{\text{NiCl}_2(\text{dppf})\ (10\ \text{mol\%}) \\ \text{THF, rt, 3 h}}]{\mathbf{11a}} \mathbf{72},\ 66\% + \mathbf{71},\ 20\% \quad(18)$$

Scheme 17. Coupling reactions of bromo-phosphonates with organometallics.

6. Beyond Borates

6.1. Limitation of Borates

In the above study using borates, the development of alkyl borates for C–C bond formation was unsuccessful. Therefore, we focused on alkyl reagents derived from (alkyl)MgX and a copper salt and found that several combinations of (alkyl)MgX, CuCN, and solvent delivered an alkyl group to cyclopentenediol monoacetate **3** to afford either **4** or **37** with high regio- and stereoselectivity (Scheme 18, Equation (19)) [30]. Subsequently, aryl, alkenyl, and benzylic reagent systems were developed as described below. We also established allylic and propargylic substitutions on acyclic secondary carbons [Equations (20) and (21)]. Acyclic allylic substitution features the use of allylic picolinate **73** to yield the *anti*-S_N2' product **74** with regio- and stereoselectivity (Equation (20)) [31]. Propargylic substitution was realized using propargylic phosphate **75** and a Cu catalyst in THF/DME to produce **76** with high stereoselectivity (Equation (21)) [32]. In contrast to allylic and propargylic substrates, the substitution of secondary *alkyl* carbons was difficult because of the lack of reaction enhancers, such as allylic and propargylic unsaturation. Nevertheless, pyridine-sulfonate **77** was invented for this purpose (Equation (22)) [33]. In the second part of this review, we briefly present the results of these reactions.

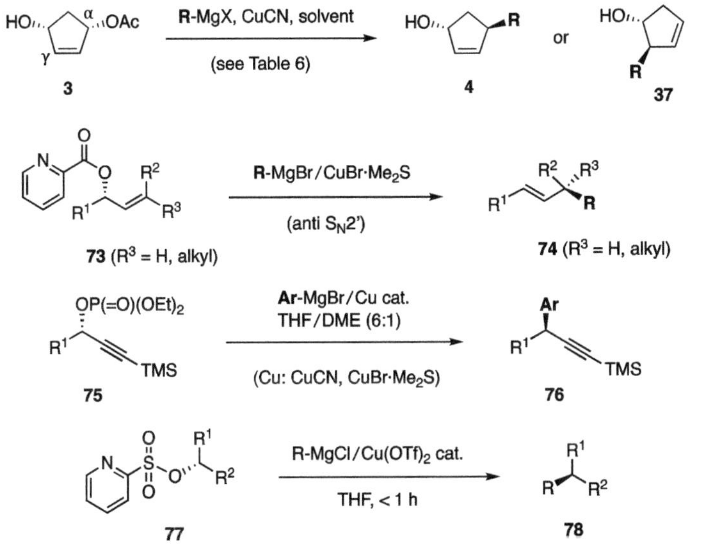

Scheme 18. C–C bond forming reactions at secondary carbons.

6.2. Substitution of Cyclopentenediol Monoacetate 3

Although copper-based reagents generally undergo preferential substitution at the γ carbon (S_N2' reaction), **3** was not used as a substrate in the previous study of allylic substitution. In a preliminary study, reagents comprising n-BuMgX (X = Cl, Br)/CuCN in ratios of 1:1, 2:1, and 10:1 in Et$_2$O and THF were systematically examined [30]. Reagents based on BuMgX (X = Cl, Br), mainly in Et$_2$O, were highly γ selective, thereby producing **37m** (Table 6). In contrast, n-BuMgCl/CuCN in 2:1 and 10:1 ratios in THF preferentially attacked the α carbon (entries 5 and 6). Surprisingly, the solvent (THF vs. Et$_2$O) was decisive for the high regioselectivity (entries 5 and 6 vs. entries 2 and 3).

Table 6. Allylic substitution with n-BuMgX/CuCN.

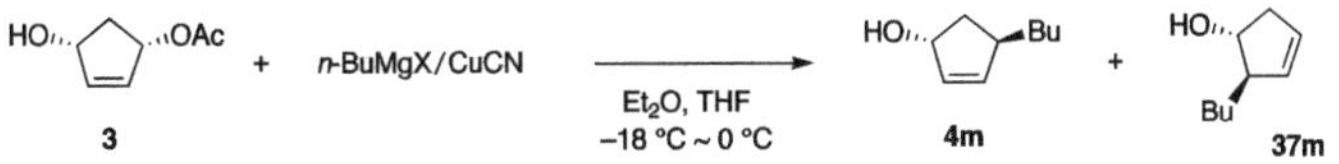

Entry	n-BuMgCl/CuCN	Solvent	4m/37m	Yield (%)
1	1:1	Et$_2$O	14:86	37
2	2:1	Et$_2$O	7:93	85
3	10:1	Et$_2$O	8:92	82
4	1:1	THF	7:93	97
5	2:1	THF	93:7	94
6	10:1	THF	94:6	100
7	1:1	Et$_2$O	4:96	77
8	2:1	Et$_2$O	6:94	88
9	10:1	Et$_2$O	5:95	94
10	1:1	THF	10:90	89
11	2:1	THF	71:29	98
12	10:1	THF	73:27	96

The combination of RMgX, CuX, and the solvent was further investigated to find aryl, alkenyl, and benzylic reagents. These protocols have been successfully used as key steps in the syntheses of plant hormones (such as 12-oxo-PDA and OPC-8:0 [34], tuberonic acid [35], isoleucine conjugate of *epi*-jasmonic acid [36]), AH-13205 (PG receptor agonist) [37], and

PGF$_{2\alpha}$ intermediates [38] (Figure 6). The ω side chain of Δ^7-PGA$_1$ methyl ester (**43**) (structure in Scheme 10) was installed to **3** with efficiency similar to that recorded with the borate [39].

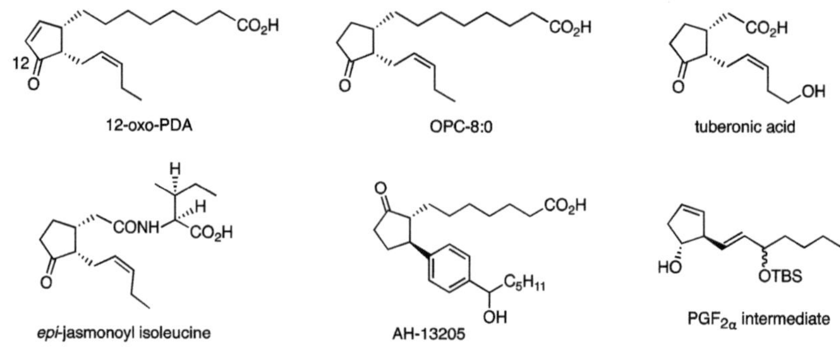

Figure 6. Cyclopentanoids synthesized by using RMgX/CuCN.

6.3. Substitution of Secondary Allylic Picolinates

The reaction in Equation (20) in Scheme 18 features the picolinoxy leaving group, and requires the *cis*-olefin geometry (R^2 ≠ H in **73**) to obtain *anti*-S$_N$2′ products with high selectivity [31]. This topic has been summarized in reviews [40,41]. In brief, substitution of **73a** with PhMgBr (2 equiv) and CuBr·Me$_2$S (0.5–1 equiv) afforded **74a** with high regioselectivity (rs) and enantiospecificity (es) (=chirality transfer C.T.) (Equation (23)). Low temperatures were necessary for high es. Allylic picolinates are chemically stable and easily prepared by the DCC-assisted esterification of *cis*-allylic alcohols with picolinic acid (2-Py-CO$_2$H). Although stable, this leaving group gains high reactivity by chelating to MgBr$_2$, which is generated in situ from RMgBr and CuBr·Me$_2$S. The substitution occurred not only with alkyl but also with aryl, alkenyl, and alkynyl reagents [42,43].

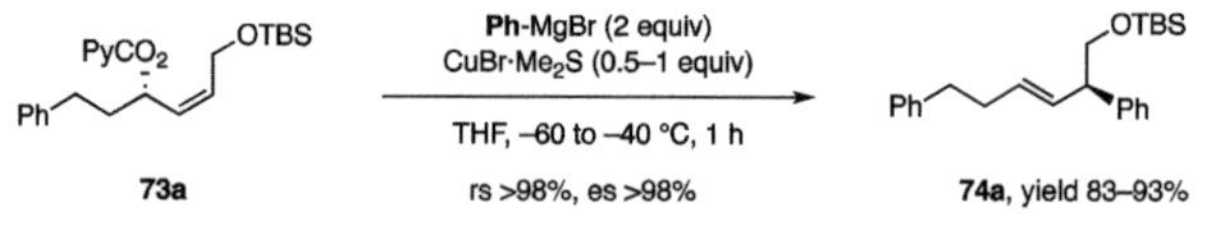

Using this substitution, the PGF$_{2\alpha}$ intermediate was synthesized again (Figure 7) [43]. Other biologically active compounds synthesized are: sesquichamaenol [44], mesembrine [45], cyclobakuchiol B [46], axenol [47], anastrephin [48], verapamil [49], and LY426965 [50] (Figure 7). The active form of anti-inflammatory loxoprofen was synthesized by performing the allylic substitution twice (Scheme 19) [51].

6.4. Substitution of Secondary Propargylic Phosphates

A reagent system derived from Grignard reagent/copper salt was next explored for the substitution of secondary propargylic alcohol derivatives (Scheme 18, Equation (21)). Although an attempted reaction of a propargylic picolinate and PhMgBr/CuCN in THF proceeded with high regioselectivity, 10–20% of the corresponding alcohol was coproduced. In contrast, phosphate **75a** and PhMgBr (2.5 equiv)/CuCN (0.25 equiv) in THF/DME (6:1) gave products **76a** (Ar = Ph) with high rs and es (Scheme 20) [32]. The use of CuCN as the catalyst and DME as the additional solvent was pivotal for producing **76b–f** with high selectivity. The performance of CuBr·Me$_2$S was as excellent as that of CuCN. The presence of TMS at the acetylene end was decisive for achieving the high regioselectivity as observed for the reaction of **75g**. The phenyl group attached to the acetylene end of **75h** also showed good regioselectivity.

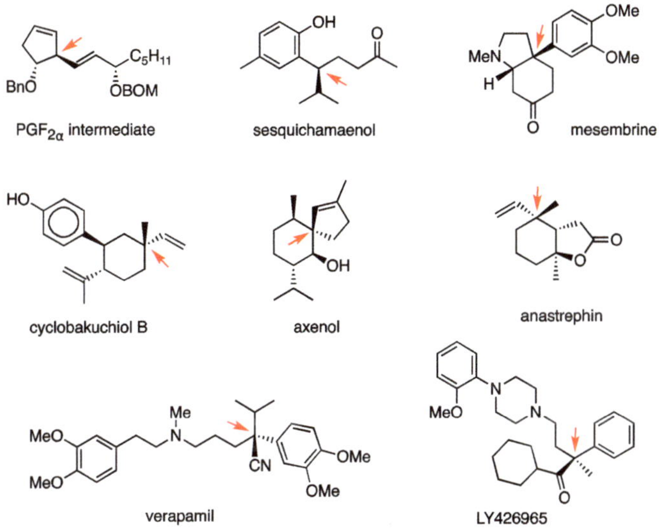

Figure 7. Biologically active compounds synthesized by using RMgX/CuCN [1]. [1] C–C bonds constructed by the *anti*-S_N2' reaction are indicated by a red arrow.

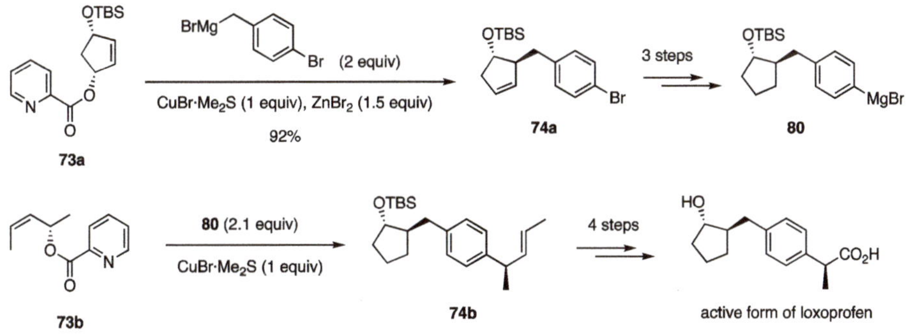

Scheme 19. Synthesis of the active form of loxoprofen using allylic substitutions twice.

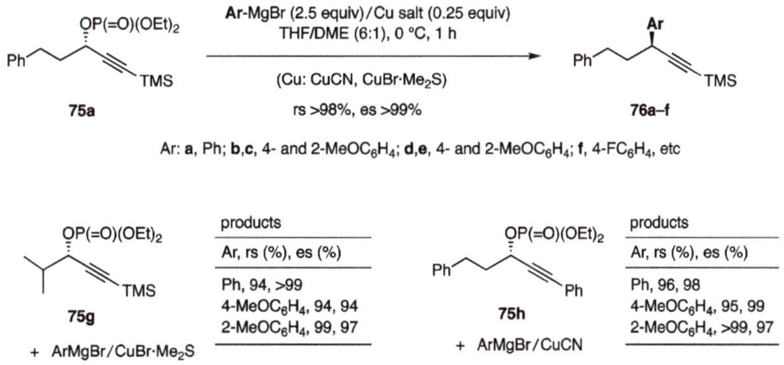

Scheme 20. Substitution at secondary propargylic carbon.

In contrast, the reagents derived from PhMgBr (3 equiv) and Cu(acac)$_2$ (1.5 equiv) gave allene **79** efficiently (Equation (24)).

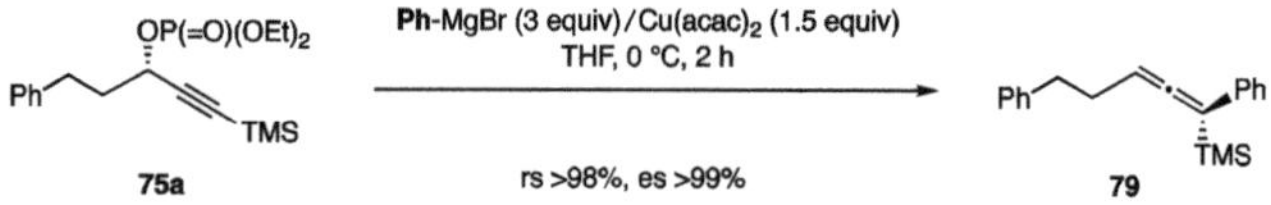

$$\text{75a} \xrightarrow[\text{THF, 0 °C, 2 h}]{\text{Ph-MgBr (3 equiv)/Cu(acac)}_2\text{ (1.5 equiv)}} \text{79} \quad (24)$$

rs >98%, es >99%

With the selective propargylic substitution in hand, the synthesis of TNF inhibitor and flurbiprofen (Figure 8) was easily designed [52], while the substitution of **80** with Grignard reagent **81** was used for the synthesis of heliannuol E [53].

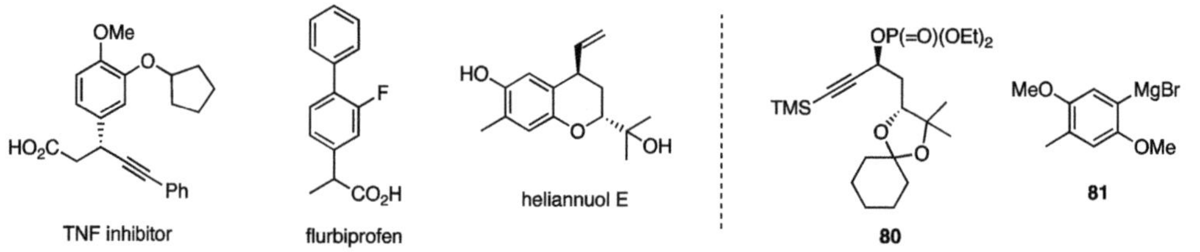

Figure 8. Compounds synthesized by using propargylic substitution [1]. [1] Phosphate **80** and Grignard reagent **81** were used for the synthesis of heliannuol E.

6.5. Substitution at Secondary Alkyl Carbons

In contrast to the allylic and propargylic substitutions, the substitution on secondary alkyl carbons generally proceeds with difficulty because of the absence of a reaction promoter, such as alkenyl and alkynyl moieties. The substitution reported by Breit relies on an ester carbonyl group [54,55]. Consequently, it was not surprising to find only one publication by Liu, who employed tosylate, CuI, and additives [56]. Negishi applied this reaction to the synthesis of phthioceranic acid [57]. Having successfully reproduced the reaction of Liu (Scheme 21, Equation (25)), MeMgCl was used in the reaction to confirm the low reactivity of MeMgCl (Equation (26)). To explore a more reactive protocol, picolinate and phosphate leaving groups (2-pyridineCO$_2$ and (EtO)$_2$PO$_2$) were examined, but unsuccessfully. As an extension of the concept using the pyridyl group, (2-pyridine)sulfonate **77b** was next examined to find sufficient reactivity toward MeMgCl in the presence of Cu(OTf)$_2$. The reaction was completed within 1 h to afford product **78b** in 77% yield (Equation (27)) [33].

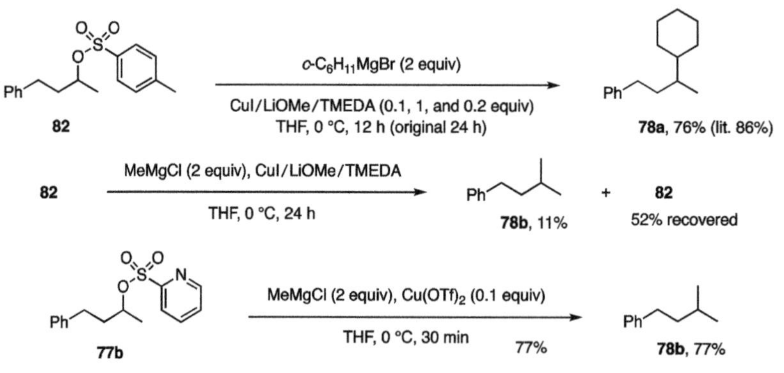

$$\text{82} \xrightarrow[\substack{\text{CuI/LiOMe/TMEDA (0.1, 1, and 0.2 equiv)} \\ \text{THF, 0 °C, 12 h (original 24 h)}}]{\text{c-C}_6\text{H}_{11}\text{MgBr (2 equiv)}} \text{78a, 76% (lit. 86%)} \quad (25)$$

$$\text{82} \xrightarrow[\text{THF, 0 °C, 24 h}]{\text{MeMgCl (2 equiv), CuI/LiOMe/TMEDA}} \text{78b, 11%} + \text{82} \quad 52\% \text{ recovered} \quad (26)$$

$$\text{77b} \xrightarrow[\text{THF, 0 °C, 30 min}]{\text{MeMgCl (2 equiv), Cu(OTf)}_2\text{ (0.1 equiv)}} \text{77\%} \quad \text{78b, 77%} \quad (27)$$

Scheme 21. Substitution at secondary alkyl carbon by Liu (Equation (25)) and by us (Equations (26) and (27)).

The substitution system shown in Equation (26) was applied to other (2-pyridine)sulfonates and RMgX to afford **78c–g** in good yields with high stereoselectivity (Scheme 22).

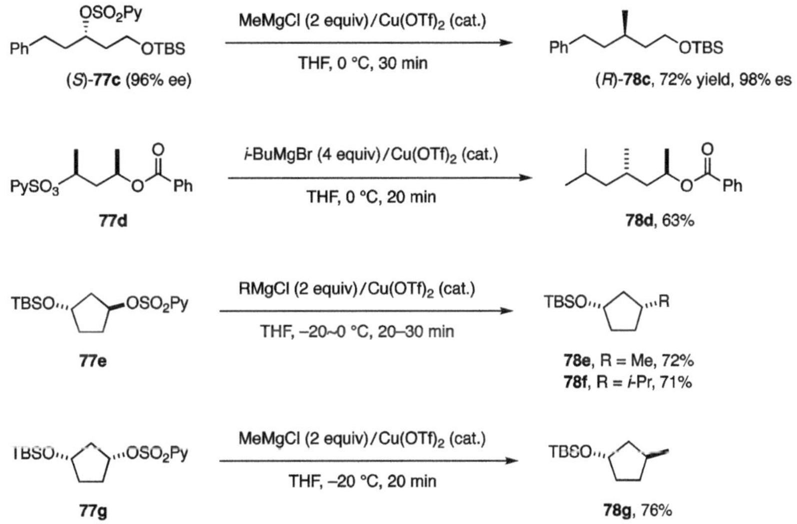

Scheme 22. Substitution on secondary alkyl carbons.

7. Summary

Allylic coupling reactions of secondary allylic carbonates and esters with aryl and alkenyl borates were found to be catalyzed by nickel complexes, such as $NiCl_2(PPh_3)$ and $NiCl_2(dppf)$, and not by palladium catalysts. The alkanediol ligand on the borates was essential for producing high reactivity, which allowed the Ni-catalyzed coupling reaction between cyclopentenediol monoacetate **3** and borates. Furthermore, carbonyl group-friendly borates were synthesized from boronate esters and MeZnCl. Alkenyl borates were used for the construction of dienes from less reactive allylic alcohol derivatives with a bromine atom at the *cis* position. In contrast to the aryl and alkenyl borates, reactive alkyl borates were not developed. Instead, substitutions on secondary carbons with RMgX under copper-based and copper-catalyzed conditions were studied to obtain the following results: The regioselectivity in the allylic substitution of cyclopentenediol monoacetate **3** was highly controlled by the ratio of RMgX/CuCN, halogen X (Cl, Br) in RMgX, and the solvent (Et_2O vs. THF). The substitution of acyclic allylic picolinates proceeded in *anti*-S_N2' mode. Propargylic phosphates underwent propargylic substitution with RMgBr and a Cu catalyst in THF-DME. In addition, the recent success in the substitution of secondary alkyl (2-pyridine)sulfonates with RMgX is presented. The reactions have been applied to the synthesis of biologically active compounds.

Funding: This review was funded by JSPS, KAKENHI Grant No. 20K05501. Fund for each research work described herein will be found in the publication.

Data Availability Statement: Not applicable.

Conflicts of Interest: The author declares no conflict of interest.

References

1. Tsuji, J. *Palladium Reagents and Catalysts: Innovations in Organic Synthesis*; Wiley: Chichester, UK, 1995; pp. 1–574, ISBN 978-0-471-97202-0.
2. Tsuji, J. *Palladium Reagents and Catalysts: New Perspectives for the 21st Century*; Wiley: Chichester, UK, 2004; pp. 1–670, ISBN 978-0-470-85032-9.

3. Stork, G.; Tsuji, J. Lithium-Ammonia Reduction of α,β-Unsaturated Ketones. II. Formation and Alkylation of A β-Carbanion Intermediate. *J. Am. Chem. Soc.* **1961**, *83*, 2783–2784. [CrossRef]
4. Stork, G.; Rosen, P.; Goldman, N.; Coombs, R.V.; Tsuji, J. Alkylation and Carbonation of Ketones by Trapping the Enolates from the Reduction of α,β-Unsaturated Ketones. *J. Am. Chem. Soc.* **1965**, *87*, 275–286. [CrossRef]
5. Miyaura, N.; Yamada, K.; Suginome, H.; Suzuki, A. Novel and convenient method for the stereo-and regiospecific synthesis of conjugated alkadienes and alkenynes via the palladium-catalyzed cross-coupling reaction of 1-alkenylboranes with bromoalkenes and bromoalkynes. *J. Am. Chem. Soc.* **1985**, *107*, 972–980. [CrossRef]
6. Pagett, A.B.; Lloyd-Jones, G.C. Suzuki–Miyaura Cross-Coupling. *Org. React.* **2019**, *100*, 547–620. [CrossRef]
7. Lennox, A.J.J.; Lloyd-Jones, G.C. Selection of boron reagents for Suzuki–Miyaura coupling. *Chem. Soc. Rev.* **2013**, *43*, 412–443. [CrossRef]
8. Ghorai, D.; Cristòfol, À.; Kleij, A.W. Nickel-Catalyzed Allylic Substitution Reactions: An Evolving Alternative. *Eur. J. Inorg. Chem.* **2021**, *2022*, e202100820. [CrossRef]
9. Kobayashi, Y. *Modern Organonickel Chemistry*; Tamaru., Y., Ed.; Wiley: Chichester, UK, 2005; Chapter 3; pp. 56–101. [CrossRef]
10. Kobayashi, Y.; Ikeda, E. Nickel-catalysed substitution reactions of allylic carbonates with aryl- and alkenyl-borates. *J. Chem. Soc. Chem. Commun.* **1994**, 1789–1790. [CrossRef]
11. Kobayashi, Y.; Mizojiri, R.; Ikeda, E. Nickel-Catalyzed Coupling Reaction of 1,3-Disubstituted Secondary Allylic Carbonates and Lithium Aryl- and Alkenylborates. *J. Org. Chem.* **1996**, *61*, 5391–5399. [CrossRef]
12. Kobayashi, Y.; Tokoro, Y.; Watatani, K. Zinc borates: Functionalized hard nucleophiles for coupling reaction with secondary allylic acetates. *Eur. J. Org. Chem.* **2000**, *2000*, 3825–3834. [CrossRef]
13. Kobayashi, Y.; Murugesh, M.G.; Nakano, M.; Takahisa, E.; Usmani, S.B.; Ainai, T. A New Method for Installation of Aryl and Alkenyl Groups onto a Cyclopentene Ring and Synthesis of Prostaglandins. *J. Org. Chem.* **2002**, *67*, 7110–7123. [CrossRef]
14. Kobayashi, Y.; Nakayama, Y.; Mizojiri, R. Nickel-catalyzed coupling reaction of sterically congested cis bromides and lithium alkenylborates. *Tetrahedron* **1998**, *54*, 1053–1062. [CrossRef]
15. Kobayashi, Y.; William, A.D.; Mizojiri, R. Scope and limitation of the nickel-catalyzed coupling reaction between lithium borates and mesylate. *J. Organometal. Chem.* **2002**, *653*, 91–97. [CrossRef]
16. Suzuki, M.; Oda, Y.; Noyori, R. Palladium(0) catalyzed reaction of 1,3-diene epoxides. A useful method for the site-specific oxygenation of 1,3-dienes. *J. Am. Chem. Soc.* **1979**, *101*, 1623–1625. [CrossRef]
17. Grisso, B.A.; Johnson, J.R.; Mackenzie, P.B. Nickel-catalyzed, chlorotrialkylsilane-assisted conjugate addition of alkenyltributyltin reagents to α,β-unsaturated aldehydes. Evidence for a [1-[(trialkylsilyl)oxy]allyl]nickel(II) mechanism. *J. Am. Chem. Soc.* **1992**, *114*, 5160–5165. [CrossRef]
18. Trost, B.M.; Verhoeven, T.R. Allylic alkylation. Palladium-catalyzed substitutions of allylic carboxylates. Stereo- and regiochemistry. *J. Am. Chem. Soc.* **1980**, *102*, 4730–4743. [CrossRef]
19. Newton, R.F.; Reynolds, D.P.; Davies, J.; Kay, P.B.; Roberts, S.M.; Wallace, T.W. Enantiocomplementary total asymmetric syntheses of prostaglandin A2. *J. Chem. Soc. Perkin Trans.* **1983**, *1*, 683–685. [CrossRef]
20. Noyori, R.; Suzuki, M. Organic Synthesis of Prostaglandins: Advancing Biology. *Science* **1993**, *259*, 44–45. [CrossRef]
21. Tokoro, Y.; Kobayashi, Y. Realisation of highly stereoselective dihydroxylation of the cyclopentene in the synthesis of (−)-aristeromycin. *Chem. Commun.* **1999**, 807–809. [CrossRef]
22. Kobayashi, Y.; Morita, M. *Cutting-Edge of Organic Synthesis and Chemical Biology of Bioactive Molecules*; Kobayashi, Y., Ed.; Springer Nature: Singapore, 2019; Chapter 9; pp. 193–231, ISBN 978-981-13-6243-9. Available online: https://link.springer.com/chapter/10.1007/978-981-13-6244-6_9 (accessed on 22 November 2022).
23. Nakayama, Y.; Kumar, G.B.; Kobayashi, Y. Synthesis of 10,11-Dihydroleukotriene B4 Metabolites via a Nickel-Catalyzed Coupling Reaction of cis-Bromides and trans-Alkenyl Borates. *J. Org. Chem.* **2000**, *65*, 707–715. [CrossRef]
24. Kobayashi, Y.; Shimazaki, T.; Taguchi, H.; Sato, F. Highly stereocontrolled total synthesis of leukotriene B4, 20-hydroxyleukotriene B4, leukotriene B3, and their analogs. *J. Org. Chem.* **1990**, *55*, 5324–5335. [CrossRef]
25. Kobayashi, Y.; Yoshida, S.; Nakayama, Y. Total synthesis of korormicin. *Eur. J. Org. Chem.* **2001**, *2001*, 1873–1881. [CrossRef]
26. Kobayashi, Y.; Yoshida, S.; Asano, M.; Takeuchi, A.; Acharya, H.P. Nickel-Catalyzed Coupling Producing (2Z)-2,4-Alkadien-1-ols, Conversion to (E)-3-Alkene-1,2,5-triol Derivatives, and Synthesis of Decarestrictine D. *J. Org. Chem.* **2007**, *72*, 1707–1716. [CrossRef] [PubMed]
27. Kobayashi, Y.; Asano, M.; Yoshida, S.; Takeuchi, A. Stereoselective Synthesis of Decarestrictine D from a Previously Inaccessible (2Z,4E)-Alkadienyl Alcohol Precursor. *Org. Lett.* **2005**, *7*, 1533–1536. [CrossRef] [PubMed]
28. Kobayashi, Y.; William, A.D. Coupling Reactions of α-Bromoalkenyl Phosphonates with Aryl Boronic Acids and Alkenyl Borates. *Org. Lett.* **2002**, *4*, 4241–4244. [CrossRef]
29. Kobayashi, Y.; William, A.D. Palladium- and Nickel-Catalyzed Coupling Reactions of α-Bromoalkenylphosphonates with Arylboronic Acids and Lithium Alkenylborates. *Adv. Synth. Catal.* **2004**, *346*, 1749–1757. [CrossRef]

30. Ito, M.; Matsuumi, M.; Murugesh, M.G.; Kobayashi, Y. Scope and Limitation of Organocuprates, and Copper or Nickel Catalyst-Modified Grignard Reagents for Installation of an Alkyl Group onto cis-4-Cyclopentene-1,3-diol Monoacetate. *J. Org. Chem.* **2001**, *66*, 5881–5889. [CrossRef] [PubMed]
31. Kiyotsuka, Y.; Acharya, H.P.; Katayama, Y.; Hyodo, T.; Kobayashi, Y. Picolinoxy Group, a New Leaving Group for anti S_N2' Selective Allylic Substitution with Aryl Anions Based on Grignard Reagents. *Org. Lett.* **2008**, *10*, 1719–1722. [CrossRef] [PubMed]
32. Kobayashi, Y.; Takashima, Y.; Motoyama, Y.; Isogawa, Y.; Katagiri, K.; Tsuboi, A.; Ogawa, N. α- and γ-Regiocontrol and Enantiospecificity in the Copper-Catalyzed Substitution Reaction of Propargylic Phosphates with Grignard Reagents. *Chem. Eur. J.* **2020**, *27*, 3779–3785. [CrossRef]
33. Shinohara, R.; Morita, M.; Ogawa, N.; Kobayashi, Y. Use of the 2-Pyridinesulfonyloxy Leaving Group for the Fast Copper-Catalyzed Coupling Reaction at Secondary Alkyl Carbons with Grignard Reagents. *Org. Lett.* **2019**, *21*, 3247–3251. [CrossRef]
34. Ainai, T.; Matsuumi, M.; Kobayashi, Y. Efficient Total Synthesis of 12-oxo-PDA and OPC-8:0. *J. Org. Chem.* **2003**, *68*, 7825–7832. [CrossRef]
35. Nonaka, H.; Ogawa, N.; Maeda, N.; Wang, Y.-G.; Kobayashi, Y. Stereoselective synthesis of epi-jasmonic acid, tuberonic acid, and 12-oxo-PDA. *Org. Biomol. Chem.* **2010**, *8*, 5212–5223. [CrossRef] [PubMed]
36. Ogawa, N.; Kobayashi, Y. Synthesis of the amino acid conjugates of epi-jasmonic acid. *Amino Acids* **2011**, *42*, 1955–1966. [CrossRef] [PubMed]
37. Kobayashi, Y.; Nakata, K.; Ainai, T. New Reagent System for Attaining High Regio-and Stereoselectivities in Allylic Displacement of 4-Cyclopentene-1,3-diol Monoacetate with Aryl- and Alkenylmagnesium Bromides. *Org. Lett.* **2004**, *7*, 183–186. [CrossRef]
38. Nakata, K.; Kiyotsuka, Y.; Kitazume, T.; Kobayashi, Y. Realization of Anti-S_N2' Selective Allylation of 4-Cyclopentene-1,3-diol Monoester with Aryl- and Alkenyl-Zinc Reagents. *Org. Lett.* **2008**, *10*, 1345–1348. [CrossRef] [PubMed]
39. Nakata, K.; Kobayashi, Y. Aryl-and Alkenyllithium Preparations and Copper-Catalyzed Reaction between the Derived Magnesium Reagents and the Monoacetate of 4-Cyclopentene-1,3-diol. *Org. Lett.* **2005**, *7*, 1319–1322. [CrossRef] [PubMed]
40. Kobayashi, Y. Alkyl Pyridinesulfonates and Allylic Pyridinecarboxylates, New Boosters for the Substitution at Secondary Carbons. *Heterocycles* **2020**, *100*, 499. [CrossRef]
41. Kobayashi, Y.; Shimoda, M. *Cutting-Edge of Organic Synthesis and Chemical Biology of Bioactive Molecules*; Kobayashi, Y., Ed.; Springer Nature: Singapore, 2019; Chapter 7; pp. 145–169, ISBN 978-981-13-6243-9. Available online: https://link.springer.com/chapter/10.1007/978-981-13-6244-6_7 (accessed on 22 November 2022).
42. Kiyotsuka, Y.; Kobayashi, Y. Formation of Chiral C(sp^3)–C(sp) Bond by Allylic Substitution of Secondary Allylic Picolinates and Alkynyl Copper Reagents. *J. Org. Chem.* **2009**, *74*, 7489–7495. [CrossRef]
43. Wang, Q.; Kobayashi, Y. Allylic Substitution on Cyclopentene and -hexene Rings with Alkynylcopper Reagents. *Org. Lett.* **2011**, *13*, 6252–6255. [CrossRef]
44. Kiyotsuka, Y.; Katayama, Y.; Acharya, H.P.; Hyodo, T.; Kobayashi, Y. New General Method for Regio- and Stereoselective Allylic Substitution with Aryl and Alkenyl Coppers Derived from Grignard Reagents. *J. Org. Chem.* **2009**, *74*, 1939–1951. [CrossRef]
45. Ozaki, T.; Kobayashi, Y. Synthesis of (−)-mesembrine using the quaternary carbon-constructing allylic substitution. *Org. Chem. Front.* **2015**, *2*, 328–335. [CrossRef]
46. Kawashima, H.; Kaneko, Y.; Sakai, M.; Kobayashi, Y. Synthesis of Cyclobakuchiols A, B, and C by Using Conformation-Controlled Stereoselective Reactions. *Chem. Eur. J.* **2013**, *20*, 272–278. [CrossRef] [PubMed]
47. Kobayashi, Y.; Ozaki, T. Concise Synthesis of (−)-Axenol by Using Stereocontrolled Allylic Substitution. *Synlett* **2015**, *26*, 1085–1088. [CrossRef]
48. Kobayashi, Y.; Wada, K.; Sakai, M.; Kawashima, H.; Ogawa, N. Efficient Synthesis of Anastrephin via the Allylic Substitution for Quaternary Carbon Construction. *Synlett* **2016**, *27*, 1428–1432. [CrossRef]
49. Kobayashi, Y.; Saeki, R.; Nanba, Y.; Suganuma, Y.; Morita, M.; Nishimura, K. Synthesis of the Verapamil Intermediate through the Quaternary Carbon-Constructing Allylic Substitution. *Synlett* **2017**, *28*, 2655–2659. [CrossRef]
50. Kobayashi, Y.; Yamaguchi, K.; Morita, M. Regio-and stereoselective SN2' reaction of an allylic picolinate in the synthesis of LY426965. *Tetrahedron* **2018**, *74*, 1826–1831. [CrossRef]
51. Hyodo, T.; Kiyotsuka, Y.; Kobayashi, Y. Synthesis of the Active Form of Loxoprofen by Using Allylic Substitutions in Two Steps. *Org. Lett.* **2009**, *11*, 1103–1106. [CrossRef]
52. Takashima, Y.; Isogawa, Y.; Tsuboi, A.; Ogawa, N.; Kobayashi, Y. Synthesis of a TNF inhibitor, flurbiprofen and an i-Pr analogue in enantioenriched forms by copper-catalyzed propargylic substitution with Grignard reagents. *Org. Biomol. Chem.* **2021**, *19*, 9906–9909. [CrossRef]
53. Ogawa, N.; Uematsu, C.; Kobayashi, Y. Stereoselective Synthesis of (−)-Heliannuol E by α-Selective Propargyl Substitution. *Synlett* **2021**, *32*, 2071–2074. [CrossRef]
54. Studte, C.; Breit, B. Zinc-Catalyzed Enantiospecific sp^3–sp^3 Cross-Coupling of α-Hydroxy Ester Triflates with Grignard Reagents. *Angew. Chem. Int. Ed.* **2008**, *47*, 5451–5455. [CrossRef]
55. Brand, G.J.; Studte, C.; Breit, B. Iterative Synthesis of (Oligo) deoxypropionates via Zinc-Catalyzed Enantiospecific sp^3–sp^3 Cross-Coupling. *Org. Lett.* **2009**, *11*, 4668–4670. [CrossRef]

56. Yang, C.-T.; Zhang, Z.-Q.; Liang, J.; Liu, J.-H.; Lu, X.-Y.; Chen, H.-H.; Liu, L. Copper-Catalyzed Cross-Coupling of Nonactivated Secondary Alkyl Halides and Tosylates with Secondary Alkyl Grignard Reagents. *J. Am. Chem. Soc.* **2012**, *134*, 11124–11127. [CrossRef] [PubMed]
57. Xu, S.; Oda, A.; Bobinski, T.; Li, H.; Matsueda, Y.; Negishi, E.-I. Highly Efficient, Convergent, and Enantioselective Synthesis of Phthioceranic Acid. *Angew. Chem. Int. Ed.* **2015**, *54*, 9319–9322. [CrossRef] [PubMed]

Disclaimer/Publisher's Note: The statements, opinions and data contained in all publications are solely those of the individual author(s) and contributor(s) and not of MDPI and/or the editor(s). MDPI and/or the editor(s) disclaim responsibility for any injury to people or property resulting from any ideas, methods, instructions or products referred to in the content.

MDPI AG
Grosspeteranlage 5
4052 Basel
Switzerland
Tel.: +41 61 683 77 34

Catalysts Editorial Office
E-mail: catalysts@mdpi.com
www.mdpi.com/journal/catalysts

Disclaimer/Publisher's Note: The title and front matter of this reprint are at the discretion of the Guest Editors. The publisher is not responsible for their content or any associated concerns. The statements, opinions and data contained in all individual articles are solely those of the individual Editors and contributors and not of MDPI. MDPI disclaims responsibility for any injury to people or property resulting from any ideas, methods, instructions or products referred to in the content.